SUSTAINABLE AGRICULTURE AND GLOBAL ENVIRONMENTAL HEALTH

Toward the Efficient Management of Resources

SUSTAINABLE AGRICULTURE AND GLOBAL ENVIRONMENTAL HEALTH

Toward the Efficient Management of Resources

Mónica L. Chávez González, PhD

Cristóbal Noé Aguilar, PhD

Juan Carlos Contreras-Esquivel, PhD

A. K. Haghi, PhD

Editors

AAP | APPLE ACADEMIC PRESS

First edition published 2025

Apple Academic Press Inc.
1265 Goldenrod Circle, NE,
Palm Bay, FL 32905 USA

760 Laurentian Drive, Unit 19,
Burlington, ON L7N 0A4, CANADA

CRC Press
2385 NW Executive Center Drive,
Suite 320, Boca Raton FL 33431

4 Park Square, Milton Park,
Abingdon, Oxon, OX14 4RN UK

Library and Archives Canada Cataloguing in Publication

Title: Sustainable agriculture and global environmental health : toward the efficient management of resources / Mónica L. Chávez González, PhD, Cristóbal Noé Aguilar, PhD, Juan Carlos Contreras-Esquivel, PhD, A.K. Haghi, PhD, editors.
Names: Chávez-González, Mónica Lizeth, 1987- editor | Aguilar, Cristóbal Noé, editor | Contreras-Esquivel, Juan Carlos, editor | Haghi, A. K., editor
Description: First edition. | Includes bibliographical references and index.
Identifiers: Canadiana (print) 20240437098 | Canadiana (ebook) 20240437128 | ISBN 9781774917183 (hardcover) | ISBN 9781774917190 (softcover) | ISBN 9781003504474 (ebook)
Subjects: LCSH: Sustainable agriculture—Technological innovations. | LCSH: Environmental health.
Classification: LCC S494.5.S86 S875 2025 | DDC 630.2/086—dc23

Library of Congress Cataloging-in-Publication Data

CIP data on file with US Library of Congress

ISBN: 978-1-77491-718-3 (hbk)
ISBN: 978-1-77491-719-0 (pbk)
ISBN: 978-1-00350-447-4 (ebk)

About the Editors

Monica Lizeth Chavez-Gonzalez, PhD
Research Scientist, School of Chemistry of the Universidad Autónoma de Coahuila, Mexico

Mónica Chávez-González, PhD, is a Research Professor at the School of Chemical Sciences of the Universidad Autónoma de Coahuila, Mexico. She is a pharmacobiological chemist. Her master's and doctoral program was in Food Science and Technology. She is the General Coordinator and founder of the Bioprocessing Mx-LATAM Network, whose objective is to promote R&D in the area of bioprocessing through interaction between universities, institutes, research centers, and relevant industries in Mexico and Latin America. She has been awarded the Latin American Women in Chemistry Award in the category of "Emerging Leader," granted by the American Chemical Society and the Latin American Federation of Chemical Associations (FLAQ). Her expertise is in the areas of fermentation processes, microbial biotransformation, enzyme production, industrial food waste valorization, extraction of bioactive compounds (antioxidants), and chemical characterization.

Cristóbal Noé Aguilar, PhD
Director, Research and Postgraduate Programs, BBG-DIA, Bioprocesses & Bioproducts Research Group, Food Research Department, Faculty of Chemical Sciences, Universidad Autonoma de Coahuila, Saltillo, Coahuila, Mexico

Cristóbal Noé Aguilar, PhD, is Director of Research and Postgraduate Programs at the Universidad Autonoma de Coahuila, Mexico. Dr. Aguilar has published more than 530 papers in indexed journals and 550 contributions in scientific meetings. He has also edited more than 30 books. He has been awarded several prizes and awards, the most important of which are the National Prize of Research (2010) from the Mexican Academy of Sciences; the Prize CCC–2008 from the Mexican Society of Biotechnology and Bioengineering; National Prize AgroBio-2005; and the Mexican Prize in Food Science and Technology. In 2019, he received the Outstanding Research

Award by the International Association of Bioprocessing (IBA). Dr. Aguilar is a member of the Mexican Academy of Science, the Central Council of the International Bioprocessing Association, the Mexican Academy of Sciences, the Mexican Society for Biotechnology and Bioengineering, and the Mexican Association for Food Science and Biotechnology. He has developed more than 41 research projects, including 25 international exchange projects.

Juan Carlos Contreras-Esquivel, PhD
Professor and Head, Laboratory of Applied Glycobiotechnology,
Faculty of Chemistry, Department of Food Research,
Universidad Autonoma de Coahuila, Saltillo, Mexico

Juan Carlos Contreras-Esquivel, PhD, is a full Professor and Head of the Laboratory of Applied Glycobiotechnology at the Food Research Department, School of Chemistry, the University of Coahuila, Mexico. He is involved in teaching undergraduate and graduate students with a special focus in food science and food bioscience. His research interests are mainly on applied glycoltechnology for polysaccharide extraction by using ecofriendly technologies (enzymatic, microwave, pulsed electric field, ultrasonic) and oligosaccharide production by degradative or biosynthetic routes. Dr. Contreras-Esquivel earned a bachelor's degree in Biological Chemistry and holds an MSc degree in Food Science and Technology, both from the Universidad Autonoma de Chihuahua, Mexico. He received his PhD from the Universidad Nacional de La Plata, Argentina.

A. K. Haghi, PhD
Research Associate, Department of Chemistry, University of Coimbra, Portugal

A. K. Haghi, PhD, is a retired professor and has written, co-written, edited or co-edited more than 1000 publications, including books, book chapters, and papers in refereed journals with over 3800 citations and h-index of 32, according to the Google Scholar database. He is currently a research associate at university of Coimbra, Portugal. Professor Haghi has received several grants, consulted for several major corporations, and is a frequent speaker to national and international audiences. He is Founder and former Editor-in-Chief of the *International Journal of Chemoinformatics and Chemical Engineering* and *Polymers Research Journal*. Professor Haghi has acted as an editorial board member of many international journals. He has served as a member of the Canadian Research and Development Center of Sciences & Cultures. He has supervised several PhD and MSc theses at the University of Guilan (UG) and co-supervised international doctoral projects. Professor Haghi holds a BSc in urban and environmental engineering from the University of North Carolina (USA) and holds two MSc, one in mechanical engineering from North Carolina State University (USA) and another one in applied mechanics, acoustics, and materials from the Université de Technologie de Compiègne (France). He was awarded a PhD in engineering sciences from the Université de Franche-Comté (France). He is a regular reviewer of leading international journals.

Contents

Contributors

Cristóbal N. Aguilar
Bioprocesses and Bioproducts Research Group, Food Research Department, School of Chemistry, Universidad Autónoma de Coahuila, Saltillo, Coahuila, Mexico; Departamento de Investigación en Alimentos, Facultad de Ciencias Químicas, Universidad Autónoma de Coahuila, Saltillo, Mexico

Miguel Á. Aguilar-González
Centro de Investigación y de Estudios Avanzados del Instituto Politécnico Nacional (CINVESTAV-IPN), Ramos Arizpe, Coahuila, Mexico

Rocio Yaneli Aguirre-Loredo
Centro de Investigación en Química Aplicada, Blvd. Enrique Reyna Hermosillo 140, San José de los Cerritos, Saltillo, Coahuila, Mexico

Elan Iñaky Laredo Alcalá
Research Center for the Conservation of Biodiversity and Ecology of Coahuila, Autonomous University of Coahuila, Cuatrociénegas, Coahuila, Mexico, ZC

Sara Andrea Álvarez-Alvarado
Facultad de Ciencias Químicas, Department de Química Analítica, Universidad Autónoma de Coahuila, Saltillo, Coahuila, Mexico

Olga B. Alvarez-Pérez
Microbiological Research and Development Department, Greencorp Biorganiks SA de CV, Blvd. Luis Donaldo Colosio, San Patricio, C.P., Saltillo, Coahuila, Mexico

Judith Amador-Hernández
Facultad de Ciencias Químicas, Department de Química Analítica, Universidad Autónoma de Coahuila, Saltillo, Coahuila, Mexico

Epifanio Castro del Ángel
Department of Parasitology, Autonomous Agrarian University Antonio Narro, Saltillo, Coahuila, Mexico, ZC

Zaida M. Aranda
Autonomous University of Coahuila, Saltillo (UAdeC), Mexico

G. Soria Argüello
Investigador por México CONACYT, Departamento de Química Macromolecular y Nanomateriales, Centro de Investigación en Química Aplicada, Saltillo, Coahuila, Mexico

Roberto Arredondo-Valdés
Facultad de Ciencias Químicas, School of Chemistry, School of Chemistry, Universidad Autónoma de Coahuila, Saltillo, Coahuila, Mexico

Abigail A. Belmares-Cerda
Departamento de Investigación en Alimentos, Facultad de Ciencias Químicas, Universidad Autónoma de Coahuila, Saltillo, Mexico

Ruth E. Belmares-Cerda
Departamento de Investigación en Alimentos, Facultad de Ciencias Químicas, Universidad Autónoma de Coahuila, Saltillo, Mexico

Julia Cecilia Anguiano Cabello
Health Sciences Area, La Salle Saltillo University, Saltillo, Coahuila, Mexico, ZC

Carlos N. Cano-Gonzalez
Centro de Investigación en Química Aplicada, Blvd. Enrique Reyna Hermosillo 140, San José de los Cerritos, Saltillo, Coahuila, Mexico

Francisco Daniel Hernández Castillo
Department of Parasitology, Autonomous Agrarian University Antonio Narro, Saltillo, Coahuila, Mexico, ZC

Juan Carlos Contreras-Esquivel
Facultad de Ciencias Químicas, Universidad Autónoma de Coahuila, Blvd. Venustiano Carranza y José Cárdenas s/n. Republica Oriente, Saltillo, Coahuila, Mexico

V. J. Cruz Delgado
Departamento de Procesos de Transformación, Centro de Investigación en Química Aplicada, Saltillo, Coahuila, Mexico

Liliana Flores-López
Center for Scientific and Technological Research and Innovation, Universidad Autónoma de Coahuila, Mexico

Cristóbal Noé Aguilar González
School of Chemistry, Autonomous University of Coahuila, Saltillo, Coahuila, Mexico, ZC

Mónica Lizeth Chávez González
School of Chemistry, Autonomous University of Coahuila, Saltillo, Coahuila, Mexico, ZC

Mayela Govea-Salas
Facultad de Ciencias Químicas, Universidad Autónoma de Coahuila, Saltillo, Coahuila, Mexico

T. E. Flores Guía
Departamento de Materiales Avanzados, Facultad de Ciencias Químicas, Universidad Autónoma de Coahuila, Saltillo, Coahuila, Mexico

A. K. Haghi
Bioprocesses and Bioproducts Research Group, Food Research Department, School of Chemistry, Universidad Autónoma de Coahuila, Saltillo, Coahuila, Mexico

José Luis Martínez Hernández
Bioprocesses and Bioproducts Group, Food Research Department, School of Chemistry, Universidad Autónoma de Coahuila, Saltillo, Coahuila, Mexico

L. M. Hernández
Bioprocesses and Bioproducts Group, Food Research Department, School of Chemistry, Universidad Autónoma de Coahuila, Saltillo, Coahuila, Mexico

María de la Luz Herrera-Estrada
Facultad de Ciencias Químicas, Universidad Autónoma de Coahuila, Blvd. Venustiano Carranza y José Cárdenas s/n. Republica Oriente, Saltillo, Coahuila, Mexico

Agustín Hernández Juarez
Department of Parasitology, Autonomous Agrarian University Antonio Narro, Saltillo, Coahuila, Mexico, ZC

Elan Laredo-Alcalá
Centro de Investigación para la Conservación de la Biodiversidad y Ecología de Coahuila (CICBEC), Universidad Autónoma de Coahuila, Saltillo, Coahuila, Mexico

Areli Lopez-Montelongo
Faculty of Architecture in Saltillo, Autonomous University of Coahuila, Mexico

Erika J. López-Zúñiga
Faculty of Nutrition and Food Sciences, University of Sciences and Arts of Chiapas, Lib. Norte Pte., Tuxtla Gutierrez, Chiapas, Mexico

Karla A. Luna
Bioprocesses and Bioproducts Group, Food Research Department, School of Chemistry, Universidad Autónoma de Coahuila, Saltillo, Coahuila, Mexico

Miriam E. Mery-Ruiz
Universidad Autónoma de Coahuila, Mexico

Víctor Navarro-Macias
Facultad de Ciencias Químicas, Universidad Autónoma de Coahuila, Saltillo, Coahuila, Mexico

Gabriela Carmona Ochoa
Autonomous University of Coahuila, Mexico

Yanhsy Hernández Portillo
Direction of Research and Graduate Studies, University of Sciences and Arts of Chiapas, Mexico

N. Ramírez-Guzman
Bioprocesses and Bioproducts Group, Food Research Department, School of Chemistry, Universidad Autónoma de Coahuila, Saltillo, Coahuila, Mexico

Nathiely Ramírez-Guzmán
Autonomous University of Coahuila, Saltillo (UAdeC), Mexico

Rodolfo Ramos-Gonzales
Facultad de Ciencias Químicas, Universidad Autónoma de Coahuila, Saltillo, Coahuila, Mexico

Karina Reyes-Acosta
Departamento de Investigación en Alimentos, Facultad de Ciencias Químicas, Universidad Autónoma de Coahuila, Saltillo, Mexico

Laihsa V. Rodríguez-González
University of Valladolid, Spain

Rosa M. Rodríguez-Jasso
Biorefinery Group, Food Research Department, School of Chemistry, Universidad Autónoma de Coahuila, Saltillo, Coahuila, Mexico

H. A. Ruiz
Bioprocesses and Bioproducts Group, Food Research Department, School of Chemistry, Universidad Autónoma de Coahuila, Saltillo, Coahuila, Mexico

Héctor A. Ruiz
Biorefinery Group, Food Research Department, School of Chemistry, Universidad Autónoma de Coahuila, Saltillo, Coahuila, Mexico

Mayela Goveza Salas
Bioprocesess and Bioproducts Group, Food Research Department, School of Chemistry, Universidad Autónoma de Coahuila, Saltillo, Coahuila, Mexico

L. F. Cano Salazar
Departamento de Materiales Avanzados, Facultad de Ciencias Químicas, Universidad Autónoma de Coahuila, Saltillo, Coahuila, Mexico

José Sandoval-Cortés
Bioprocesses and Bioproducts Research Group, Food Research Department, School of Chemistry, Universidad Autónoma de Coahuila, Saltillo, Coahuila, Mexico; Departamento de Investigación en Alimentos, Facultad de Ciencias Químicas, Universidad Autónoma de Coahuila, Saltillo, Mexico

Lluvia de Abril Alexandra Soriano-Melgar
Facultad de Ciencias Químicas, Universidad Autónoma de Coahuila, Blvd. Venustiano Carranza y José Cárdenas s/n. Republica Oriente, Saltillo, Coahuila, Mexico

Veymar G. Tacias-Pascacio
Faculty of Nutrition and Food Sciences, University of Sciences and Arts of Chiapas, Lib. Norte Pte., Tuxtla Gutierrez, Chiapas, Mexico

Cristian Torres-León
Reaserch Center and Ethnobiological Garden (CIJE-UAdeC), Universidad Autonoma de Coahuila, Unidad Torreón, Viesca, Coahuila, Mexico

Marco Antonio Tucuch-Pérez
School of Chemistry, Autonomous University of Coahuila, Saltillo, Coahuila, Mexico, ZC

Gilber Vela-Gutiérrez
Faculty of Nutrition and Food Sciences, University of Sciences and Arts of Chiapas, Lib. Norte Pte., Tuxtla Gutierrez, Chiapas, Mexico

Miguel Velázquez-Manzanares
Facultad de Ciencias Químicas, Department de Química Analítica, Universidad Autónoma de Coahuila, Saltillo, Coahuila, Mexico

Héctor S. Villada-Carrillo
Department of Food Research, Universidad Autónoma de Coahuila, Mexico

Preface

New technologies have the potential to transform the agri-food industry by improving agricultural productivity. Recent innovations in this sector are making a useful contribution to environmental sustainability as well. This new book presents advanced methods and modern technologies in agri-food sectors to document the improvements in environmental health and food security considering a global scale. This new volume focuses on prioritizing sustainable agriculture and global environmental health. It explores the principles of environmental sustainability and explains how these principles can be applied in practice in agri-food systems. It also provides an accessible framework for the effective management of resources for the reduction of chronic disease in humans and mitigating the degradation of the environment.

This book:

- Describes the key concepts of sustainable food production for scientists interested in the environmental impacts of agriculture.
- Reviews sustainable options to the industrial production methods used today to adopt healthier foods and lifestyles.
- Reports on sustainable resources for the food industry along with sustainable processes for food scientists of all levels, both in academia and industry.
- Explores technical innovations to connect the gap between a healthy environment and sustainable agriculture using the theories and principles set out in each chapter to address real-world problems.
- Presents case studies related to our growing demand for food from the perspectives of disciplines ranging from environmental science to public health.

CHAPTER 1

An Introduction to the Environmental Sustainability of Biofuels

HÉCTOR A. RUIZ,[1] ROSA M. RODRÍGUEZ-JASSO,[1]
JOSÉ SANDOVAL-CORTÉS,[2] A. K. HAGHI,[2] and CRISTÓBAL N. AGUILAR[2]

[1]*Biorefinery Group, Food Research Department, School of Chemistry, Universidad Autónoma de Coahuila, Saltillo, Coahuila, Mexico*

[2]*Bioprocesses and Bioproducts Research Group, Food Research Department, School of Chemistry, Universidad Autónoma de Coahuila, Saltillo, Coahuila, Mexico*

ABSTRACT

Population growth and its inevitable associated industrialization have accelerated the global energy demand never seen in past times, which now has become a major concern worldwide. There is already a considerably huge energy crisis due to two major factors: on the one hand, the limited availability of conventional energy resources, and on the other hand, the great consumption of non-renewable fossil fuels. In addition, there is a great necessity to explore alternative renewable fuels that can meet the energy needs of the growing population and overcome daunting environmental problems to be solved, among others greenhouse effect, warming at the global level, air pollution, etc. Exploration of in-depth knowledge of food waste and details of the processing, as well as opportunities to utilize it for mass-scale biofuel production, is a necessity. Particularly, green, non-expensive, and easy-to-produce options describing advantages, mechanisms, strategies, drawbacks, and future scope. In this chapter, some achievements and progress made for biofuel production, including new possibilities derived from the use of food waste as a potential, alternative, and reachable feedstock, are analyzed and discussed.

Sustainable Agriculture and Global Environmental Health: Toward the Efficient Management of Resources. Mónica L. Chávez González, Cristóbal Noé Aguilar, Juan Carlos Contreras-Esquivel, and A. K. Haghi (Eds.)

1.1 INTRODUCTION

Biomass is the most abundant resource on Earth, which is integrated with living organisms, including plants, animals, and microorganisms (Rosero-Chasoy et al., 2023). The agro-industrial wastes have a relevant role in biofuels production because of their wide availability, practically no cost, and mainly composed of materials such as cellulose, hemicellulose, starch, and lignin (Romero-Cedillo et al., 2017; Escamilla-Alvarado et al., 2017).

Biofuels are obtained from biomass through physical, chemical, or biological processes. Fermentation and digestion are the processes most used in the production of biofuels such as ethanol and biogas because many microorganisms can convert the sugars inside agro-industrial waste into biofuels after pretreatment on the biomass (Barrera et al., 2016).

The Food and Agriculture Organization (FAO) reports that food waste is a major global challenge that is becoming increasingly urgent to be addressed because one-third of all food is wasted globally each year. This waste not only results in economic losses and environmental impacts but also has social consequences, contributing to issues such as food insecurity and malnutrition. In recent years, there has been growing interest in using food waste as a feedstock for biofuel production, due to its high organic content and potential to generate clean energy.

To decrease the elevated accumulation of greenhouse gases in the environment, it is necessary to substitute traditional fuels with the new generation of biofuels to cover the worldwide requirements. The obtaining of efficient and green renewable energy fuels is the main challenge for the entire world to combat adverse climatic problems. Increasing attention has been focused on biomass use as a new resource to obtain energy.

Biofuels are an attractive alternative to conventional fuels because they are renewable, emit fewer greenhouse gas emissions, and are more environmentally friendly. One potential feedstock for biofuel production is food waste.

Table 1.1 shows a general overview of different biofuel options along with their advantages and drawbacks. Each of these biofuel options has its advantages and drawbacks. The key to achieving sustainable and economically feasible biofuels is a combination of policies and technological advancements that support their production, distribution, and use. While there is still room for improvement, the future scope of biofuels is promising, given the increasing demand for renewable and low-carbon fuels.

TABLE 1.1 General Overview of Different Biofuel Options

Biofuel	Advantages and Drawbacks
Bioethanol	It is made by fermenting sugars from crops such as corn, sugarcane, and wheat. It is blended with gasoline to create a more sustainable fuel for cars. This alcohol has the potential to reduce greenhouse gas emissions and is economically feasible due to its renewable nature. However, it requires large amounts of land and water and may lead to deforestation and soil degradation.
Biodiesel	It is produced by processing vegetable oil, animal fat, or algae oil. It is a cleaner alternative to diesel fuel and has lower emissions compared to traditional diesel, making it an attractive option for transportation. However, it is more costly to produce than traditional diesel, and the feedstocks required for producing biodiesel may compete with food production.
Biogas	It is produced by breaking down organic matter such as animal manure, food waste, and sewage through a process called anaerobic digestion. It is used for heating and producing electricity. Biogas has the potential to reduce greenhouse gas emissions, and its production can also support waste management. However, it requires a constant supply of organic matter, and the process of anaerobic digestion requires a significant upfront investment
Hydrogen	can be produced from biomass or other renewable sources through a process called electrolysis. It can be used as fuel in fuel cell vehicles, producing only water as a byproduct. Hydrogen has the potential to significantly reduce emissions and has a range of applications beyond transportation. However, it is expensive to produce, and the infrastructure required to distribute hydrogen is not yet in place.
Aviation fuels	Sustainable aviation fuels are derived from renewable feedstocks such as plant oils, waste oils, and algae. They have the potential to decarbonize the aviation industry, which is a significant contributor to greenhouse gas emissions. However, the cost of production is currently high, and scaling up production to meet the demand of the aviation industry remains a challenge.

1.2 PROCESSING FOOD WASTE FOR BIOFUEL PRODUCTION

The process of converting food waste into biofuels typically involves several steps. First, the waste is collected and sorted to remove any non-organic components, such as plastic or metal. Then, it is treated in a pre-processing step to break down any physical barriers to decomposition, such as cellulose or lignin, and to reduce particle size. This can be achieved through mechanical, chemical, or biological treatments, such as shredding, grinding, heating, or enzymes. Once the waste is pre-processed, it is ready for anaerobic digestion (AD), which is a process that breaks down organic matter in the absence of oxygen, producing biogas as a result. The biogas can be used directly as fuel or further processed into biomethane, a purified form of biogas with higher energy content. The remaining digested residue can also be used as a fertilizer or soil conditioner, providing additional economic and environmental benefits.

Alternatively, food waste can be converted into biofuels through biochemical or thermochemical processes. In the biochemical process, enzymes or microorganisms are used to break down the organic matter into sugars, which are then fermented into ethanol or other biofuels. In the thermochemical process, high temperatures and pressure are used to chemically break down the waste into gas or liquid fuels.

1.3 OPPORTUNITIES FOR MASS SCALE BIOFUELS PRODUCTION

The use of food waste as a feedstock for biofuel production presents several opportunities for mass-scale implementation. First, it can help to diminish global warming gas emissions by diverting waste from landfills, where it produces methane, a potent greenhouse gas. Secondly, it can help to meet the increasing demand for new energy sources, particularly in urban areas with large amounts of waste streams. Thirdly, it can provide economic benefits by reducing waste disposal costs and generating revenue from the sale of biofuels and other by-products.

Several countries have implemented policies and programs to promote the use of food waste for biofuel production. For example, in the UK, the "Renewable Transport Fuel Obligation" requires fuel producers to ensure that a certain percentage of their products are made from renewable sources, including waste-based biofuels. Similarly, in the US, the Renewable Fuel Standard (RFS) program mandates the use of renewable fuels, including those derived from waste, in transportation fuels.

Food waste is a significant global challenge, but it also presents an opportunity to produce clean energy through mass-scale biofuel production. By diverting organic waste from landfills and converting it into biogas, biomethane, or other biofuels, we can reduce greenhouse gas emissions, provide renewable energy, and generate economic benefits. With the support of policies and incentives, the use of food waste for biofuel production could become a significant contributor to the transition from a high-carbon economy to green fuels. Some of the recent achievements and progress made in biofuels production using food waste are discussed in succeeding sections.

1.4 BIO-ALCOHOL FEEDSTOCKS

Various wastes from processing agricultural materials have been demonstrated to have a high potential to be converted into new biofuels to diminish the use of conventional fuels. Among these alternative fuels include two important bioalcohols: biobutanol, bioethanol, and biomethanol. Bioalcohols are an alternative to old fuels, and it is well known that butanol is an attractive alternative to gasoline; this alcohol has better features than other biological alcohols when used as biofuel, mainly due to its high-octane number, higher energy content, lower corrosivity, and poor water solubility. Biobutanol is obtained from dilute acid hydrolysis of agricultural waste to produce fermentable sugars. Three species of *Clostridium* are involved, including *C. acetobutylicum, C. beijerinckii*, and *C. saccharoperbutylacetonicum*.

The most used agricultural wastes have been wheat starch liquid, cauliflower waste, de-oiled rice bran, rice straw, mango peels, corn fiber, wood pulp, brown seaweed, pineapple waste juice, palm oil mill waste, and crude sugarcane bagasse. The higher concentration of produced biobutanol has been found in wood pulp through alkali extraction, acidic hydrolysis, and fermentation using adsorption detoxification as a separation method.

The oversupply of bread products contributes significantly to food waste generation, particularly bread waste, which is incinerated or dumped. In both cases, the energy requirements are highly inefficient and harmful to the environment. During the last decades, the production of biofuels using non-expensive substrates has been focused on lignocellulosic biomass as the key substrate. However, many limitations have been found, including low yields and high costs of the bioprocess. Bread waste has gained attention as an alternative substrate for the production of biobutanol via fermentation. The bioprocess is characterized by a shorter acidogenesis phase and high bacterial and enzyme activity compared to when glucose is the substrate. This corresponds with great butanol production yields and higher Acetone-Butanol-Ethanol productivity. In conclusion, it is possible to consider that bread waste is viable for butanol production, providing a higher efficiency method for energy production to cover energy requirements, and establishing a safer alternative to treat bread waste (Joseph et al., 2021).

Milk whey is another substrate to produce biofuels because it is a main by-product in the production of cheese and casein. This waste should be disposed of or processed in a greener way. An advantage of obtaining biofuels is linked to the substantial increase in cheese production, generating high whey amounts. The treatment of dairy waste leads to minimizing the biochemical oxygen demand, and the bioethanol produced is used as fuel for automobiles and power generation (Rajarajan et al., 2016).

Excellent biomasses are represented by distillery and brewery industrial waste, however, in some of them, the use of additional enzymes or pretreatments is needed.

1.5 BIODIESEL FEEDSTOCKS

The fat and oil content of animal by-products can be converted into biodiesel through a process called transesterification, which is also preferred over vegetable oils because biodiesel can have a superior flashpoint, higher cetane number, more stability against oxidation, lower CO production during the combustion process, lower levels of nitrated compounds, less particulate matter, and lower unburned hydrocarbon or aromatic compounds (Andreo-Martínez et al., 2022).

Biodiesel production from meat waste is also beneficial to the environment because this kind of waste is biologically unstable if not properly managed due to its negative impact on soil, water, and air. In addition, meat waste is a really good biomass resource due to the absence of costs like cultivation, harvesting, and land area necessary to obtain traditional oil resources for transesterification (Okoro, Sun, & Birch, 2019).

A clear example of the use of meat waste is the Polish Eco-Industrial Park in Śmiłowo in the Notéc Valley region, which processes 300,000 tons of meat waste per year to produce 110,000 tons of meat bone meal biofuel, reaching a production of 460,000 GJ of bioenergy and managing to eliminate 92,000 tons of CO_2 emissions (Kowalski et al., 2023).

To improve the traditional biodiesel production from palm oil, which yields a product extremely similar to normal diesel, palm oil can be combined with animal fat such as fat from colored tropical goats. The oil extracted from goat fat reaches 56.81%. Then, the ultrasound-assisted process allows for the production of biodiesel with a 78.5% conversion rate and 98.69% purity (Kumoro & Saeed, 2022).

Protein hydrolysis and subsequent fermentation can be carried out to obtain highly energetic compounds such as methane or hydrogen, intermediates in biofuel production, or other interesting biochemicals. The initial step in the biofuel production process is hydrolysis. This reaction, which aims to obtain free amino acids or peptides to be converted into volatile fatty acids or methane, depends on the pH value of the medium. The optimal pH value is

7.0, showing that a neutral pH allows for the faster production of free amino acids compared to acidic conditions, by a factor of 5. This is significant to consider given that the protein content in wastewater ranges from 20% to 40% in domestic effluents and from 60% to 90% in food industry wastewater (Duong et al., 2019).

For the anaerobic degradation of proteins, prior solubilization is necessary. Protein solubilization from chicken breast waste can be facilitated by an electric pulse technique developed by Ghosh et al. This technique involves applying high voltage, short pulses initially, followed by low voltage, long pulses, with a total energy input of 38.4±1.2 Jg^{-1}, resulting in the release of approximately 12% of the chicken waste biomass (Ghosh et al., 2019).

The use of anaerobic bacteria to do protein catabolism for direct biofuel production is an option to produce molecules like hydrogen, ethanol, butanol, and butyrate since amino acids. A good example of this is *Clostridium sticklandii* which can catabolize these kinds of molecules resulting in a really good metabolic model at the genome scale to determine if *Clostridium* species are an excellent candidate for the production of biofuel at an industrial scale (Sangavai & Chellapandi, 2017).

1.5.1 MICROALGAL WASTE

The oleochemical industry demands better oil resources than animal fats as a feedstock to produce biofuels. Oleaginous microbes are a really good option to achieve this due to their oil content (20–70%). Microalgae are considered good oil producers that can grow in autotrophic or heterotrophic conditions. This is good because, under autotrophic conditions, it is necessary to have large areas for cultivation, which is why it is better to install heterotrophic processes. However, in these conditions, there is still a problem related to the high cost (60% of total production cost) of the substrate to feed the microalgae (Patel et al., 2022). To solve this problem, the use of wastewater from food, agricultural, or beverage industries is suggested. This is not just a good strategy to produce biofuel but also helps with environmental improvement through the bioremediation of wastewater with microalgae. Wastewater from the agricultural or food industry is especially rich in nutrients like potassium, phosphorus, nitrogen, and organic carbon, providing good conditions for the growth of microalgae. Additionally, microalgae are microorganisms with high CO_2 usage efficiency, a short lifespan, and a high growth rate (Ali et al., 2022).

Microalgae like *Arthrospira platensis* can be used directly in biofuel production due to their content of carbohydrates (10–20%) and lipids (6–13%). Additionally, there is a protein fraction that can be utilized for human or animal nutrition. Furthermore, biodiesel production can be enhanced by directly using the lipid fraction for transesterification. As a result, a significant amount of deoiled and deproteinized residue is generated from the biofuel and food supplements industry. This residue can be further processed to obtain fuel liquid, gas, and char through pyrolysis, resulting in superior bio-oils compared to those obtained from lignocellulosic residues. This is attributed to their high stability, heating value, and lower oxygen content (Casazza et al., 2020).

1.5.2 SPENT COFFEE GROUNDS

Coffee consumption as a beverage demands around 170 million 60 kg bags per year (Lee et al., 2021; Durán-Aranguren et al., 2021). The SCG is a contaminant because its decomposition produces large amounts of CO_2 and methane (Lauberts et al., 2023); its use as a feedstock for biorefineries can diminish the environmental impact, and its value can be maximized by converting biomass into various types of valuable products, such as biofuels, biochemicals, and bioplastics (Lee et al., 2021), achieving a sustainable utilization of waste (Banu et al., 2020).

Spent coffee grounds have a high content of organic matter, including cellulose 8.6–15.3%, hemicellulose 31.7–41.7%, and lignin 22.2–33.6% (Lee et al., 2021). These compounds can be converted into a variety of valuable products through various biorefinery processes, such as fermentation, hydrolysis, and pyrolysis.

Spent coffee grounds (SCG) have been increasingly recognized as a promising feedstock for biorefineries due to their low-cost, non-edible waste, and high amounts generated annually (Araujo et al., 2022). SCG oil content is about 15%, of which 85.85% consists of fatty acids, mainly linoleic, oleic, stearic, and palmitic acids (Kafková et al., 2023; Lauberts et al., 2023).

1.6 GLOBAL AND MEXICAN ACHIEVEMENTS AND PROGRESS

In recent years, the biofuel production agenda has been of great interest in developed countries. According to the International Energy Agency (https://www.iea.org/), biofuels are an important strategy in the decarbonization of the transport sector. However, it is necessary to increase the production of liquid, gaseous, and solid biofuels to move towards the Zero Net Emissions Scenario by 2050. The Renewable Fuels Association (RFA) in this year 2023 has published the Ethanol Industry Outlook (https://ethanolrfa.org/) and currently, 199 biorefineries are operating for the production of biofuels first-generation ethanol from corn, with a production of 15,400 million gallons in the year 2022. Ethanol production in the United States is based mainly on corn starch (94%); cellulosic biomass/starch (4%), corn/sorghum/wheat starch (2%), and waste sugars/starch (<1%). It is important to mention that the production of this biofuel in 2022 had a significant social/economic impact in the United States (Direct Jobs 78,802). Without a doubt, this ethanol industry will be able to achieve net-zero carbon emissions, seeking as an important strategy the development of biorefineries and employment of renewable energies, with cogeneration in the processes themselves (https://ethanolrfa.org/). Related to the second country with the highest production worldwide, Brazil produces bioethanol from sugarcane (7,420 million gallons in 2022). There are currently 269 first-generation ethanol production plants operating (https://apps.fas.usda.gov/). The other main biofuel producers are the European Union: 1,330 (5%), India: 1,090 (4%), China: 920 (3%), Canada: 460 (2%), Thailand: 380 (1%), Argentina: 290 (1%), Rest of World: (870) 3% [million gallons, (world percentage)] (https://ethanolrfa.org/).

On the other hand, the use of advanced raw materials should be an important alternative in the production of biofuels, in this case, agro-industrial residues (lignocellulosic biomasses), also known as second-generation biorefineries and third-generation aquatic biomass. In addition, Food Waste (FW) becomes an important alternative in the production of liquid or gas biofuels such as bioethanol, biodiesel, and biogas. In a recent work, Arya et al. (2022) reported that 33% of food produced is wasted. These organic materials such as FW are mainly composed of proteins and carbohydrates, so the amount of this biomass and chemical composition produced makes it a promising material for the production of biofuels. Figure 1.1 shows alternative ways to obtain biofuels in liquid or gaseous form from FW, as well as the production of enzymes.

The production of biofuels at an industrial level still has important limitations, from economic viability and biomass availability, but the employment of several new raw materials as an FW will lead to rethinking the design of more robust processes for the development of advanced biofuels: first, second, and third generation for the benefit of society and decarbonization reduction.

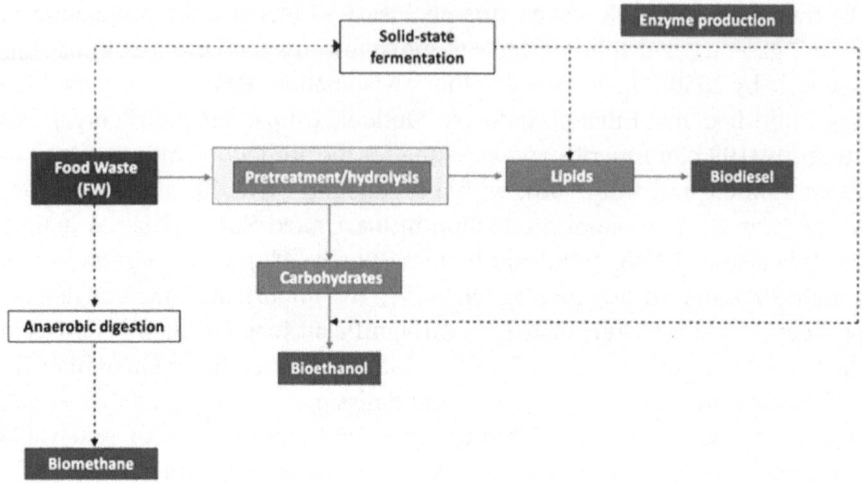

FIGURE 1.1 Production of bioethanol, biodiesel, and biomethane from food waste.

The interest in biofuels in Mexico is very recent (2016) and is mainly due to an effort to reduce CO_2 emissions into the atmosphere. The Mexican government approved in 2008 the Bioenergetics Act that promotes the use of bioethanol and biodiesel in fuels sold by Petróleos Mexicanos (PEMEX), the most important state-owned industry that explores, produces, refines, and transforms crude oil (Barrera et al., 2016).

Ethanol production in Mexico in 2019 was 144 million liters, and only 37% were from biofuels primarily from sugar cane fermentation (CEDRSSA, 2023). Ethanol production from first-generation feedstocks such as sugar cane, corn, and sorghum has several disadvantages due to ethanol competing with food production since corn crops and sugarcane are also used for food production. Increased demand for these crops for ethanol production can drive up food prices and create competition between food and fuel production (Escamilla-Alvarado et al., 2017). Also, large-scale production of first-generation feedstocks can lead to land use change, deforestation, and other environmental impacts. This can contribute to habitat loss, biodiversity loss, and greenhouse gas emissions (Escamilla-Alvarado et al., 2017).

Therefore, biofuel production from agro-industrial waste is a great challenge for ensuring food security. Mexico is the fifth country in the search for biofuels from agro-industrial waste as a sustainable form of energy. Of the 72 documents published in Scopus from the search "agro-industrial waste" and "biofuels," 46 documents are articles. Around 17% of the documents are focused on producing hydrogen, 20% on producing methane, and 13% on producing ethanol.

The national outlook on projects in methane generation from municipal waters is encouraging. Currently, there are 25 wastewater treatment plants involved in the energy recovery of sludge and 12 landfills able to recover biogas to generate power and supply in a few sites (Mediavilla, 2018). It is important to mention that municipal wastes have a distribution where only 46% are organic wastes suitable for methane production. These consist of 29% food waste, 14% yard trimmings, and 4% other organic waste (Mediavilla, 2018).

The utilization of organic wastes to produce biofuels presents an appealing option due to its potential to address waste management challenges, reduce environmental pollution, and offer an alternative energy source that can promote sustainable development (Pagani, 2021). However, this approach is not without challenges. One significant hurdle is the production cost, which can be higher compared to biofuels derived from first-generation feedstocks. Additionally, the process of biofuel production from organic wastes demands considerable energy and water resources, which can be a limiting factor in some areas (Dufosse et al., 2017). In summary, producing biofuels from organic wastes is a promising solution for sustainable energy production and waste management, but its economic, social, and environmental impacts should be carefully considered to ensure its long-term viability and sustainability.

In 2018, the global biofuels market was valued at $68.18 billion and is expected to reach $218.7 billion by 2025. According to a report by the World Bioenergy Association, biofuel production from waste biomass has the potential to contribute up to 17% of the global transportation fuel demand by 2050. The European Union (EU) has set a target to increase the share of renewable energy in transport to 10% by 2020, and biodiesel production from waste oil and fats is one of the ways to achieve this target. The United States (US) Renewable Fuel Standard Program mandates the use of a minimum volume of biofuels, including advanced biofuels, in the transportation sector. China is the world's largest ethanol producer, and it has recently introduced a policy to promote the use of biofuels in the transportation sector. In 2020, the Mexican Government announced that it would invest $4.5 billion in the development of biofuels, including ethanol and biodiesel, over the next two decades.

In 2018, the Mexican Government launched an initiative to promote the use of renewable energy, including biofuels, in the transport sector. Several

Mexican companies, such as Biofield and Beta Renewables, are investing in the development of biofuels from food waste, such as corn stover and sugarcane bagasse. The Mexican Government has set a target to increase the share of biofuels in the transport sector to 10% by 2024. The Mexican state of Jalisco has launched a program to convert food waste into biofuels, aiming to promote environmental sustainability and create new job opportunities.

It is important to note that using food waste as a feedstock for biofuels has great potential for reducing waste, promoting sustainability, and mitigating climate change. The global and local achievements and progress in this field demonstrate the growing interest and commitment to biofuel production from food waste.

KEYWORDS

- **biofuels**
- **climate change**
- **energy resources**
- **environmental sustainability**
- **food waste**
- **methane**
- **renewable energy**
- **renewable fuel standard program**
- **sugarcane bagasse**

REFERENCES

Ali, S. S., Mastropetros, S. G., Schagerl, M., Sakarika, M., Elsamahy, T., El-Sheekh, M., Sun, J., & Kornaros, M. (2022). Recent advances in wastewater microalgae-based biofuels production: A state-of-the-art review. *Energy Reports, 8*, 13253–13280. https://doi.org/10.1016/j.egyr.2022.09.143.

Andreo-Martínez, P., Ortiz-Martínez, V. M., Salar-García, M. J., Veiga-del-Baño, J. M., Chica, A., & Quesada-Medina, J. (2022). Waste animal fats as feedstock for biodiesel production using non-catalytic supercritical alcohol transesterification: A perspective by the PRISMA methodology. *Energy for Sustainable Development, 69*, 150–163. https://doi.org/10.1016/j.esd.2022.06.004.

Araujo, M. N., dos Santos, K. C., Diniz, N. C., de Carvalho, J. C., & Corazza, M. L. (2022). A biorefinery approach for spent coffee grounds valorization using pressurized fluid extraction to produce oil and bioproducts: A systematic review. *Bioresource Technology Reports, 18*, 101013. https://doi.org/10.1016/j.biteb.2022.101013.

Arya, P. S., Yagnik, S. M., Rajput, K. N., Panchal, R. R., & Raval, V. H. (2022). Valorization of agro-food wastes: Ease of concomitant-enzymes production with application in food and biofuel industries. *Bioresource Technology, 361*, 127738. https://doi.org/10.1016/j.biortech.2022.127738.

Banu, J. R., Kavitha, S., Kannah, R. Y., Kumar, M. D., Preethi, Atabani, A. E., & Kumar, G. (2020). Biorefinery of spent coffee grounds waste: Viable pathway towards circular bioeconomy. *Bioresource Technology, 302*, 122821.

Barrera, I., Amezcua-Allieri, M. A., Estupiñan, L., Martínez, T., & Aburto, J. (2016). Technical and economical evaluation of bioethanol production from lignocellulosic residues in Mexico: Case of sugarcane and blue agave bagasse. *Chemical Engineering Research and Design, 107*, 91–101.

Biofuels Annual. (n.d.). Retrieved from: https://apps.fas.usda.gov/ (accessed on 28 May 2024).

Casazza, A. A., Spennati, E., Converti, A., & Busca, G. (2020). Production of carbon-based biofuels by pyrolysis of exhausted Arthrospira platensis biomass after protein or lipid recovery. *Fuel Processing Technology, 201*, 106336. https://doi.org/10.1016/j.fuproc.2020.106336.

Centro de Estudios para el Desarrollo Rural Sustentable y la Soberanía Alimentaria (CEDRSSA). (2023). La producción y el comercio de los biocombustibles en México y en el mundo.

Dufosse, K., Aoun, W. B., & Gabrielle, B. (2017). Life-cycle assessment of agricultural feedstock for biorefineries. In *Life-Cycle Assessment of Biorefineries*.

Duong, T. H., Grolle, K., Nga, T. T. V., Zeeman, G., Temmink, H., & Van Eekert, M. (2019). Protein hydrolysis and fermentation under methanogenic and acidifying conditions. *Biotechnology for Biofuels, 12*(1), 1–10. https://doi.org/10.1186/s13068-019-1592-7.

Durán-Aranguren, D. D., Robledo, S., Gomez-Restrepo, E., Arboleda Valencia, J. W., & Tarazona, N. A. (2021). Scientometric overview of coffee by-products and their applications. *Molecules, 26*(24), 1–25.

Escamilla-Alvarado, C., Pérez-Pimienta, J. A., Ponce-Noyola, T., & Poggi-Varaldo, H. M. (2017). An overview of the enzyme potential in bioenergy-producing biorefineries. *Journal of Chemical Technology & Biotechnology, 92*(5), 906–924.

Ethanol Industry Outlook. (n.d.). Retrieved from: https://ethanolrfa.org/ (accessed on 28 May 2024).

Ghosh, S., Gillis, A., Sheviryov, J., Levkov, K., & Golberg, A. (2019). Towards waste meat biorefinery: Extraction of proteins from waste chicken meat with non-thermal pulsed electric fields and mechanical pressing. *Journal of Cleaner Production, 208*, 220–231. https://doi.org/10.1016/j.jclepro.2018.10.037.

International Energy Agency. (n.d.). Retrieved from: https://www.iea.org/ (accessed on 28 May 2024).

Joseph, Q. S., Justin, K. W., Jovian, H. Y., Guandong, S., & Jianzhong, H. (2021). Biobutanol biofuel production via ABE fermentation from bread waste for an energy-sustainable Singapore.

Kafková, V., Kubinec, R., Mikulec, J., Variny, M., Ondrejíčková, P., Ház, A., & Brisudová, A. (2023). Integrated approach to spent coffee grounds valorization in biodiesel biorefinery. *Sustainability, 15*(5612), 1–17.

Kowalski, Z., Kulczycka, J., Makara, A., Mondello, G., & Salomone, R. (2023). Industrial symbiosis for sustainable management of meat waste: The case of Smiłowo. *International Journal of Environmental Research and Public Health, 20*(5162), 1–22. https://doi.org/10.3390/ijerph20065162.

Kumoro, A. C., & Saeed, M. T. M. N. (2022). Ultrasound-assisted transesterification of tropical goat fat – palm oil blend for biodiesel synthesis. *Energy Conversion and Management: X, 14*(January), 100213. https://doi.org/10.1016/j.ecmx.2022.100213.

Lauberts, M., Mierina, I., Pals, M., Abdul Latheef, M. A., & Shishkin, A. (2023). Spent coffee grounds valorization in biorefinery context to approaches and solvents. *Plants, 12*(30), 1–14.

Lee, K. H., Jang, Y., Lee, J., Ki, S., Park, C., & Yoo, H. Y. (2021). Statistical optimization of alkali pretreatment to improve sugars recovery from spent coffee grounds and utilization in lactic acid fermentation. *Processes, 9*(494), 1–15.

Mediavilla Merchan, V. (2018). *Waste-to-energy in Mexico*. Technical potential for biogas production and greenhouse gas mitigation from the anaerobic digestion of municipal solid waste (Master's thesis). Universiteit Utrecht, Alemania.

Okoro, O. V., Sun, Z., & Birch, J. (2019). Techno-economic assessment of a scaled-up meat waste biorefinery system: A simulation study. *Materials, 12*, 1–27. https://doi.org/10.3390/ma12071030.

Pagani, D. (2021). The importance of green mobility. In *Distributed Renewable Energies for Off-Grid Communities* (pp. 345–391). Elsevier.

Patel, A., Krikigianni, E., Rova, U., Christakopoulos, P., & Matsakas, L. (2022). Bioprocessing of volatile fatty acids by oleaginous freshwater microalgae and their potential for biofuel and protein production. *Chemical Engineering Journal, 438*, 135529. https://doi.org/10.1016/j.cej.2022.135529.

Rajarajan, G., Irshad, A., Raghunath, B. V., Mahesh Kumar, G., & Punnagaiarasi, A. (2016). Utilization of cheese industry whey for biofuel–ethanol production. In M. Prashanthi & R. Sundaram (Eds.), *Integrated Waste Management in India* (pp. 121–142). Springer. https://doi.org/10.1007/978-3-319-27228-3_6.

Romero-Cedillo, L., Poggi-Varaldo, H. M., Ponce-Noyola, T., Ríos-Leal, E., Ramos-Valdivia, A. C., Cerda-García Rojas, C. M., & Tapia-Ramírez, J. (2017). A review of the potential of pretreated solids to improve gas biofuels production in the context of an OFMSW biorefinery. *Journal of Chemical Technology & Biotechnology, 92*(5), 937–958.

Rosero-Chasoy, G., Rodríguez-Jasso, R. M., Aguilar, C. N., Buitrón, G., Chairez, I., & Ruiz, H. A. (2023). Hydrothermal kinetic modeling for microalgae biomass under subcritical condition cultivated in a close bubble tubular photobioreactor. *Fuel, 334*, 126585.

Sangavai, C., & Chellapandi, P. (2017). Amino acid catabolism-directed biofuel production in *Clostridium sticklandii*: An insight into model-driven systems engineering. *Biotechnology Reports, 16*, 32–43. https://doi.org/10.1016/j.btre.2017.11.002.

CHAPTER 2

Advanced Oxidation Processes in the Treatment of Effluents Contaminated with Dyes

SARA ANDREA ÁLVAREZ-ALVARADO, JUDITH AMADOR-HERNÁNDEZ, and MIGUEL VELÁZQUEZ-MANZANARES

Facultad de Ciencias Químicas, Department de Química Analítica, Universidad Autónoma de Coahuila, Saltillo, Coahuila, Mexico

ABSTRACT

The pollution of water by the textile industry dyes is an increasing problem because of its accelerated growth over time. Therefore, the treatment of effluent is necessary. Among these, advanced oxidation processes (AOPs) are commonly used alternatives for dye elimination because they prioritize the environment and healthcare. However, dyes possess high chemical stability, making them persistent pollutants. Thus, methodologies for the treatment of these effluents are subjected to continuous development and innovation to improve their operational costs and efficiency. Unfortunately, most conventional methods for dye degradation have room for improvement, as they have several drawbacks. In contrast, the research and development focused on the improvement of AOPs while maintaining sustainability has increased in recent years. In this review, the utility of four AOP methodologies – ozonation, the Fenton process, electrochemical oxidation, and electrocoagulation – is discussed in terms of their operational conditions, advantages, disadvantages, and recent applications in the treatment of wastewater effluents from the textile industry.

Sustainable Agriculture and Global Environmental Health: Toward the Efficient Management of Resources. Mónica L. Chávez González, Cristóbal Noé Aguilar, Juan Carlos Contreras-Esquivel, and A. K. Haghi (Eds.)

2.1 INTRODUCTION

Currently, the textile industry is one of the most polluting industries; as part of its activities, it releases wastewater containing numerous types of dyes. This industry includes clothing manufacturing. The discharge of this water into municipal drainage impacts the environment and human health. Therefore, wastewater treatment is of great importance because it minimizes the consequences. From the physicochemical point of view, these waters present high levels of chemical oxygen demand (COD), biochemical oxygen demand (BOD), total organic carbon (TOC), and dissolved solids, and it is considered that water with a concentration of 1 mg/L of dye can no longer be reused for other purposes (Nippatlapalli & Philip, 2021). Recently, Islam et al. (2023) addressed with great concern the impact of wastewater discharge containing dyes from certain industrial activities on health and the imbalance in ecosystems. Undoubtedly, the increase in industrial activity of this type of company is linked to population growth, which demands more consumer goods, particularly clothing and other types of household goods.

Unfortunately, in fabric dyeing processes, dyes are not fully utilized, generating wastewater with high concentrations of these pollutants. The treatment of effluents contaminated with these dyes is a complex procedure because of the chemical stability of these organic molecules; therefore, even after treatment, a percentage of the dye remains in the wastewater. Certainly, this varies depending on the method used; some turn out to be better than others, and the results of treatment training can be beneficial. However, at the industrial level, this does not depend solely on the applied procedure; costs, energy, equipment, supplies, and many other important variables must be considered. Biological processes are not capable of degrading dyes, and physicochemical methods are not efficient enough to be considered a good option; advanced oxidation processes are the ones that are best cataloged in this type of application (Fadzli et al., 2022; Nippatlapalli & Philip, 2021; Senthil & Saravanan, 2017).

Advanced oxidation processes (AOPs) involve the generation of hydroxyl radicals ($OH^{\bullet}$) that react with organic molecules to obtain inorganic ions, CO_2, and water at the end of the treatment. AOPs are primarily used to remove organic compounds; thus, wastewater of different industrial natures can be treated. In fact, when any of these processes are used for pretreatment, the aim is to condition the effluents so that the next step is more efficient. However, when implemented as a second or third treatment, hardly any biodegradable contaminants can be eliminated using traditional methods (Korpe & Rao, 2021; Oturan & Aaron, 2014).

Various authors have discussed the advantages and disadvantages of traditional techniques for treating effluents with dyes (Siddiqui et al., 2019; Zhou et al., 2019). These techniques include the use of adsorbents (zeolites, clays, and biosorbents such as chitosan), which have been shown to be options for the treatment of this type of contaminant (Zhou et al., 2019). In this type of technology, it is important to consider several parameters that can affect efficiency, such as pH and temperature (Katheresan, 2018). A diagram of the general advantages and disadvantages of AOPs is shown in Figure 2.1.

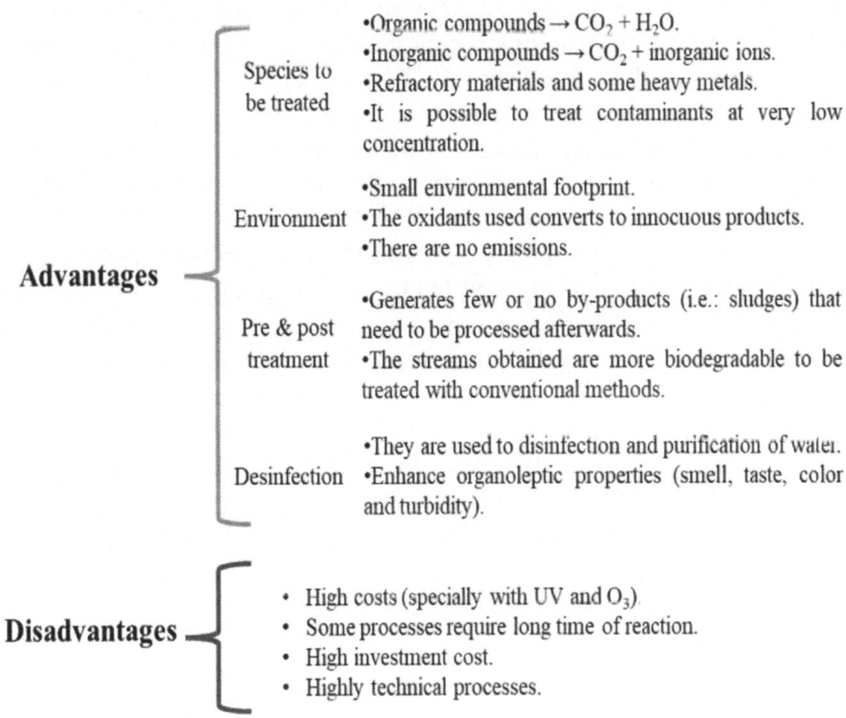

FIGURE 2.1 Advantages and disadvantages of AOPs.

Source: Adapted from Cong, 2016; Korpe & Rao, 2021; Monge et al., 2018

This review discusses recent studies regarding the treatment of water contaminated with dyes, as well as textile effluents treated by advanced oxidation processes, specifically electrochemical oxidation, ozonation, the Fenton process, and electrocoagulation.

2.2 METHODS TO TREAT CONTAMINATED EFFLUENTS

2.2.1 OZONATION

The strongest oxidants are hydroxyl radicals and ozone (Korpe & Rao, 2021). Through ozonation, organic molecules can be degraded and transformed into smaller ones. Chemically, ozone acts on functional groups rich in electrons (Fadzli et al., 2022; Nippatlapalli & Philip, 2021). Therefore, this methodology is widely used to destroy structurally stable organic contaminants such as azo dyes. The reaction with O_3 can occur directly with the organic substances of interest, generally at acidic pH, or indirectly due to its instability, generating $OH^{\bullet}$ radicals that attack organic matter, triggering the reaction, which is observed at basic pH. However, the ozonation method is commonly used as part of a tertiary treatment for the destruction of dyes due to its high cost. This is owing to the fact that the half-life of ozone is short, which continuously requires a higher concentration of this precursor in comparison with other methodologies (Fadzli et al., 2022).

To enhance the degradation capacity of O_3, it is combined with catalysts such as Fe, Co, TiO_2, Al_2O_3, and H_2O_2, as well as the use of UV radiation to increase its efficiency (Korpe & Rao, 2021; Nippatlapalli & Philip, 2021). Its applications in the preliminary phases are mainly disinfection, detoxification, and sludge reduction. As in most processes, the use and concentration of O_3 depend on what is to be treated. In the case of dyes, the chemical structure and COD levels of the effluents determine the amount of ozone to be used. During the course of the process, O_3 decomposes in an aqueous medium to generate $OH^{\bullet}$ free radicals; the reactions that occur are shown below:

$$2O_3 + H_2O \rightarrow OH^* + O_2 + HO_2 \quad (1)$$

$$O_3 + HO_2^- \rightarrow HO_2 + O_3^- \quad (2)$$

$$O_3^- +.H_2O \rightarrow OH^* + OH^- + O_2^* \quad (3)$$

Table 2.1 shows some examples of the ozonation treatments of dyes and/or textile effluents.

TABLE 2.1 Ozonation in the Degradation of Textile Dyes

Technique	Dye	Ci (mg/L)	pH	Tt (min.)	COLR (%)	CODR (%)	References
O_3/anaerobic treatment	Reactive black 5	1,500	–	25	94	90	Venkatesh et al. (2017)
O_3/UV	Acid red 17	100	11	25	100	–	El Nemr et al. (2018)
O_3 (biological post-treatment)	Reactive red 239	50	7	20	100	62	Dias et al. (2019)
O_3 and O_3/Ni	Methyl orange	500	9	60	96	30 and 70	El Hassani et al. (2019)
O_3/$MnFe_2O_4$	Reactive Blue 19	200	–	10	100	60	Hu et al. (2019)
O_3	Reactive orange 16, reactive red 120, direct red 80	500	11	30	100	–	Muniyasamy et al. (2020)
O_3/activated carbon catalyst	Reactive black 5	842	11.26	30	–	39.5	Bilinska et al. (2020)
O_3/zeolites	Reactive black 5	1,000	10	30	90.31	–	Ikhlaq et al. (2021)
UV/O_3/SBR	Reactive red 198	50–300	5–9	60	94	–	Mahdizadeh et al. (2021)
O_3/H_2O_2	Procian blue	60	10	–	92	–	Bhad et al. (2022)
O_3/UV (125 W)	Azo red-60	100	5–10	10–60	100	–	Shahamat et al. (2022)

Source: Initial concentration (C_i): treatment time (T_t); color removal (COL_R); and chemical oxygen demand (COD_R).

A possible disadvantage of the technique is the rate of reaction; COD reduction is slow and can take several hours for the process to take place. Furthermore, according to Nippatlapalli & Philip (2021), ozonation is not useful for all types of organic dyes because, in many cases, the mineralization of the pollutant is not reached; in cases where they are completely degraded, a considerably high value of pollutant removal is achieved, as well as a reduction in TOC; otherwise, the procedure takes a long time. In a critical review of ozonation applications, Lim et al. (2022) discussed the mechanistic processes of the ability of ozone to degrade different types of organic compounds, demonstrating the importance of this technique in the treatment of polluted water. Similarly, Cong et al. (2016) found in laboratory studies that the combination of ozonation with a bioreactor can make the treatment of water contaminated with highly recalcitrant compounds more efficient as the required O_3 concentration is reduced.

Thus, the treatment of textile effluents by ozonation can be expensive because of the high levels of ozone consumed; however, coupling it with other techniques or catalysts can make it cheaper and improve its performance. This is the case for the procedure described by El-Nemr et al. (2018), where the treatment for a reactive dye improves when O_3/UV is applied simultaneously, unlike with ozone alone, thereby increasing the rate of discoloration and reducing the toxicity of the effluent to the zooplankton. It has also been combined with hydrogen peroxide (O_3/H_2O_2) to increase the degradation capacity for a large number of dyes and other contaminants in aqueous media (Fadzli et al., 2022; Merle et al., 2017).

2.2.2 *FENTON PROCESS*

The Fenton process has been proposed for many years (Fenton, 1984); thus, it is now widely known. In this process, the formation of hydroxyl radicals is also involved; the redox reactions that are observed in this process are detailed in the following equations (Ameta et al., 2018; Nippatlapalli & Philip, 2021):

$$Fe^{2+} + H_2O_2 \rightarrow Fe^{3+} + HO_2 + H^+ \tag{4}$$

$$Fe^{2+} + H_2O_2 \rightarrow Fe^{3+} + OH^* + OH^- \tag{5}$$

This process has been found to be efficient under acidic pH conditions. Similarly, in the laboratory study of the discoloration of Drimaren X-6BN with hematite, it was found that the efficiency of the Fenton process depended to a great extent on the acidity of the medium (Araujo et al., 2011).

The Fenton process coincides with ozonation in the following ways: a) it is necessary to combine it with solid catalysts to increase the surface area and improve adsorption, and b) it is coupled with UV radiation, ultrasound (US), or other energy sources to improve the degradation performance. The Fenton process without a combination of solid catalysts is inefficient, resulting in high energy costs. The ferrous ion (Fe^{2+}) is a crucial element in Fenton processes and related techniques because it acts as a catalyst for the generation of hydroxyl radicals (Araujo et al., 2011). An increase in its concentration generates a higher rate of pollutant degradation; however, an excess of Fe^{2+} can be counterproductive because of the excess sludge that is produced, in addition to the fact that the degradation rate tends to decrease (Aoudj et al., 2021). Therefore, it is believed that the catalysts contain Fe ions on their surfaces, among which hematite, goethite, and magnetite are highlighted (Araujo et al., 2011; Scheres et al., 2018).

In addition, it has been found that the Fenton and photo-Fenton processes have better efficiencies than other techniques, such as US, UV, and O_3. Some of their main advantages are that they are environmentally friendly methodologies, Fe^{2+} and Fe^{3+} ions are not considered toxic, and a smaller amount of sludge is generated compared to conventional treatment owing to the oxidation and coagulation in the process (Goi & Trapido, 2002; Korpe & Rao, 2021). Unfortunately, it is difficult to treat, and pH-dependent sludge is generated, leading to an increase in its overall cost.

With this technique, an average of 90% of the COD can be reduced in 90 min during Erythrosine B treatment, which is considered a process with rapid kinetics (Clematis & Panizza, 2021). However, similar to ozonation, the degradation kinetics of the Fenton process can vary depending on the physicochemical properties of the contaminants. Other variants of the Fenton process include the photo-Fenton, electro-Fenton (EF), photoelectro-Fenton (PEF), solar photoelectro-Fenton (SPEF), and sonoelectro-Fenton (SEF) processes. All of them follow the same principle as the Fenton reaction and agree on the disadvantage of requiring an acidic medium; acidification (pretreatment) and neutralization (post-treatment) must be taken into account. In contrast, EF and its different versions have the advantage of electrochemically regenerating the catalyst owing to the reduction of Fe^{3+} in the cathode, thus avoiding the addition of a large amount of Fe^{2+} into the system, unlike the conventional Fenton process (Aoudj et al., 2021; Clematis & Panizza, 2021; He & Zhou, 2017). Table 2.2 lists some studies carried out by other researchers that used the Fenton process for the degradation of dyes and/or textile effluents.

Some observations to note from the Fenton process are:

- The performance improves when the hydrogen peroxide concentration is increased, although this is not as noticeable.
- The use of iron oxides causes the uncontrolled leaching of Fe from the catalyst surface, favoring sludge formation.

TABLE 2.2 Fenton Process in the Degradation of Textile Dyes

Technique	Dye or Effluent	Ci (mg/L)	pH	Tt (min)	COLR (%)	CODR (%)	References
Electro-Fenton: carbon fiber (cathode) and stainless steel. (anode)	Real textile effluent	–	2	180	85	55	Ramirez-Pereda et al. (2020)
Fe^{2+} (0.5 mM)/ H_2O_2 (50 mg/L)	Blue 19 red 195 yellow 145	75	–	20	22–52	–	Tavares et al. (2020)
Fe^{2+} (3.6 mM/L)/ H_2O_2 (194 mM/L)	Acid red 14	1	3	20	100	–	Wakrim et al. (2021)
Electro-Fenton, graphite electrodes, Fe^{2+} (2 mM).	Textile industry wastewater	–	3	30–60	89	93	Kuleyin et al. (2021)
Fe^{2+} (150 mg/L)/ H_2O_2 (100 mg/L)	Sudan black B	50	3	30	85	–	Suhan et al. (2021)
Magnetite (0.2 g/L)/ H_2O_2 (0.5 mM)	Reactive Yellow 3	50	3.5	40	95.23	–	Wang et al. (2022)
Fenton-H_3NOH^+	Basic fuchsin	–	3–5	5	80–95	–	Merouani et al. (2022)
Fenton	Reactive yellow 145	–	3	30	90–95	66	Islam et al. (2019)
Sono-Fenton, 80 W power	Congo red dye	25	3.5	60	89	83	Nawaz et al. (2022)
Photo-Fenton/bacterial cellulose/Fe_3O_4	Mixture of textile dyes	–	5	–	92.19	83.24	da Rocha et al. (2023)

- More focused research has been conducted on modified Fenton catalysts and processes.
- Photo-Fenton processes are considered economical options for treating highly contaminated effluents.

2.2.3 *ELECTROCHEMICAL OXIDATION*

The concepts of electrochemical oxidation (EO), electrodegradation, electrochemical degradation, and electrochemical incineration are identical. In all cases, the aim is to oxidize organic molecules and try to reach the mineralization of the compounds (CO_2 + H_2O) or else obtain innocuous molecules.

Electrochemical advanced oxidation processes (EAOP) are widely used to degrade organic dyes present in wastewater. Similar to the other techniques, this process involves the generation of OH• radicals. This is of great interest to the scientific community and industry because it is relatively easy to control operating and automation conditions (Nian et al., 2022; Nidheesh et al., 2020; Santos et al., 2020). In addition to the characteristics of the contaminant, the influential variables in this process are effluent pH, electrolyte concentration, and electrode material. The physicochemical properties of the chemical species to be degraded are determining factors in the choice of the electrode material, generally the anode. For example, PbO_2, boron-doped diamond (BDD), and other materials can provide higher electrocatalytic capacity to electrodes (Abdelhay et al., 2021; Wang et al., 2020; Zhou et al., 2022). Figure 2.2 shows some electrodes commonly used as anodes, their basic characteristics, and the importance of electrolytes.

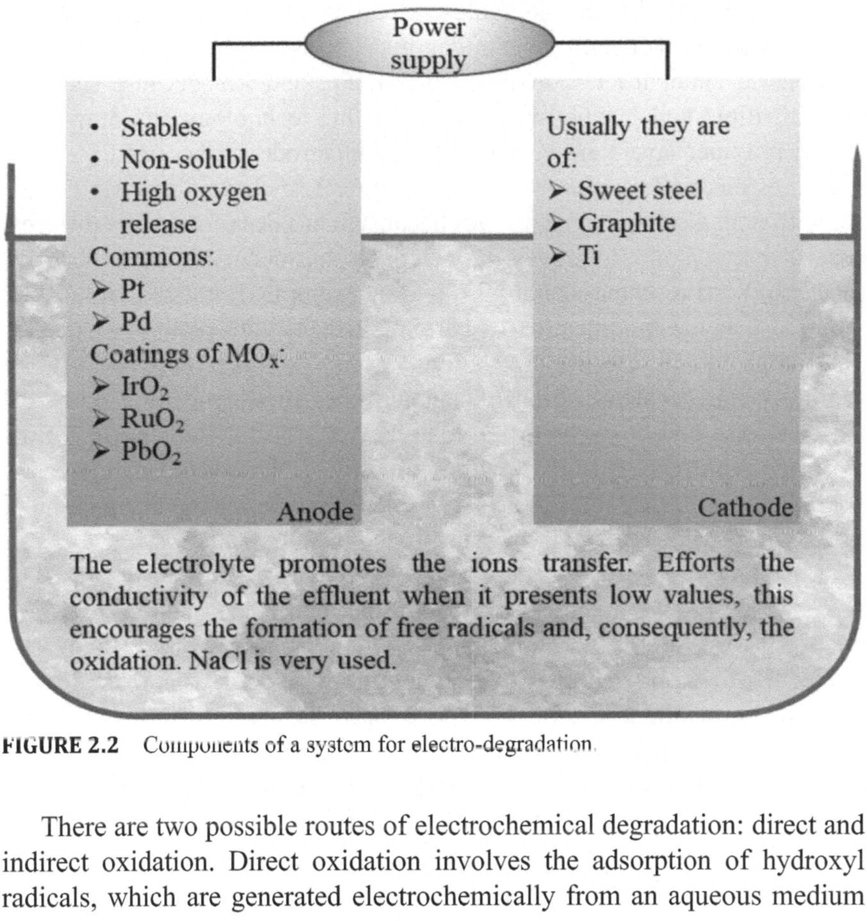

FIGURE 2.2 Components of a system for electro-degradation.

There are two possible routes of electrochemical degradation: direct and indirect oxidation. Direct oxidation involves the adsorption of hydroxyl radicals, which are generated electrochemically from an aqueous medium on the anode surface, and oxidize organic contaminants. In contrast, in indirect oxidation, external oxidizing agents are produced, which may originate in the electrolyte (it is common to find active chlorine, ozone, persulfates, etc.), facilitating the transfer of electrons between organic substances and electrodes (Aoudj et al., 2021; Yasri & Gunasekaran, 2017).

According to Brillas & Martínez-Huitle (2015), there are two possible results for the degradation of organic molecules in EAOP: (a) when they become biodegradable substances; and (b) when complete mineralization is reached (CO_2, inorganic ions). In both cases, the difference in the potential of the electrochemical cell is decisive, such that high values of cell potential improve oxidation, whereas low values diminish it because the phenomenon of adsorption occurs on the surface of the electrode, thus hindering the treatment of the effluent.

Currently, there is considerable interest in the chemical modification of the surfaces of metallic electrodes that act as anodes, such as the use of metallic nanoparticles, mainly Co, Ni, Fe, and Zn, because they are biocompatible and inexpensive. However, this technology has drawbacks in that polymer layers are coated onto the electrode surface, and an effect known as electrode surface poisoning occurs. Tests have also been carried out with doped carbon nanoparticles because it has been found that the presence of heteroatoms improves the reactions that occur on these electrode surfaces (Yasri & Gunasekaran, 2017). For example, Hamous et al. (2021) reported that the modification of carbon surfaces with platinum nanoparticles improved the electrodegradation capacity of orange G dye. Thus, the synthesis and characterization of nanoparticles to chemically modify the flat electrode surface seeks to increase the electrocatalytic capacity of the electrode with respect to that of the unmodified electrode (Belding et al., 2010; Falola, 2022). Table 2.3 lists examples of studies on the degradation of dyes and/or textile effluents using electrochemical oxidation.

Abdelhay et al. (2021) studied the effectiveness of EAOP using BDD at the anode for the treatment of dyes in the textile and paint sectors. The treatment was effective for both effluents, where the use of Na_2SO_4 and NaCl as electrolytes improved the COD reduction in both effluents, from 84% to 94% for the textile effluent and from 21% to 71% and 85% for the paint effluent with Na_2SO_4 and NaCl, respectively, whereas the discoloration of textile effluents was 84% to 96%. Therefore, it is expected that the introduction of a supporting electrolyte improves this type of treatment, although this may vary depending on the characteristics of the sample and/or the chemical properties of the electroactive species.

TABLE 2.3 Electrochemical Oxidation in the Degradation of Textile Dyes

Electrode	Sample	j (mA/cm2)	Ci (mg/L)	pH	tE (min)	COLR (%)	CODR (%)	References
BDD/flow reactor	Ruby F-2B	8.8	20	3	420	94.9	86.3	Pieczyńska et al. (2019)
Graphene-based electrode materials	Reactive black 5	2.5	5	6.6	30	95	–	Li et al. (2019)
BDD/flow reactor	Blue D-5RN	8.8	20	3	120	98.9	99.1	Pieczyńska et al. (2019)
Activated carbon cloth	Textile dye bath wastewater	10	–	10	90	95.5	95.5	Isik et al. (2020)
$BDD_{(anode)}$; $Pt_{(cathode)}$	Reactive brilliant yellow X-6G	100	100	3	120	100	–	Tang et al. (2020)
Ru	Acid orange 74	40	50	–	120	65.7	28.2	Li et al. (2021)
IrO_2						73.8	43.5	
PbO_2						85.1	68.2	
BDD						100	84.3	
Ti/TiO_2, IrO_2, RuO_2 (anode)	Rhodamine-B	15	100	3–5	10	100	–	Kothari et al. (2020)
BDD, photoelectro-Fenton	Mixed of dyes	20	60	3	60	80–97	58–95	Medrano-Rodriguez et al. (2020)
BDD Na_2SO_4 (electrolyte)	Dye-containing wastewater	2.8	–	5.4	360	84	84	Abdelhay et al. (2021)
Pt/KCl (electrolyte)	Reactive blue 21	100	–	3	20	99	–	El Aggadi et al. (2021)
Ti/IrO_2 (anode) graphite (cathode)	Basic yellow 28 (Na_2SO_4 and NaCl electrolyte)	30	8.66	2.5 6.5	15	92.90 93.30	26.8 46.2	Belal et al. (2021)

Note: Current density (j) and electrolysis time (t_E).

However, the use of salts with chloride as electrolytes can induce the reaction of this anion with organic molecules present in the effluents and those generated throughout the process, giving rise to organochlorine compounds. It is important to emphasize that, as in other AOPs, the chemical substances to be destroyed highly influence the operating conditions; therefore, the suppression of electrolytes such as sodium or potassium chloride cannot be generalized to all electrochemical oxidation practices. For example, it has been observed that the use of NaCl has increased the electrodegradation of certain organic compounds (Carneiro et al., 2019, 2020; Yao et al., 2022).

On the contrary, Santos et al. (2020) demonstrated that the degradation of the RB19, RR195, and RY145 dyes is possible by electrochemical oxidation using dimensionally stable anodes (DSA®) and that the incorporation of NaCl in the system improved the degradation efficiency. However, after treatment, regardless of the addition of salt, they observed high levels of toxicity in the sample; therefore, this methodology could be used as a pretreatment. Additionally, to approximate a real industrial effluent, they prepared a mixture of these dyes in which the degradation was lower than the results obtained individually. However, when evaluating the toxicity of the mixture, they observed a reduction in toxicity with increasing dye concentrations. The authors suggested that a low concentration of dyes in the mixture promoted synergy between the dyes and the intermediate products, preventing degradation. Similarly, they stated that the importance of the presence of chlorides obtained from the electrolyte goes into the background because a greater number of species in the medium contribute to the formation of parallel reactions, preferring direct reactions on the surface of the electrode because, in theory, the oxidation of dyes is prioritized. The effluents generated after treatment, both from the samples of the isolated dyes and from the mixture, showed significant toxicity. They measured the germination and growth rates of lettuce (*Lactuca sativa*) roots. They reported that when diluting the treated effluent of the mixture to 25%, the relative germination rate was satisfactory. Although other parameters presented an inhibitory effect, it was concluded that this was due to the sensitivity of this specific seed to the species generated during this procedure.

Finally, it should be noted that in a laboratory study, it is possible to eliminate the chlorinated compounds that are produced along the EAOP and that are adsorbed on the electrode surface by polishing it with activated carbon, thereby avoiding the generation of these toxic compounds under conditions that are too complex for later removal (Yasri & Gunasekaran, 2017).

2.2.4 ELECTROCOAGULATION

Electrocoagulation (EC) is a simple and efficient technique widely used in the treatment of dyes and textile effluents, as well as the removal of solids present in the water to be treated (Sahu et al., 2014). It has been shown to be reliable and profitable; thus, its application and research have increased, and it has become popular (Tahreen et al., 2002). One of its main advantages is the change in the coagulation agents used compared with conventional chemical coagulation. EC came to supplant these classical techniques in the early 1900s (Tchamango & Darchen, 2018), which were considered highly expensive; however, with the passage of time and improvements, this drawback has been eliminated.

This technique is useful for eliminating COD and BOD in industrial and urban wastewater. In addition, this process is attractive for the treatment of natural organic matter and microbiological contaminants and is very effective in removing dissolved phosphorus (Elazzouzi et al., 2018). Recent laboratory studies have shown that increasing the Fe^{2+} concentration in the electrocoagulation process and using Fe anodes favor the kinetics of methyl orange removal under acidic conditions (Akter & Islam, 2022). In contrast, Łuba et al. (2020) treated wastewater with Red 167 dye using Cu–Al anodes and achieved the removal of approximately 99% of the dye. Likewise, other types of alloys have been used, among which the metals Al or Fe prevail, which are the ones that actively participate in the removal of contaminants, such as Cu–Fe anodes (Kuczynski et al., 2021).

EC consists of the *in-situ* generation of flocs or clots (Nippatlapalli & Philip, 2021). These clots are formed by applying an electric current that promotes the oxidation of the anode, which is why its ions enter the solution and neutralize the charges of the colloids present in the effluents (Tchamango & Darchen, 2018). Simultaneously, the ions coming from the anode react with the hydroxyl ions produced by the cathode to obtain metal hydroxides that favor the formation of flocs. In these processes, destabilized pollutant particles are adsorbed and separated by settling or flotation (Elazzouzi et al., 2018). The reactions that occur at the anode and cathode are shown in Figure 2.3, providing an analogy in which the contaminants can be considered ligands.

The materials used in the electrodes vary. At small scales, BDD, Ti/SiO_2, and Ti/PbO_2 can be used because of their high cost (Elazzouzi et al., 2018; Sahu et al., 2014). The most commonly used materials are Fe and Al (Kobya et al., 2003). In addition, the development of new reactors with different electrode arrangements has been investigated. The simplest reactor has only a pair of electrodes and is generally used for the development and/or modification of the method. However, some reactors have more than two electrodes in their configuration, thereby increasing the capacity and efficiency of the process. Therefore, the search for new configurations continues with the aim of reducing the main disadvantages of this type of electrochemical cells

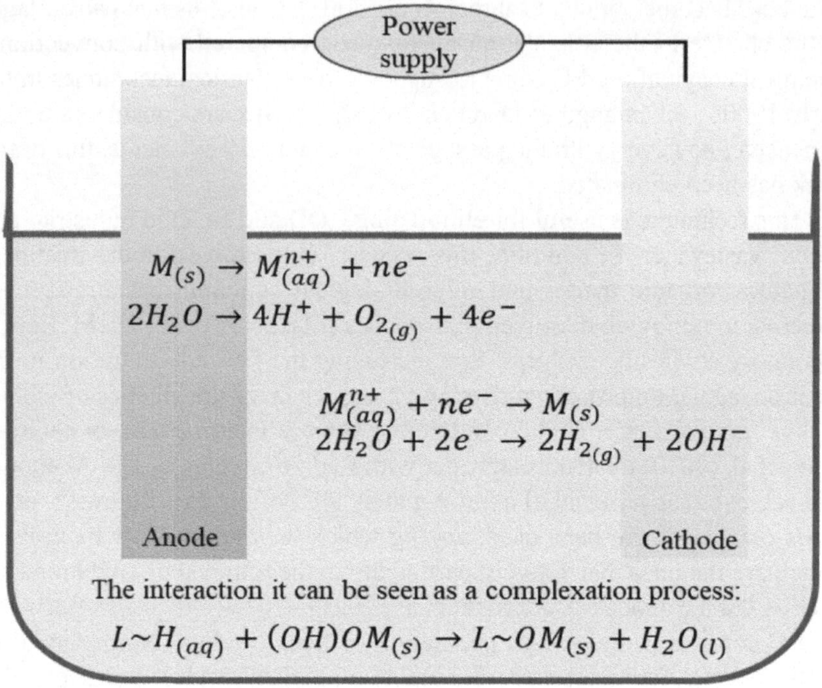

FIGURE 2.3 Reactions present in electrocoagulation.

– very large electrodes, inefficient effluent circulation, and excessive energy consumption – because a poorly directed flow increases the heterogeneous wear of the electrodes (Tchamango & Darchen, 2018). An interesting proposal that has been made in the field of sustainability is the removal of textile dyes by electrocoagulation powered by solar panels (Khemila et al., 2018), which represents a decrease in operating costs from an energy perspective.

The electrodes used in EC can also be chemically modified to improve their performance. Elazzouzi et al. (2018) compared a reactor with an insulator on its electrodes with a conventional reactor and found that under some conditions, the insulated edges helped to better distribute the current density, which led to a higher COD and phosphorus reduction percentage in urban wastewater. They also observed that energy consumption was reduced, and therefore, the cost of operation, although this decrease was not significant. The most remarkable aspect of their work was that they obtained effluent suitable for discharge into natural areas in accordance with Moroccan standards. This is important because, in the treatment of textile, urban, or industrial effluents, the final disposal of water must be considered.

Thus, it can be expected that EC assists efficiently in the degradation of dyes at the expense of Fe and Al anodes, although this often implies a high operational cost (Akter & Islam, 2022; Kobya et al., 2003). In such a scenario, Fe electrodes turn out to be better than Al electrodes because they consume less energy, not to mention they also have lower cost, in addition to the fact that the use of $Fe(OH)_3$ has presented a higher percentage of removal of carbon dioxide contaminant (Kobya et al., 2003). In addition, the use of Al creates larger bubbles that prevent flocculation and break up the floccules formed by the roughness of its surface. Furthermore, the iron anodes participate in both flocculation and electrochemical oxidation. Table 2.4 shows examples of the use of electrodegradation for the treatment of dyes and/or textile effluents.

TABLE 2.4 Electrocoagulation in the Degradation of Textile Dyes

Electrode	Sample	j (mA/cm2)	Ci (mg/L)	pH	tE (min)	COLR (%)	CODR (%)	References
SAE 1005 steel – Al	Residual textile effluent	8	–	7	10	86	59	Núñez et al. (2019)
Al–Ti	Methylene blue	5	30	7	30	99.37	–	Liu & Wu (2019)
Fe	Dyeing process of the textile industry	8	–	7.1	10	86	59	Núñez et al. (2019)
Al	Mixed industrial wastewater	1.6	–	10	60	69	~35	Nidheesh et al. (2020)
Al	Reactive Blue 198 Reactive Yellow 145 Reactive Blue 19	1.04	50	–	20	98.8 95.8 98.2	–	Keyikoglu & Can (2021)
Al	Eriochrome black	–	200	2–7	55	98.5	–	Cestarolli et al. (2019)
Fe	Congo red dye	23.8	1,000	3	30	97	89	Akhtar et al. (2020)
Fe (stainless steel 304)	Methyl orange	0.4	200	4–5	32	93.1	88.5	Akter & Islam (2022)
Fe Al	Reactive blue 2	16.6	5	3	6	95 80	–	Yaqub et al. (2023)
Mg (anode) Fe (cathode)	Real textile wastewater	144.7 A/m^2	–	4.5	29	87	93	Zaldivar-Díaz et al. (2023)

However, the increase in the current density in these systems must be managed carefully because it can be counterproductive. Although the rate of degradation of organic substances increases with the current density, it is possible that an excessive increase leads to accelerated wear of the electrode. Although this wear is expected, it is preferable to establish operating conditions that balance the electrode loss with the degradation rate and power consumption. This is the case in the work carried out by Keyikoglu & Can (2021), who adjusted this parameter to 10.42 A/m^2 and observed high energy consumption and a loss of 1.65 kg of Al for each kilogram of dye treated.

Many of these contaminants have been studied for their use in dye treatment. According to Khandegar & Saroha (2013), there is a slight preference for reactive and acidic dyes. Regardless of the material of the electrodes involved, most dyes have a removal percentage close to 94%, and it is expected that subsequent technological innovations will reach 100%.

Finally, it must be recognized that EC has certain disadvantages. On the one hand, subsequent treatment is necessary to eliminate the coagulants formed because high volumes of precipitates are generated (Yasri & Gunasekaran, 2017). On the other hand, owing to the sacrificial anodes in the electrochemical process, their continuous renewal is necessary, which could be costly. Therefore, it is constantly sought to reduce the periodicity of the change of the anodes by means of modified materials or the adjustment of the operating conditions.

2.3 CONCLUSION

Currently, wastewater effluents containing dyes from the textile industry present a serious environmental problem because of their large volume, alarmingly high COD and BOD parameters, high concentrations of substances with great chemical stability, and persistence in the environment, most of which are toxic and highly bioaccumulative. Owing to the inefficiency of their removal by conventional techniques, new processes in which AOPs play a leading role, either for their applications or in combination with other methodologies, are being actively sought.

Among these techniques, electrochemical oxidation and electrocoagulation are quite similar processes, both in their development and achievable results. Although it has been reported that electrodegradation presents slightly better results, the choice between the two depends on the chemical composition of the effluent to be treated and the nature of the dye, for which electrodes of various materials and different cell configurations are available to achieve operating conditions that make the process more efficient.

Continuous study of AOP continues to be necessary since there are various areas of opportunity at the chemical and engineering levels that lead to addressing problems inherent to our society, which are highly relevant in terms of health and the environment. Thus, work is being conducted on innovations that lead to practical, efficient, profitable, automatable, and scalable solutions at the industrial level, as well as sustainable solutions, which is why their application is commonly used in conjunction with other techniques.

ACKNOWLEDGMENT

SAAA thanks CONACyT for the scholarship awarded (No. 1178094) for postgraduate studies. Likewise, COECyT-Coahuila is thanked for financing through project COAH-2022-C19-C064.

KEYWORDS

- **advanced oxidation processes**
- **dyes**
- **electrodegradation**
- **iron anodes**
- **ozonation**
- **sustainable solutions**
- **wastewater effluents**
- **water treatment**

REFERENCES

Abdelhay, A., Jum'h, I., Albsoul, A., Abu Arideh, D., & Qatanani, B. (2021). Performance of electrochemical oxidation over BDD anode for the treatment of different industrial dye-containing wastewater effluents. *Journal of Water Reuse and Desalination, 11*(1), 110–121. https://doi.org/10.2166/wrd.2020.064.

Akhtar, A., Aslam, Z., Asghar, A., Bello, M. M., & Raman, A. A. A. (2020). Electrocoagulation of Congo Red dye-containing wastewater: Optimization of operational parameters and process mechanism. *Journal of Environmental Chemical Engineering, 8*(5), 104055. https://doi.org/10.1016/j.jece.2020.104055.

Akter, S., & Islam, M. (2022). Effect of additional Fe^{2+} salt on electrocoagulation process for the degradation of methyl orange dye: An optimization and kinetic study. *Heliyon, 8*, e10176. https://doi.org/10.1016/j.heliyon.2022.e10176.

Ameta, R., Chohadia, A. K., Jain, A., & Punjabi, P. B. (2018). Fenton and Photo-Fenton Processes. In S. C. Ameta & R. Ameta (Eds.), *Advanced Oxidation Processes for Wastewater Treatment* (pp. 49–87). Academic Press. https://doi.org/10.1016/B978-0-12-810499-6.00003-6.

Aoudj, S., Bahloul, K., & Khelifa, A. (2021). Degradation of dyes by electrochemical advanced oxidation processes. In S. S. Muthu & A. Khadir (Eds.), *Advanced Removal Techniques for Dye-containing Wastewaters, Sustainable Textiles: Production, Processing, Manufacturing & Chemistry* (pp. 129–174). Singapore: Springer. https://doi.org/10.1007/978-981-16-3164-1_5.

Araujo, F. V., Yokoyama, L., Teixeira, L. A., & Campos, J. C. (2011). Heterogeneous Fenton process using the mineral hematite for the discoloration of a reactive dye solution. *Brazilian Journal of Chemical Engineering, 28*(4), 605–616. https://doi.org/10.1590/S0104-66322011000400006.

Belal, R. M., Zayed, M. A., El-Sherif, R. M., & Abdel Ghany, N. A. (2021). Advanced electrochemical degradation of basic yellow 28 textile dye using IrO_2/Ti meshed electrode in different supporting electrolytes. *Journal of Electroanalytical Chemistry, 882*, 114979. https://doi.org/10.1016/j.jelechem.2021.114979.

Belding, S. R., Campbell, F. W., Dickinson, E. J., & Compton, R. G. (2010). Nanoparticle-modified electrodes. *Physical Chemistry Chemical Physics, 12*(37), 11208–11221. https://doi.org/10.1039/C0CP00233J.

Bhad, R. M., Das, A., & Kodape, S. M. (2022). Ozonation of Procion Blue Reactive Dye and its Kinetics Study. *Pollution, 8*(2), 529–541. doi: 10.22059/POLL.2021.330871.1191.

Bilinska, L., Blus, K., Foszpanczyk, M., Gmurek, M., & Ledakowicz, S. (2020). Catalytic ozonation of textile wastewater as a polishing step after industrial scale electrocoagulation. *Journal of Environmental Management, 265*, 110502. https://doi.org/10.1016/j.jenvman.2020.110502.

Brillas, E., & Martínez-Huitle, C. A. (2015). Decontamination of wastewaters containing synthetic organic dyes by electrochemical methods: An updated review. *Applied Catalysis B: Environmental, 166–167*, 603–643. https://doi.org/10.1016/j.apcatb.2014.11.016.

Carneiro, J. F., Aquino, J. M., Silva, A. J., Barreiro, J. C., Cass, Q. B., & Rocha-Filho, R. C. (2019). Corrigendum to "The effect of the supporting electrolyte on the electrooxidation of enrofloxacin using a flow cell with a BDD anode: Kinetics and follow-up of oxidation intermediates and antimicrobial activity." *Chemosphere, 223*, 759. https://doi.org/10.1016/j.chemosphere.2019.03.016.

Carneiro, J. F., Aquino, J. M., Silva, B. F., Silva, A. J., & Rocha-Filho, R. C. (2020). Comparing the electrochemical degradation of the fluoroquinolone antibiotics norfloxacin and ciprofloxacin using distinct electrolytes and a BDD anode: evolution of main oxidation byproducts and toxicity. *Journal of Environmental Chemical Engineering, 8*(6), 104433. https://doi.org/10.1016/j.jece.2020.104433.

Cestarolli, D. T., de Oliveira, A., Das G., & Guerra, E. M. (2019). Removal of Eriochrome Black textile dye from aqueous solution by combined electrocoagulation–electrofotation methodology. *Applied Water Science, 9*, 101. https://doi.org/10.1007/s13201-019-0985-x.

Clematis, D., & Panizza, M. (2021). Electro-Fenton, solar photoelectro-Fenton and UVA photoelectro-Fenton: Degradation of erythrosine B dye solution. *Chemosphere, 270*, 129480. https://doi.org/10.1016/j.chemosphere.2020.129480.

da Rocha, S. R. M., Napoleão, D. C., Rodriguez-Diaz, J. M., Gomes de, M. R. K., Silva, G. M., Estolano de Lima, V. M., de Melo Neto, A. A., Vinhas, G. M., & Duarte, B. M. M. M. (2023). Efficient microbial cellulose/Fe_3O_4 nanocomposite for photocatalytic degradation by advanced oxidation process of textile dyes. *Chemosphere, 326*, 138453. https://doi.org/10.1016/j.chemosphere.2023.138453.

Dias, N. C., Bassin, J. P., Sant'Anna, G. L., & Dezotti, M. (2019). Ozonation of the dye Reactive Red 239 and biodegradation of ozonation products in a moving-bed biofilm reactor: Revealing reaction products and degradation pathways. *International Biodeterioration & Biodegradation, 144*, 104742. doi: 10.1016/j.ibiod.2019.104742.

El Aggadi, S., El Abbassi, Z., & El Hourch, A. (2021). Color removal from dye-containing aqueous solutions by electrooxidation. *Desalination and Water Treatment, 215*, 232–236. doi: 10.5004/dwt.2021.26766.

El Hassani, K., Kalnina, D., Turks, M., Beakou, B. H., & Anouar, A. (2019). Enhanced degradation of an azo dye by catalytic ozonation over Nicontaining layered double hydroxide nanocatalyst. *Separation and Purification Technology, 210*, 764–774. doi: 10.1016/j.seppur.2018.08.074.

El Nemr, A., Hassaan, M. A., & Madkour, F. F. (2018). Advanced oxidation process (AOP) for detoxification of acid red 17 dye solution and degradation mechanism. *Environmental Processes, 5*, 95–113. doi: 10.1007/s40710-018-0284-9.

Elazzouzi, M., El Kasmi, A., Haboubi, K., & Elyoubi, M. S. (2018). A novel electrocoagulation process using insulated edges of Al electrodes for enhancement of urban wastewater treatment: Techno-economic study. *Process Safety and Environment Protection, 116*, 506–515. doi: 10.1016/j.psep.2018.03.006.

Fadzli, J., Ku Hamid, K. H., Nik Him, N. R., & Puasa, S. W. (2022). A critical review on the treatment of reactive dye wastewater. *Desalination and Water Treatment, 257*, 185–203. doi: 10.5004/dwt.2022.28028.

Falola, T. O. (2022). Nanoparticles Modified Electrodes: Synthesis, Modification, and Characterization—A Review. *World Journal of Nano Science and Engineering, 12*, 29–62. https://doi.org/10.4236/wjnse.2022.123003.

Fenton, H. (1984). Oxidation of tartaric acid in presence of iron. *Journal of the Chemical Society, Transactions, 65*, 899–910. doi: 10.1039/CT8946500899.

Goi, A., & Trapido, M. (2002). Hydrogen peroxide photolysis, fenton reagent and photo-fenton for the degradation of nitrophenols: a comparative study. *Chemosphere, 46*(6), 913–922. doi: 10.1016/s0045-6535(01)00203-x.

Hamous, H., Khenifi, A., Orts, F., Bonastre, J., & Cases, F. (2021). Carbon textiles electrodes modified with RGO and Pt nanoparticles used for electrochemical treatment of azo dye. *Journal of Electroanalytical Chemistry, 887*, 115154. doi: 10.1016/j.jelechem.2021.115154.

He, H., & Zhou, Z. (2017). Electro-fenton process for water and wastewater treatment. *Critical Reviews In Environmental Science and Technology, 47*(21), 2100–2131. doi: 10.1080/10643389.2017.1405673.

Hu E., Shang S., & Chiu K. L. (2019). Removal of Reactive Dyes in Textile Effluents by Catalytic Ozonation Pursuing on-Site Effluent Recycling. *Molecules, 24*(15), 755. https://doi.org/10.3390/molecules24152755.

Ikhlaq A., Zafar M., Javed F., Yasar A., Akramd A., Shabbira S., & Qi F. (2021). Catalytic ozonation for the removal of reactive black 5 (RB-5) dye using zeolites modified with $CuMn_2O_4/gC_3N_4$ in a synergic electro flocculation-catalytic ozonation process. *Water Science & Technology, 84*(8), 1943–1953. https://doi.org/10.2166/wst.2021.404.

Isik, Z., Ozbey-Unal, B., Karagunduz, A., Keskinler, B., & Dizge, N. (2020). Electrochemical Treatment of Textile Dye Bath Wastewater Using Activated Carbon Cloth Electrodes. *Avicenna Journal of Environmental Health Engineering, 7*(1), 47–52. doi: 10.34172/ajehe.2020.07.

Islam S., Shaikh I. A., Firdous N., Ali A. & Sadef Y. (2019). A new approach for the removal of unfixed dyes from reactive dyed cotton by Fenton oxidation. *Journal of Water Reuse and Desalination, 9*(2), 133–141. doi: 10.2166/wrd.2019.011.

Islam, T., Repon, M. R., Islam, T., Sarwar, Z., & Rahman, M. M. (2023). Impact of textile dyes on health and ecosystem: a review of structure, causes, and potential solutions. *Environmental Science and Pollution Research, 30*(4), 9207–9242. doi: 10.1007/s11356-022-24398-3.

Katheresan, V., Kansedo, J., & Lau, S. Y. (2018). Efficiency of various recent wastewater dye removal methods: A review. *Journal of Environmental Chemical Engineering, 6*(4), 4676–4697. doi: 10.1016/j.jece.2018.06.060.

Keyikoglu, R., & Can, O. T. (2021). The role of dye molecular weight on the decolorization performance of the electrocoagulation. *Environment, Development and Sustainability, 23*, 3917–3928. doi: 10.1007/s10668-020-00749-3.

Khandegar, V., & Saroha, A. K. (2013). Electrocoagulation for the treatment of textile industry effluent – A review. *Journal of Environmental Management, 128*, 949–963. doi: 10.1016/j.jenvman.2013.06.043.

Khemila, B., Merzouk, B., Chouder, A., Zidelkhir, R., Leclerc, J., & Lapicque, F. (2018). Removal of a textile dye using photovoltaic electrocoagulation. *Sustainable Chemistry and Pharmacy, 7*, 27–35. doi: 10.1016/J.SCP.2017.11.004.

Kobya, M., Can, O. T., & Bayramoglu, M. (2003). Treatment of textile wastewaters by electrocoagulation using iron and aluminum electrodes. *Journal of Hazardous Materials, 100*(1-3), 163–178. doi: 10.1016/S0304-3894(03)00102-X.

Korpe, S., & Rao, P. V. (2021). Application of advanced oxidation processes and cavitation techniques for treatment of tannery wastewater—A review. *Journal of Environmental Chemical Engineering, 9*(105234). doi: 10.1016/j.jece.2021.105234.

Kothari M. S. & Shah K. A. (2020). Electrochemical advanced oxidation process for synthetic dye wastewater treatment. *Pollution Research, 39*(Nov Suppl. Issue): S221–S224.

Kuczynski, M., Łuba, M., Mikołajczyk, T., Pierożyński, B., Jasiecka-Mikołajczyk, A., Smoczyński, L., & Wojtacha, P. (2021). Electrodegradation of Acid Mixture Dye through the Employment of Cu/Fe Macro-Corrosion Galvanic Cell in Na_2SO_4 Synthetic Wastewater. *Molecules, 26*(15), 4580. doi: 10.3390/molecules26154580.

Kuleyin, A., Gök, A., & Akbal, F. (2021). Treatment of textile industry wastewater by electro-Fenton process using graphite electrodes in batch and continuous mode. *Journal of Environmental Chemical Engineering, 9*, 104782. doi: 10.1016/j.jece.2020.104782.

Li, A., Weng, J., Yan, X., Li, H., Shi, H., & Wu, X. (2021). Electrochemical oxidation of acid orange 74 using Ru, IrO2, PbO2 and BDD anodes: Direct and indirect oxidation. *Journal of Electroanalytical Chemistry, 898*, 115622. doi: 10.1016/j.jelechem.2021.115622.

Li, J., Guan, Q., Hong, J., & Chang C. T. (2019). Electrochemical Oxidation of Azo Dye Wastewater Using Graphene-Based Electrode Materials. *Journal of Nanoscience and Nanotechnology, 19*(11), 7308–7314. doi: 10.1166/jnn.2019.16656.

Lim, S., Shi, J. L., Gunten, U. V., & McCurry, D. L. (2022). Ozonation of organic compounds in water and wastewater: A critical review. *Water Research, 213*, 118053. doi: 10.1016/j.watres.2022.118053.

Liu, N., & Wu, Y. (2019). Removal of methylene blue by electrocoagulation: a study of the effect of operational parameters and mechanism. *Ionics, 25*, 3953–3960. doi: 10.1007/s11581-019-02915-8.

Łuba, M., Mikołajczyk, T., Pierożyński, B., Smoczyński, L., Wojtacha, P., & Kuczyński, M. (2020). Electrochemical Degradation of Industrial Dyes in Wastewater through the Dissolution of Aluminum Sacrificial Anode of Cu/Al Macro-Corrosion Galvanic Cell. *Molecules, 25*(18), 4108. doi: 10.3390/molecules25184108.

Mahdizadeh, H., Dadban Shahamat, Y., & Rodríguez-Couto, S. (2021). Discoloration and mineralization of a textile azo dye using a hybrid UV/O3/SBR process. *Applied Water Science, 11*, 159. doi: 10.1007/s13201-021-01479-1.

Medrano-Rodríguez F., Picos-Benítez A., Brillas E., Bandala E. R., Pérez T. & Peralta-Hernández J. M. (2020). Electrochemical advanced oxidation discoloration and removal of three brown diazo dyes used in the tannery industry. *Journal of Electroanalytical Chemistry, 873*, 114360. doi: 10.1016/j.jelechem.2020.114360.

Merle, T., Pronk, W., & Gunten, U. V. (2017). MEMBRO3X, a Novel Combination of a Membrane Contactor with Advanced Oxidation (O_3/H_2O_2) for Simultaneous Micropollutant Abatement and Bromate Minimization. *Environmental Science & Technology Letters, 4*(5), 180–185. doi: 10.1021/acs.estlett.7b00061.

Merouani, S., Dehane, A., Belghit, A., Hamdaoui, O., Boussalem, N. E-H., & Daif, H. (2022). Removal of persistent textile dyes from wastewater by Fe(ii)/H_2O_2/H_3NOH^+ integrated system: process performance and limitations. *Environmental Science: Advances, 1*, 192–207. doi: 10.1039/D2VA00011C.

Monge, S., Silva, A., & Bengoa, C. (2018). Manual Técnico sobre Procesos de Oxidación Avanzada aplicados al Tratamiento de Aguas Residuales Industriales. España: CYTED. doi: 978-84-09-08637-5.

Muniyasamy, A., Sivaporul, G., Gopinath, A., Lakshmanan, R., Altaee, A., Achary, A., & Padmanaban, V. C. (2020). Process development for the degradation of textile azo dyes (mono-, di-, poly-) by advanced oxidation process – Ozonation: Experimental & partial derivative modelling approach. *Journal of Environmental Management, 265*, 110397. doi: 10.1016/j.jenvman.2020.110397.

Nawaz S., Siddique M., Khan R., Riaz N., Waheed U., Shahzadi I., & Ali A. (2022). Ultrasound-Assisted Hydrogen Peroxide and Iron Sulfate Mediated Fenton Process as an Efficient Advanced Oxidation Process for the Removal of Congo Red Dye. *Polish Journal of Environmental Studies, 31*(3), 1–13. doi: 10.15244/pjoes/144298.

Nian, C., Jingyi, H., & Yingru, W. (2022). Establishment of Electrochemical Treatment Method to Dye Wastewater and Its Application to Real Samples. *Main Group Chemistry, 21*(2), 523–537. doi: 10.3233/mgc-210148.

Nidheesh, P. V., Kumar, A., Syam Babu, D., Scaria, J., & Kumar, M. S. (2020). Treatment of mixed industrial wastewater by electrocoagulation and indirect electrochemical oxidation. *Chemosphere, 251*, 126437. doi: 10.1016/j.chemosphere.2020.126437.

Nippatlapalli, N., & Philip, L. (2021). Advanced Oxidation Processes for Dye Removal. En S. S. Muthu, & A. Khadir, Advanced Removal Techniques for Dye-containing Wastewaters, Sustainable Textiles: Production, Processing, Manufacturing & Chemistry (pp. 71–128). Singapur: Springer. doi: 10.1007/978-981-16-3164-1_4.

Núñez, J., Yeber, M., Cisternas, N., Thibaut, R., Medina, P., & Carrasco, C. (2019). Application of electrocoagulation for the efficient pollutants removal to reuse the treated wastewater in the dyeing process of the textile industry. *Journal of Hazardous Materials, 371*, 705–711. doi: 10.1016/j.jhazmat.2019.03.030.

Oturan, M. A., & Aaron, J. J. (2014). Advanced Oxidation Processes in Water/Wastewater Treatment: Principles and Applications. A Review. *Critical Reviews in Environmental Science and Technology, 44*(23), 2577–2641. doi: 10.1080/10643389.2013.829765.

Pieczyńska A., Ossowski T., Bogdanowicz R. & Siedlecka E. (2019). Electrochemical degradation of textile dyes in a flow reactor: effect of operating conditions and dyes chemical structure. *International Journal of Environmental Science and Technology, 16*, 929–942. doi: 10.1007/s13762-018-1704-0.

Ramirez-Pereda, B., Álvarez-Gallegos, A., Bustos-Terrones, Y. A., Silva-Martínez, S., & Hernández-Pérez, A. (2020). Effective Electro-Fenton Treatment for a Real Textile Effluent: A Case Study. *Journal of Water Process Engineering, 37*, 101434. doi: 10.1016/j.jwpe.2020.101434.

Sahu, O., Mazumdar, B., & Chaudhari, P. (2014). Treatment of wastewater by electrocoagulation: A review. *Environmental Science and Pollution Research, 21*, 2397–2413. doi: 10.1007/s11356-013-2208-6.

Santos, D. H., Duarte, J. L., Tavares, M. G., Friedrich, L. C., Meili, L., Meili, L., Pimentel, W. R. O., Tonholo, J., & Zanta, C. L. P. S. (2020). Electrochemical degradation and toxicity evaluation of reactive dyes mixture and real textile effluent over DSA® electrodes. *Chemical Engineering & Processing: Process Intensification, 153*, 107940. doi: 10.1016/j.cep.2020.107940.

Scheres F., D., Rocha Ribeiro, R., de Liz, M. V., & Peralta-Zamora, P. (2018). Investigations on iron leaching from oxides and its relevance for radical generation during Fenton-like catalysis. *Environmental Earth Sciences, 77*(1009–1020). doi: 10.1007/s12665-018-7274-0.

Senthil K., P., & Saravanan, A. (2017). Sustainable wastewater treatments in textile sector. En S. S. Muthu, The Textile Institute Book Series, *Sustainable Fibres and Textiles* (pp. 323–346). Woodhead Publishing. doi: 10.1016/B978-0-08-102041-8.00011-1.

Shahamat Y. D., Masihpour M., Borghei P., Hoda R. S. (2022). Removal of azo red-60 dye by advanced oxidation process O3/UV from textile wastewaters using Box-Behnken design. *Inorganic Chemistry Communications, 143*, 1097785. doi: 10.1016/j.inoche.2022.109785.

Siddiqui, S. I., Fatima, B., Tara, N., Rathi, G., & Chaudhry, S. A. (2019). Recent advances in remediation of synthetic dyes from wastewaters using sustainable and low-cost adsorbents. En S. Islam, & B. S. Butola, The Textile Institute Book Series, *The Impact and Prospects of Green Chemistry for Textile Technology* (pp. 471–507). Woodhead Publishing. doi: 10.1016/B978-0-08-102491-1.00015-0.

Suhan, M. K., Mahtab, S. T., Aziz, W., Akter, S., & Islam, M. (2021). Sudan black B dye degradation in aqueous solution by Fenton oxidation process: Kinetics and. *Case Studies in Chemical and Environmental Engineering, 4*, 100126. doi: 10.1016/j.cscee.2021.100126.

Tahreen, A., Jami, M. S., & Ali, F. (2002). Role of electrocoagulation in wastewater treatment: A developmental review. *Journal of Water Process Engineering, 37*, 101440. doi: 10.1016/j.jwpe.2020.101440.

Tang, Y., He, D., Guo, Y., Qu, W., Shang, J., Zhou, L., Pan, R., & Dong, W. (2020). Electrochemical oxidative degradation of X-6G dye by boron-doped diamond anodes: Effect of operating parameters. *Chemosphere, 258*, 127368. doi: 10.1016/j.chemosphere.2020.127368.

Tavares, M. G. R., Santos, D. H. S., Tavares, M. G., Duarte, J. L. S., Meili, L., Pimentel, W. R. O., Tonholo, J., & Zanta, C. L. P. S. (2020). Removal of Reactive Dyes from Aqueous Solution by Fenton Reaction: Kinetic Study and Phytotoxicity Tests. *Water Air Soil Pollut., 231*, 82. https://doi.org/10.1007/s11270-020-4465-6.

Tchamango, S. R., & Darchen, A. (2018). Investigation and optimization of a new electrocoagulation reactor with horizontal bipolar electrodes: Effect of electrode structure on the reactor performances. *Journal of Environmental Chemical Engineering, 6*(4), 4546–4554. doi: 10.1016/j.jece.2018.06.044.

Venkatesh, S., Venkatesh, K., & Quaff, A. (2019). Dye decomposition by combined ozonation and anaerobic treatment: Cost effective technology. *Journal of Applied Research and Technology, 15*(4). https://doi.org/10.1016/j.jart.2017.02.006.

Vo Huu Cong, Sakakibara, Y., Komori, M., Kishimoto, N., Watanabe, T., Mishima, I., Ihara, I., Tanaka, T., Yoshida, Y., & Ozaki, H. (2016). Recent Developments in Electrochemical Technology for Water and Wastewater Treatments. *Journal of Water and Environmental Technology, 14*(2), 25–36. doi: 10.2965/jwet.15-029.

Wakrim, A., Dassaa, A., Zaroual, Z., El Ghachtouli, S., Jamal-Eddine, J., & Azzi, M. (2021). Mechanistic study of carmoisine dye degradation in aqueous solution by Fenton process. *Materials Today: Proceedings, 37*(3), 3847–3853. doi: 10.1016/j.matpr.2020.08.405.

Wang, W. M., Li, X., Du, X., & Wu, Q. Y. (2022). A feasible approach for azo dye degradation using natural magnetite in heterogeneous Fenton oxidation. *Water Cycle, 3*, 100–105. doi: 10.1016/j.watcyc.2022.06.002.

Wang, W., Duan, X., Sui, X., Wang, Q., Xu, F., & Chang, L. (2020). Surface characterization and electrochemical properties of PbO2/SnO2 composite anodes for electrocatalytic oxidation of m-nitrophenol. *Electrochimica Acta, 335*, 135649. doi: 10.1016/j.electacta.2020.135649.

Yao, Y., Chen, Q., & Zhou, J. (2022). Influence of typical electrolytes on electrooxidation of bio-refractory reactive dye. *International Journal of Environmental Science and Technology, 19*(3), 1799–1810. doi: 10.1007/s13762-021-03184-7.

Yaqub, A., Raza, H., Ajab, H., Shah, S. H., Shad, A., & Bhatti, Z. A. (2023). Decolorization of reactive blue-2 dye in aqueous solution by electrocoagulation process using aluminum and steel electrodes. *Journal of Hazardous Materials Advances, 9*, 100248. doi: 10.1016/j.hazadv.2023.100248.

Yasri, N. G., & Gunasekaran, S. (2017). Electrochemical Technologies for Environmental Remediation. En N. Anjum, S. Gill, & N. Tuteja (Eds.), *Enhancing Cleanup of Environmental Pollutants* (pp. 5–73). Springer, Cham. doi: 10.1007/978-3-319-55423-5_2.

Zaldivar-Díaz, J. M., Martínez-Miranda, V., Castillo-Suárez L. A., Linares-Hernández, I., Ríos S. M. J., & Alcántara-Valladolid A. E. (2023). Synergistic electrocoagulation–precipitation process using magnesium electrodes for denim wastewater treatment: Bifunctional support electrolyte effect. *Journal of Water Process Engineering, 51*, 103369. https://doi.org/10.1016/j.jwpe.2022.103369.

Zhou, Q., Zhou, X., Zheng, R., Liu, Z., & Wang, J. (2022). Application of lead oxide electrodes in wastewater treatment: A review. *Science of The Total Environment, 806*(1), 150088. doi: 10.1016/j.scitotenv.2021.150088.

Zhou, Y., Lu, J., Zhou, Y., & Liu, Y. (2019). Recent advances for dyes removal using novel adsorbents: A review. *Environmental Pollution, 252*(A), 352–365. doi: 10.1016/j.envpol.2019.05.072.

CHAPTER 3

Built Environment Impact on Health, Water, Air, and Sustainable Urban Agriculture: A Case Study of Alameda Zaragoza in Saltillo, Mexico

MIRIAM E. MERY-RUIZ,[1] GABRIELA CARMONA OCHOA,[2] and ARELI LOPEZ-MONTELONGO[2]

[1]*PhD Student, Facultad de Arquitectura, Universidad Autónoma de Coahuila, Unidad Saltillo, Saltillo, Coahuila, Mexico*

[2]*Professor-Investigator, Facultad de Arquitectura, Universidad Autónoma de Coahuila, Unidad Saltillo, Saltillo, Coahuila, Mexico*

ABSTRACT

The sustainability of the built environment has become an urgent concern, and urban design plays a crucial role in shaping sustainable cities. Urban morphology, in particular, significantly impacts how people perceive and interact with their surroundings, affecting their quality of life. This study investigates the relationship between urban morphology and sustainability in the built environment, focusing on the case study of Alameda Zaragoza in Saltillo, Mexico. The ethnographic methodology was used to conduct qualitative social media research. Images and text were collected through Microsoft Bing word searches and analyzed visually using *nubedepalabras* software to identify semantic networks. The findings confirm that online communities' urban imaginaries offer valuable insights into sustainability strategies in various contexts. The recommendation is to expand the analysis to other case studies and include search results in different languages to comprehensively

Sustainable Agriculture and Global Environmental Health: Toward the Efficient Management of Resources. Mónica L. Chávez González, Cristóbal Noé Aguilar, Juan Carlos Contreras-Esquivel, and A. K. Haghi (Eds.)

understand global urban sustainability perceptions. By exploring sustainable urban design practices in Alameda Zaragoza, the contribution is to better understand the relationship between urban morphology and sustainability in the built environment.

3.1 INTRODUCTION

Sustainability in the built environment has emerged as a vital concern recently, driven by a growing awareness of the impacts of climate change, resource scarcity, and rapid urbanization (Bibri & Krogstie, 2020). The discipline of urban design plays a pivotal role in shaping sustainable cities by integrating ecological, social, and economic considerations (Rosa, 2021). In this context, urban morphology, which refers to studying urban spaces' physical form and structure (Moudon, 2019), has become increasingly significant. Understanding urban morphology can provide valuable insights into how people perceive and interact with their surroundings, affecting their quality of life and overall sustainability (Bibri & Krogstie, 2020).

This study investigates the relationship between urban morphology and sustainability in the built environment, focusing on a case study of Alameda Zaragoza in Saltillo, Mexico. This typology of location presents a unique opportunity to examine sustainable urban design practices in a rapidly developing city (García-Palomares et al., 2019).

The research employs netnographic methodology, a qualitative approach that explores the online representation and discussion of urban spaces through social media and other digital platforms (Kozinets, 2020).

By analyzing the visual and textual content found in these online sources, this study seeks to understand how urban morphology influences sustainability and provides recommendations for future research.

3.1.1 EXPLORING THE CONNECTION BETWEEN URBAN MORPHOLOGY, SUSTAINABILITY, AND PERCEPTION IN THE BUILT ENVIRONMENT

Urban morphology is the study of urban spaces' physical form and structure, including the layout of streets, buildings, and open spaces and their evolution over time (Moudon, 2019; Levy, 2017). It is an essential aspect of urban design as it directly affects urban environments' overall functionality and aesthetics (Ye, Yue, & Li, 2020).

Research in urban morphology contributes to understanding the relationship between the built environment and various aspects of urban life, such as social interactions, economic development, and environmental sustainability (Whitehand & Gu, 2019; Madanipour, 2011; Gehl, 2010).

Land use is one of the critical components of urban morphology, which refers to how land is utilized within a city. It includes residential, commercial, industrial, and open spaces. Land use arrangement influences urban areas' vitality and functionality (Gehl, 2010). According to Jacobs (1961), a mix of land uses within a city encourages pedestrian activity and social interaction, making the city more vibrant and lively.

Building design is another component of urban morphology, referring to the shape, size, and form of buildings within a city. It affects the city's aesthetic appeal and how people experience it (Madanipour, 2011). Buildings can be designed to promote energy efficiency, sustainability, and well-being, making them a critical element in creating a sustainable urban environment (Chen, 2023).

Street patterns and public spaces also play a critical role in urban morphology. They shape how people move within a city, interact, and live (Alnaim, 2023; Gehl, 2010). A well-designed street pattern can encourage walking, cycling, and public transportation, reducing reliance on private vehicles and promoting sustainability (Alsaffar & Alobaydi, 2023; Calthorpe, 2011).

Public spaces such as parks, squares, and plazas provide opportunities for social interaction, recreation, and relaxation (Faggi & Breuste, 2023). Transportation systems are also an essential component of urban morphology. They include roads, highways, public transportation, and cycling infrastructure. A well-designed transportation system can improve mobility, reduce congestion, and promote sustainability (Saunders et al., 2023).

Urban morphology is critical in creating sustainable cities; its components are interrelated and interdependent. A sustainable urban environment requires careful consideration of all these components and their relationship to each other (Jayasinghe et al., 2023). It is recognized as a significant factor in determining the sustainability of the built environment (Wang et al., 2023). The concept of urban morphology refers to the physical form of urban areas, including the arrangement of buildings, public spaces, streets, and transportation networks (Galal, 2023). It has been argued that sustainable urban development requires understanding how urban form can impact the environment, economy, and society (Elfversson et al., 2023).

The relationship between urban morphology and sustainability in the built environment is multifaceted. One aspect is the influence of urban form on transportation systems and mobility. Compact, mixed-use development is often associated with higher levels of walking, cycling, and public transportation use, which can reduce greenhouse gas emissions and promote healthy living (Rice et al., 2020).

In contrast, low-density, sprawling development promotes automobile use, significantly contributing to air pollution and traffic congestion (Vanhanen, 2023).

Another aspect of the relationship between urban morphology and sustainability is the built environment's impact on energy consumption. Urban form can affect the energy needed for heating, cooling, and lighting buildings and the efficiency of energy distribution networks (Liu et al., 2023). For example, compact development with high building densities can reduce the distance between buildings, reducing the energy needed to heat or cool buildings.

Additionally, walkable neighborhoods with mixed-use development can encourage the use of shared energy infrastructure, such as district heating and cooling systems.

Furthermore, urban morphology can impact the availability and quality of natural resources, including water and air quality.

Dense urban environments can generate more heat and pollution, contributing to local air quality problems. However, compact development can also promote green infrastructure, such as parks and green roofs, mitigating the urban heat island effect and improving air quality (Fallmann et al., 2023).

The urban form is also fundamental in managing water resources, including stormwater runoff and recharge. The design of buildings and public spaces can affect the quantity and quality of runoff. Compact development can promote using rain gardens and other green infrastructure to manage runoff and improve water quality (Lojko et al., 2023).

In addition to the physical characteristics of urban morphology, social and economic factors contribute to the sustainability of the built environment. For example, mixed-use developments incorporating residential, commercial, and office spaces can reduce the need for car travel, decreasing emissions and promoting sustainability (Khavarian-Garmsir et al., 2023).

Green spaces and parks within cities have positively impacted residents' mental health and well-being (Stepansky et al., 2023). Therapeutic Impact of Engagement in Green Spaces.). Access to public transportation and active transportation options, such as walking and cycling, also contribute to a sustainable urban environment (Bao et al., 2023).

Perception is critical in shaping urban sustainability by influencing how people interact with and experience their surroundings (Bibri & Krogstie, 2020). People's perceptions of the built environment can directly affect their behavior, preferences, and attitudes toward sustainable urban practices (Zhao et al., 2021). For instance, the perception of safety, accessibility, and attractiveness of public spaces can encourage more active transportation modes, such as walking or cycling, contributing to a more sustainable urban lifestyle (Rosa, 2021).

Understanding people's perceptions of urban spaces can provide valuable insights for planners and designers to create more sustainable and livable environments (Zhao et al., 2021). Several studies have explored the relationship between urban design and sustainability, focusing on various aspects such as energy efficiency, transportation, and public space (Bibri & Krogstie, 2020; García-Palomares et al., 2019; Rosa, 2021).

These studies highlight the importance of integrating sustainability principles into urban design practices to create more environmentally friendly, socially inclusive, and economically viable urban environments (Bibri & Krogstie, 2020). However, further research is still needed to understand the complex relationship between urban morphology and sustainability, particularly in rapidly developing cities and diverse cultural settings (Ye et al., 2020).

3.1.2 THE "ALAMEDA ZARAGOZA" IN SALTILLO

This space is essential to Saltillo's history and culture; it is a historical park in the Historical Center of Saltillo, Coahuila, Mexico (Recio, 2017). It is located in the west-central part of the city as shown in Figure 3.1. It spans around 10 acres and is renowned for its beautiful green spaces, trees, and fountains.

Initially established as a cemetery in the early 19th century, the park was transformed into a public space in 1827 (Flores Verduzco, 2021). Over time, it has undergone several renovations and upgrades, and today it is a popular destination for locals and tourists. The park also features several facilities and buildings, including a children's playground, two libraries, restrooms, gardens, and sitting areas, see Figure 3.2.

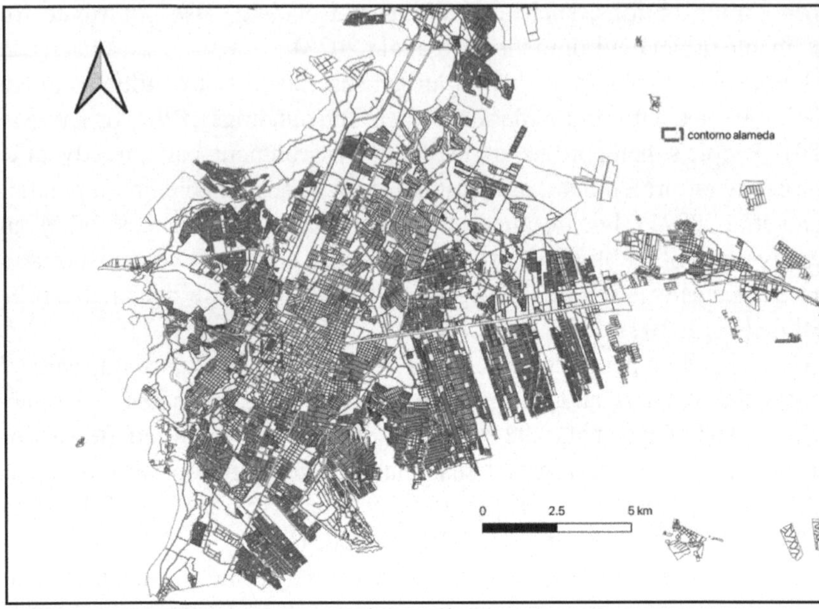

FIGURE 3.1 Localization of the Alameda in Saltillo.

Source: Mery-Ruiz (2023).

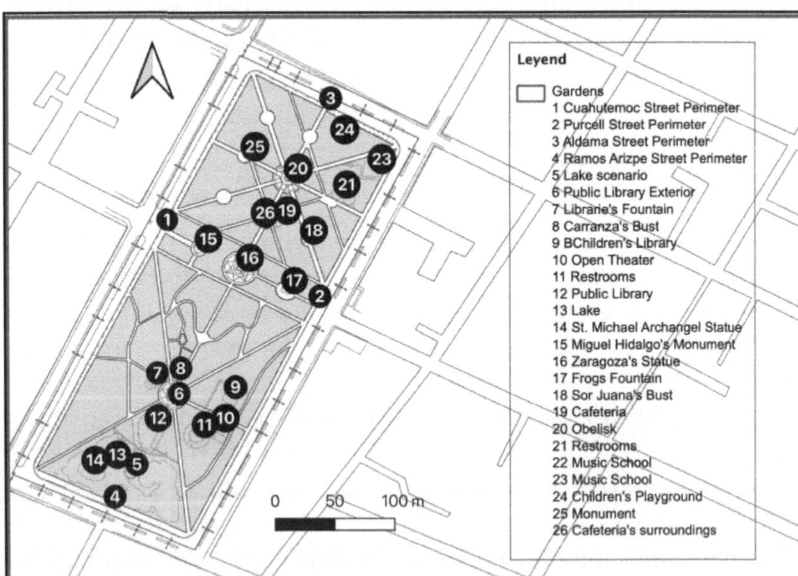

FIGURE 3.2 Localization SIG map of the Alameda Zaragoza in Saltillo.

Source: Mery-Ruiz (2023).

3.1.3 URBAN IMAGINARIES: A FUNDAMENTAL ELEMENT IN UNDERSTANDING THE PLACE

Understanding the concept of urban imaginaries is fundamental to comprehending the essence of a place. The urban imaginary is a mental process through which people gather and interpret information from various sources and types, allowing them to form a representation of their perception of a place. This abstract process is aided by memory and enables people to relate to and interpret past and current information, experiences, emotions, feelings, and more, creating a new reality.

This research seeks to explore sustainability strategies implemented in the urban morphology of public space, using the understanding of the urban imagination of people about the place.

Zaldívar (2007) explains that exploring the imaginary can reveal the cultural universe of people or a specific group since it has a shared understanding of the theme or event.

The information used to create the urban imaginary can come from various sources, including education, family, media, mass communication, and the Internet. Carmona (2015) notes the increasing use of information technologies and digital media to search for information, communicate, conduct business, and more. Thus, the internet's power to provide vast information from a group on a particular topic is essential in creating the urban imagination. This research is based on the conscious or unconscious images that internet users have left on the Microsoft Bing search engine, which aid in establishing the urban imaginary.

Lindon (2007) describes urban imaginaries as networks or plots with special meanings recognized by society, giving cities and their places qualities. An imaginary cannot be formed outside a context or historical process but within it. Thus, recognizing and understanding the urban imaginary of a place is essential since qualitative information, such as the urban imaginary, provides data that quantitative means cannot access. Failing to understand the urban imagination of a place leaves a gap in the total comprehension of the topic being studied.

3.1.4 APPLIED NETNOGRAPHY FOR SUSTAINABILITY AND MORPHOLOGICAL DESIGN IN ARCHITECTURE AND URBAN APPROACHES

Netnography is a qualitative research approach that combines elements of ethnography and online data collection to explore the cultural practices, behaviors, and beliefs of online communities (Narváez & Carmona, 2022; Kozinets, 2020).

The primary data collection method in netnography is the analysis of online interactions, such as social media posts, forum discussions, and blogs. The advantage of netnography is that it enables researchers to observe and analyze social interactions in real-time in online communities.

In architecture and urban design, netnography can be a valuable tool for exploring how people perceive and discuss urban spaces and their sustainability through social media and other digital platforms (Zhao et al., 2021).

Between urban morphology and sustainability and identify potential areas for improvement in urban design practices (García-Palomares et al., 2019).

Applying netnography in architecture and urban design allows for a more profound understanding of how online communities perceive sustainability and morphological design in urban environments. It also provides an innovative approach to studying the transformations in daily life due to the internet and technological advances (Discetti & Anderson, 2023).

Using netnography, researchers can uncover new insights that can inform sustainable design practices and contribute to developing more livable and environmentally conscious urban spaces.

3.1.5 MICROSOFT BING SEARCH ENGINE FOR NETNOGRAPHIC STUDIES ABOUT ARCHITECTURAL AND URBAN PUBLIC SPACES

Microsoft Bing search engine plays a relevant role in the netnography approach, especially when researching online communities and their behaviors. With its vast database of websites and online content, search engines provide researchers with a convenient and efficient means of gathering data about the online communities they want to study (Carmona, 2015).

By using relevant keywords and search terms, researchers can use search engines to locate online communities relevant to their research topic. This can include forums, blogs, social media platforms, and other digital spaces where people gather to discuss topics related to sustainability, morphological design, and urban approaches (Carmona, 2015).

Once researchers have identified relevant online communities, they can use the Bing engine to collect data from these platforms by analyzing the textual and visual content posted by members of these communities. This data can then be analyzed using netnography techniques to gain insights into the online community's cultural practices, behaviors, and beliefs (Carmona, 2015).

This allows for a more in-depth understanding of how people interact and communicate online and can inform sustainable design practices and contribute to the development of more livable and environmentally conscious urban spaces (Narváez & Carmona, 2022).

Using search engines, researchers can gain valuable insights into how people perceive and discuss sustainability, morphological design, and urban approaches in online communities. This, in turn, can inform the development of sustainable and environmentally conscious urban spaces that meet the needs and expectations of communities (Villarreal et al., 2022; Carmona, 2017).

3.1.6 SEMANTIC NETWORKS FOR URBAN DESIGN

Semantic networks are helpful for urban designers to analyze and represent complex relationships between urban elements. As Carmona (2015) notes, semantic networks are a visual representation of concepts connected by a series of links representing relationships between them.

In urban design, a semantic network can represent various city elements, such as streets, buildings, public spaces, and transportation systems. This allows designers to understand how these elements interact with one another and how they can be modified or improved to serve the needs of residents better.

By analyzing the connections between different elements in the network, designers can identify potential bottlenecks or areas where congestion is likely to occur (Carmona, 2015). Semantic networks can also be used to model different scenarios and test the impact of various design interventions, such as introducing new bike lanes or reconfiguring a public square (Kozinets, 2020).

García-Palomares et al. (2019) also use digital methods, specifically Twitter, to study public discourse around urban sustainability. By analyzing the content of tweets, they can identify key themes and trends in how people talk about issues such as transportation, green spaces, and climate change. This information can inform the design of sustainable and resilient urban environments, which can be represented in semantic networks.

Narváez & Carmona (2022) discuss the use of ethnography and netnography in the study of imaginaries or the shared cultural representations and symbols that shape our understanding of the world. The chapter provides a theoretical framework for the relationship between digital and physical environments and how they shape our perceptions of the world.

Semantic networks can also represent these shared cultural representations and symbols, providing insights into the underlying patterns and structures that govern how people perceive and interact with urban environments.

3.2 METHODOLOGY

For this study, netnographic analysis was conducted on Microsoft Bing, focusing on urban design, morphology, and sustainability. Specifically, the study focused on posts and discussions about the Alameda in Saltillo as a sustainable and livable urban environment, urban design, and morphology.

The data collection used keywords related to the study's research questions to identify relevant online communities and posts and was limited to English content.

The search criteria included keywords related to sustainable urban environments, such as "Alameda de Saltillo" and "sustainability." The keyword related to urban design was "urban." Finally, the keyword related to morphology included the term "form."

The collected data was analyzed using the visual analysis methodology proposed by Acaso (2009) and applied by (Villarreal et al., 2022; Carmona, 2015). It included four steps: classification of the visual product by support and function, analysis of the content through pre-iconographic and iconographic analysis, the study of the context including the author(s), places, and moments, and enunciation, which involved identifying both the explicit or manifest message and the implicit or latent message.

After the visual analysis of the collected data, semantic network analysis was performed, focusing on the relationships between words and concepts (Piselli et al., 2022; Carmona, 2017), with the use of the software nubedepalabras.es, which is a qualitative data analysis software that allows researchers to analyze data through semantic networks.

This study's netnographic methodology through visual and semantic network analysis provided an in-depth understanding of the interplay between urban design, morphology, and sustainability in creating sustainable and livable urban environments.

3.3 RESULTS AND DISCUSSION

3.3.1 MICROSOFT BING WORD SEARCH

As part of the image-centered observation conducted on April 22, 2023, at 07:26:52 p.m. using Microsoft Bing, it can be appreciated that the images generated by the search engine using keywords such as Alameda de Saltillo, sustainability, urban, and form, consist of 15 pictures from 15 websites as shown in Figure 3.3.

3.3.2 *VISUAL ANALYSIS*

The images obtained through the Bing search engine were examined. Each image was clicked to access the associated webpage and analyze its content.

One must analyze the manifest and latent messages to fully understand a visual representation. The manifest message refers to explicit information that the viewer believes they are receiving. In contrast, the latent message is implicit information the viewer receives but is often unaware of (Acaso, 2009).

3.3.2.1 *VISUAL CONTENT CLASSIFICATION*

The initial step in understanding the message is to categorize the visual product obtained from the search: Alameda de Saltillo, sustai-ability, urban, form–Bing images at https://www.bing.com/images/search?q=Alameda%20de%20Saltillo%2C%20%20sustainability%2C%20urban%2C%20form&qs=n&form=QBIR&sp=-1&lq=0&pq=alameda%20de%20saltillo%2C%20%20sustainability%2C%20urban%2C%20form&sc=0–49&cvid=EEDEF4775D51485A86F26131E305B1C5&ghsh=0&ghacc=0&first=1.

3.3.2.1.1 *Physical Features*

First, the pictures were labeled with a number in order of appearance in the mosaic, from left to right, and from top to bottom as shown in Figure 3.4.

3.3.2.1.2 *Function*

The content of these posts primarily features images, descriptions, and personal experiences related to the Alameda Zaragoza Park in Saltillo, Coahuila, Mexico.

FIGURE 3.3 Alameda de Saltillo, sustainability, urban form.

Source: Print screen from Microsoft Bing search.

FIGURE 3.4 Picture classification by order of appearance.

Source: Mery-Ruiz (2023) based on a print screen from Microsoft Bing search.

These posts emphasize different aspects of the park, such as its lake, greenery, surrounding areas, and various events like snowfall. The sources include a variety of platforms like Twitter, Pinterest, Zocalo, TripAdvisor, Reddit, Blogspot, Minube, Destinos México, and ZonaTuristica.

Many of these posts have a touristic emphasis, as they aim to showcase the park's attractions and unique features, making it appealing to potential visitors. This focus on the touristic aspects of Alameda Zaragoza can be observed through the visual content, personal experiences shared by users, and the various travel review websites included in the sources.

By exploring these links, you can gain a more profound understanding of the park from diverse perspectives, which can be useful for research purposes, planning a visit, or learning more about Alameda Zaragoza. Additionally, the touristic emphasis provides insight into the site's popularity and the experiences it offers to visitors.

3.3.2.2 Contents Study

Analyzing visual representations' content involves conducting pre-iconographic and iconographic analysis. The pre-iconographic analysis examines narrative elements and visual language tools, while iconographic analysis focuses on the symbolic meaning of the elements (Acaso, 2009). Applying this approach to the pictures found in posts about Alameda Zaragoza in Saltillo, Coahuila, Mexico shown in Figure 3.3 leads to Table 3.1.

Table 3.1 provides a detailed analysis of various photographs that depict Alameda Zaragoza, a prominent park located in Saltillo, Mexico. Table 3.1 is divided into two sections: pre-iconographic analysis, which describes the content of each image, and iconographic analysis, which interprets the meaning and significance of each photograph. These analyses showcase the park's unique features, such as its lake, trees, and architectural elements, as well as highlight its role in promoting environmental awareness, urban sustainability, and cultural significance within the community. Table 3.1 serves as a comprehensive guide to understanding the various aspects of Alameda Zaragoza and its importance as a must-visit destination in Saltillo.

The presented table organizes an insightful analysis of 15 different photographs of Saltillo, Mexico, specifically highlighting Alameda Zaragoza. By breaking down each picture into pre-iconographic and iconographic analysis,

TABLE 3.1 Pre-Iconographic and Iconographic Analyzes of the Pictures

Picture	Pre-Iconographic Analysis	Iconographic Analysis
1.	Reveals an aerial view of the lake at Alameda Zaragoza, with its particular shape of the Mexican Republic and trees surrounding it.	Suggesting the lake represents a peaceful escape within the city and the importance of clean water. Emphasizing the park's tranquility and beauty.
2.	Highlights the park's perimetral paths, shaded spaces, parking spots, and trees.	Implies a connection between nature and well-being.
3.	Shows a comprehensive view of the park, focusing on the lake, the scenario, its greenery, and architecture.	The image communicates the importance of green spaces and artificial lakes with clean water in urban environments.
4.	Presents the frontal view of the lake, focusing on an architectural element that provides shade.	Proposes visiting the lake, highlights the importance of water, and represents a shaded escape within the city, and a gathering spot.
5.	Emphasizes the park's obelisk and the public vendors, as well as the products they offer, such as colorful balloons and pink cotton candy.	Highlights the park's role as a family-friendly atmosphere and recreational offerings, and a gathering place for the community.
6.	Captures a park's aerial view during a snowfall.	Underscores the park's unique beauty and adaptability to changing seasons and weather conditions, as well as its dimensions, morphology, lake, and the urban element of tree predominance.
7.	Offers a frontal view of the lake, focusing on an architectural element that provides shade and another big shaded space.	Positions the park as a significant landmark within the city by showing the lake, architectural elements, and trees.
8.	Shows the park's lake surrounded by lush greenery and its scenario.	Reinforces the importance of architecture, nature, and water in urban settings.
9.	Focuses on the park's lake, trees, and an architectural white element.	Proposes that the lake is a central feature that shapes the park's identity.
10.	Provides a view of the children's playground dinosaur slide space.	Underlines the park's children's playground appeal to tourists and locals alike by emphasizing the dinosaur slides space, remarking that Saltillo is the land of dinosaurs.
11.	Presents an overview of the park's landscape by placing emphasis on the shaded paths and the tall trees.	Emphasizes the park's role as a symbol of the city's natural beauty by showing the importance of having a shaded public space.

TABLE 3.1 *(Continued)*

Picture	Pre-Iconographic Analysis	Iconographic Analysis
12.	Offers information on Saltillo by providing a mosaic composed of three pictures, one large and two smaller ones. The first one shows Alameda's frontal view of the lake, focusing on an architectural element that provides shade and another large shaded space. The second picture displays the monument in Nueva Tlaxcala's plaza in front of the Government Palace, and the third picture shows the front view of Garcia Carrillo's theater.	Highlights the significance of the Alameda and its lake as a must-visit destination in Saltillo. Conveys the park's multi-faceted appeal as fundamental to the city's tourism offerings within governmental and cultural places.
13.	Provides a view of the paths showing their way to Carranza's monument and its location space.	Emphasizes the cultural significance of Alameda Zaragoza within the community through monuments.
14.	Focuses on the park's frontal view of the lake.	Remarks on the cultural significance of the lake at Alameda Zaragoza within the community.
15.	Showcases the park's lake, its green spaces, and architecture.	Underscores the park's role in promoting environmental awareness and urban sustainability.

Source: Mery-Ruiz (2023).

the table allows us to better understand the various elements and themes captured in the images.

Pre-iconographic analysis, the first column, describes the content of each photograph, while the second column, iconographic analysis, delves deeper into the meaning and symbolism of the visual elements. By examining these interpretations, we gain insights into the significance of the park and its various components, such as the lake, the surrounding greenery, and the architectural elements.

The table also emphasizes the park's role as a gathering place for the community, a family-friendly recreational space, and its importance as a symbol of the city's natural beauty. Moreover, it highlights the significance of Alameda Zaragoza's lake and its relation to the city's tourism offerings within governmental and cultural places.

By exploring the table, readers can better understand and appreciate the unique features and essence of Saltillo's Alameda Zaragoza Park, as well as its cultural and environmental impact on the community.

In Saltillo, Coahuila, it is possible to appreciate the importance of the observer's role in identifying the culture, wisdom, and creativity of the individual, as mentioned by Carmona (2017). A profound meaning of the city and its cultural richness can be obtained through various visual components.

Table 3.2 offers a systematic analysis of the various units of meaning found in photographs of Saltillo, Mexico. By categorizing these elements and counting their appearances, we can better understand the significance of each aspect in defining the city's character and charm. This comprehensive overview serves as a valuable guide for those interested in learning about Saltillo's rich history, culture, and urban landscape (Table 3.2).

Table 3.2 presents a comprehensive breakdown of various units of meaning found in photographs depicting the city of Saltillo, Mexico. Table 3.2 is organized into three columns: the unit of meaning, the number of appearances of that unit, and the meaning associated with it. The units of meaning encompass a wide range of elements, from the city's landmarks, such as the lake and the obelisk, to its natural environment, including vegetation and water elements.

It highlights important aspects of the city's history, culture, and identity, as well as the significance of architectural elements and the urban landscape. It also sheds light on the city's diverse population and their daily lives, as represented in photographs featuring people. Furthermore, Table 3.2 accentuates the importance of recreational spaces for children and the presence of street vendors who contribute to the local economy and culture.

TABLE 3.2 Units of Meaning, Appearances, and Meaning

Unit of Meaning	Number of Appearances	Meaning
The lake	10	An artificial space, which is a landmark, where the inhabitants can enjoy the beauty and serenity of the water and appreciate the local flora and fauna.
Obelisk	1	This monument, another landmark, stands out in the urban landscape and symbolizes the history and progress of Saltillo.
Government Palace	1	An emblematic building that represents the authority and governance in the region, as well as its historical architecture.
Teatro García Carrillo	1	This cultural space is the epicenter of artistic and cultural events, showcasing creativity and local talent.
Saltillo	15	The city itself, with its rich history and traditions, reflects the identity and pride of its inhabitants.
Vegetation	15	The local flora shows the diversity and beauty of the region's natural environment.
Photographs	17	Captured images that reflect emblematic moments, people, and places in Saltillo.
Images with people	3	Photographs representing the city's inhabitants, showing their diversity and daily life.
Water element	10	The water resource in the city gives life and sustenance to its inhabitants and the natural environment.
Aerial views	2	Perspectives from the sky offer a panoramic view of the city and its surroundings.
Architectural elements	13	Saltillo's various styles and buildings reflect the history and urban evolution.
Monuments	5	The structures and sculptures pay homage to important historical figures and events in the city.
Shade	12	Shady spaces are generated by architecture and vegetation, which provide comfort and shelter to the inhabitants.
Paths	9	The routes that connect the spaces, facilitate movement and interaction.
Street vendors	1	Typical characters of urban life who offer their products and services, representing the informal economy and local culture.
Turquoise color	11	A tone that can be present in architectural and visual elements such as the traditional Saltillo sarape and the official baseball team "Los Saraperos" logo, adding vitality and freshness, as well as an identity to the urban landscape and culture of Saltillo.
Children's games	1	Spaces for recreation and fun for children, where socialization and physical and emotional development are encouraged.

Source: Mery-Ruiz (2023).

By examining the frequency of appearances of each unit of meaning, we can gain a deeper understanding of the prominence of these elements in Saltillo's visual narrative. This analysis enables us to appreciate the city's unique features, such as the turquoise color, which is present in various architectural and visual elements and contributes to the city's vibrant culture and identity.

3.3.2.3 CONTEXTUAL STUDY

The third step in analyzing visual representations involves examining the context in which they were created. This includes considering the author(s), locations, and moments captured. Understanding the context can offer insights into the meaning of the visual representation (Acaso, 2009), as shown in Table 3.3.

Table 3.3 provides an overview of 15 photographs featuring various elements of Saltillo, Mexico, with a focus on Alameda Zaragoza. Table 3.3 includes information about each photograph, such as the picture number, author, date, link, location, captured moment, colors, and the website where it was found. These photographs are sourced from various websites like Twitter, Pinterest, Zocalo, TripAdvisor, Reddit, Blogspot, Minube, and Destinos Mexico.

The images depict a range of captivating moments within Alameda Zaragoza, from aerial views of the lake and park to intimate scenes of people enjoying its recreational spaces. Table 3.3 also details the colors present in each photograph, revealing the vibrant and diverse hues that characterize Saltillo's urban landscape.

3.3.2.4 ENUNCIATION

In this step, the manifest and latent messages are identified in the visual representations related to Alameda Zaragoza in Saltillo, Coahuila, Mexico. By analyzing both types of messages, we can gain a more in-depth understanding of the meaning of these visual representations.

Manifest messages are the obvious and direct meanings conveyed by the images. They are typically easy to identify and understand, as they represent the literal content of the images (Acaso, 2009). The manifest messages found in the pictures are the beauty and tranquility of the park, the presence of lush greenery and natural elements within the park, the importance of the

TABLE 3.3 Contextual Study of the Pictures

Website	Picture Number	Author	Date	Link	Location	Captured Moment	Colors
Twitter	1	Adrián de la Peña	May 25, 2018	https://twitter.com/Adrian_delaPV/status/987029222295089153	The lake	Aerial picture of Alameda's Lake form during the day.	Turquoise, green, brick red, gray, white.
Pinterest	2	Rodrigo Escobar	August 24, 2017	https://www.pinterest.com.mx/pin/384635624398894174/	Paths	A path during a sunny day, shaded by trees intermittent spaces, and parked cars.	Green, red brick, sky blue, gray, white.
Pinterest	3	Unknown	December 14, 2013	https://www.pinterest.com.mx/pin/256775616244652792/	The front of the lake	Cloudy day at the solitary lake.	Turquoise, green, gray, white, red brick.
Zocalo	4	Unknown	July 12, 2020	http://zocalo.com.mx/new_site/articulo/las-dos-alamedas-de-saltillo-sobre-el-cementerio	The front of the lake	The lake is an architectural element that provides shade, shaded spaces, and the scenario.	Turquoise, green, red brick, sky blue, white.
Trip Advisor	5	Unknown	2020	https://www.tripadvisor.com.mx/Attraction_Review-g150785-d8790543-Reviews-or10-Alameda_Zaragoza-Saltillo_Northern_Mexico.html	The obelisk	A walk-in winter at the obelisk surrounded by trees, shade, the cafeteria, three public vendors' products, a father and his child walking, an ice cream cart, balloons, pink cotton candy, corn chips and corn, and a garbage bin.	Green, turquoise, blue, gray, pink, yellow, fluorescent green, and pink.
Reddit	6	Unknown	April 23, 2018	https://www.reddit.com/r/mexico/comments/7iz0hf/la_alameda_de_saltillo_en_la_nevada/	The Alameda	An aerial view of the Alameda and its surroundings during a snowfall.	White, green, gray, turquoise, brick red.
Blogspot	7	Erikson, Erick	Unknown	https://eriksonerick.blogspot.com/2010/10/saltillo-description.html	The lake	The lake is an architectural element that provides shade, shaded spaces, and the scenario.	Turquoise, green, red brick, sky blue, white.

TABLE 3.3 *(Continued)*

Website	Picture Number	Author	Date	Link	Location	Captured Moment	Colors
Minube	8	Unknown	October 4, 2017	https://www.minube.com.mx/rincon/le-lac-de-la-alameda-a2218881	The lake	The lake, the scenario during the morning.	Turquoise, green, green, sky blue, white.
Blogspot	9	Erikson, Erick	July 27, 2019	https://conocesaltillo.blogspot.com/2018/09/el-lago-de-la-republica-en-la-alameda.html	The lake	The lake, trees, two ducks, and an architectonic white element.	Green, turquoise, white.
TripAdvisor	10	Unknown	March 27, 2016	https://www.tripadvisor.com.mx/Attraction_Review-g150785-d8790543-Reviews-Alameda_Zaragoza-Saltillo_Northern_Mexico.html	Children's playground	An evening at the children's playground, two children inside and 2 persons walking outside. A bench, the dinosaur slides, and the concrete big slide.	Sky blue, green, red, gray, white.
TripAdvisor	11	Unknown	Unknown	https://www.tripadvisor.com/Attraction_Review-g150785-d8790543-Reviews-Alameda_Zaragoza-Saltillo_Northern_Mexico.html	Paths	Four persons walk through the shaded pathway on a sunny day.	Green, brick red.
Destinos Mexico	12	Unknown	Unknown	https://programadestinosmexico.com/que-ver/destinos-mexico/destinos-mexico-saltillo.html	The lake	A sunny day at the lake without water, trees, and a lamp.	Green, turquoise, sky blue, white, brick red.
TripAdvisor	13	Unknown	January 17, 2017	https://www.tripadvisor.com.mx/Attraction_Review-g150785-d8790543-Reviews-Alameda_Zaragoza-Saltillo_Northern_Mexico.html	The monument to Carranza	A cloudy day on the brick-red paths, one man walking.	Brick red, green, white.

TABLE 3.3 *(Continued)*

Website	Picture Number	Author	Date	Link	Location	Captured Moment	Colors
Pinterest	14	Unknown	September 11, 2011	https://www.tripadvisor.com.mx/Attraction_Review-g150785-d8790543-Reviews-Alameda_Zaragoza-Saltillo_Northern_Mexico.html	The lake	A cloudy day at the brick-red paths in front of the lake.	Brick red, turquoise, green, white.
Pinterest	15	Unknown	November 3, 2018	https://www.tripadvisor.com.mx/Attraction_Review-g150785-d8790543-Reviews-Alameda_Zaragoza-Saltillo_Northern_Mexico.html	The lake	A sunny day at the lake, the trees, and the blue sky.	Sky blue, turquoise, green, white.

Source: Mery-Ruiz (2023). Based on data from Pinterest, TripAdvisor, Twitter, Zocalo Saltillo, Reddit, Blogspot, Minube, and Destino Mexico.

park as a recreational and leisure space for the community, the park as a gathering place for families and friends, and the architectural and historical significance of the park.

Latent messages are the underlying meanings that may not be immediately apparent but can be inferred by analyzing the images further. These messages often convey more profound meanings, emotions, or cultural values related to visual representations (Acaso, 2009). The latent messages found in the pictures are the importance of preserving and promoting green spaces within urban environments to foster well-being and environmental awareness, the transformative power of nature and public spaces in revitalizing areas with a solemn history, such as those built over cemeteries, the park as a symbol of local pride and cultural identity for the people of Saltillo, the adaptability of the park to changing seasons and weather conditions, representing the resilience and versatility of nature in urban settings, and the role of the park in promoting social connections and fostering a sense of community among locals and visitors alike.

By analyzing both manifest and latent messages, we can better understand the various layers of meaning present in the visual representations of Alameda Zaragoza. The images showcase the park's beauty and recreational offerings and highlight its cultural significance and role in promoting well-being, environmental awareness, and community connections.

3.3.3 SEMANTIC NETWORK ANALYSIS

The text from the previous content's studio analysis was analyzed through nubedepalabras software. Prepositions and articles were removed to emphasize the main elements presented in the content. Only words with five or more repetitions were used. The result is shown in Figure 3.5.

The semantic network shown in Figure 3.5 presents the frequency of various words and themes found in a collection of images and descriptions about Alameda Zaragoza, a park in Saltillo. Words such as "history," "identity," "landscape," "local," and "natural" all appeared six times, highlighting the park's connection to Saltillo's past and its significance in the community. The park's cultural elements are represented by words like "culture" (7), "monument" (10), and "cultural" (12). Esthetics and the environment are also prominently featured, with words such as "beauty" (8), "trees" (11), "vegetation" (17), "shade" (18), and "water" (18) appearing frequently. The color palette is evident with "blue" (9), "sky" (9), "brick-red" (12), "white"

FIGURE 3.5 Semantic network from content studio text.

Source: Mery-Ruiz (2023).

(15), "turquoise" (23), and "green" (24) as recurring elements. The urban setting is emphasized through terms like "urban" (14), "city" (20), and "architectural" (26). Lastly, the sources of the images, "Pinterest" (6) and "TripAdvisor" (6), demonstrate the platform where people share their experiences of the park. Overall, the table underlines the multifaceted nature of Alameda Zaragoza, showcasing its historical, cultural, esthetic, and natural significance in Saltillo.

After obtaining the semantic network from the content studio, the text from each webpage hosting the Bing images was extracted and compiled into a text document. Prepositions and articles were taken out to highlight the main elements of the content. The word search was performed in English. Only words with five or more repetitions were used. The results showed the word "city" as the only word in English among the other 19 in Spanish.

The analysis of the text revealed the frequency of word repetitions, indicating the prominence of specific topics and themes within the content. The results are as follows: Words such as "family," "people," "July," "Pinterest," and "square" were repeated five times each, suggesting that these elements are present but not the primary focus of the content. "September" and "couple" were repeated six times, while "Mexico," "November," and "visit" appeared seven times, indicating a moderate emphasis on these topics. The phrase "public walking" was repeated eight times, showing a stronger connection to the content. The word "lake" was mentioned 10 times and "pleasant" 13 times, highlighting their significance in the material. With 14 repetitions, "Zaragoza" is an important location within the text.

"Place" was repeated 20 times, "city" 11 times, "Ciudad" 11 times, and "Alameda" 23 times, emphasizing the centrality of these concepts to the content. "Saltillo" was mentioned 49 times, indicating a critical topic within the analyzed material. To conclude, "TripAdvisor" appeared 60 times, demonstrating the travel platform's prominent role in the context of the analyzed content.

The words in Spanish were subsequently translated into English, while the word "city" was translated into Spanish. All the words were then used to create a semantic network using nubedepalabras.es, a free online software tool, as shown in Figure 3.6.

FIGURE 3.6 Semantic network from content studio text.

Source: Mery-Ruiz (2023).

Overall, these results provide insight into the recurring themes and subjects in the text, such as locations, popular travel platforms, and various activities or experiences associated with the content.

3.3.3.1 COMPARATIVE ANALYSIS OF THE SEMANTIC NETWORKS OF THE STUDY

Both sets of results present the frequency of words and themes found in descriptions and images of Alameda Zaragoza, a park in Saltillo. The first set of results highlights the park's connection to Saltillo's history, culture, and environment. It showcases various aspects of the park, such as its aesthetics, color palette, urban setting, and natural elements. The second set of results also focuses on the park, emphasizing the prominence of words such as "Zaragoza," "place," "city," "Ciudad," "Alameda," and "Saltillo."

However, the second set of results has a different emphasis, with a stronger focus on the park as a destination for visitors. It mentions the frequency of words related to visiting the park, such as "family," "couple," "visit," and "TripAdvisor." It also highlights the temporal aspect, with months like "July," "September," and "November" being mentioned multiple times. In contrast, the first set of results delves deeper into the park's features, such as its monuments, vegetation, and shade.

While both sets of results underline the multifaceted nature of Alameda Zaragoza and its significance in Saltillo, they emphasize different aspects of the park. The first set of results is more focused on the park's cultural, historical, and natural elements, whereas the second set of results highlights its role as a destination for visitors and the experiences they share on platforms like TripAdvisor.

3.4 CONCLUSION

In conclusion, this study highlights the importance of understanding the relationship between urban morphology and sustainability in shaping more sustainable, livable, and attractive urban environments. The use of netnographic methodology to explore people's perceptions and imaginaries of urban spaces through social media and other digital platforms, such as Bing Images, Twitter, Pinterest, Zocalo, TripAdvisor, Reddit, Blogspot, Minube, Destinos México, and ZonaTuristica, provides valuable insights into the complex relationship between urban imaginaries, urban morphology, and sustainability.

The findings of this study, which focused on a case study of Alameda Zaragoza in Saltillo, Mexico, reveal the beauty, recreational offerings, and cultural significance of the park, which contributes to well-being, environmental awareness, and community connections among locals and visitors. Urban planners and designers can use this knowledge to create more sustainable and livable environments that cater to diverse cultural settings and the needs of rapidly developing cities.

The listed references, such as Acaso (2009); Carmona Ochoa (2015, 2017); Calthorpe (2011); Gehl (2010); Jacobs (1961); Madanipour (2011); and Whitehand & Gu (2019), among others, provide a theoretical framework for understanding the complex relationship between urban morphology and sustainability. Other studies, such as Alnaim & Noaime (2023); Alsaffar & Alobaydi (2023); Galal (2023); and Jayasinghe et al. (2023), offer insights into the evaluation of public spaces, street configurations, sustainable urban mobility, and interdependencies between urban transport, water, and solid waste infrastructure systems, respectively.

The use of netnographic methodology in this study, which was discussed in studies such as Carmona Ochoa (2015); Discetti & Anderson (2023); Kozinets (2020); and Piselli et al. (2022), among others, demonstrates the potential of this research method for understanding the relationship between urban imaginaries, urban morphology, and sustainability.

Future research should continue to explore this complex relationship in different urban contexts and integrate various qualitative and quantitative methods, as suggested in studies such as Bibri & Krogstie (2020); Chen (2023); Elfversson et al. (2023); Khavarian-Garmsir et al. (2023); and Vanhanen (2023), among others. Ultimately, the development of comprehensive and context-specific strategies for sustainable urban design will improve urban dwellers' quality of life and enhance our built environments' overall sustainability.

KEYWORDS

- **architecture**
- **digital platforms**
- **netnographic methodology**
- **netnography**
- **perception**
- **semantic network analysis**
- **sustainability**
- **urban imaginaries**
- **urban morphology**

REFERENCES

Acaso, M. (2009). El lenguaje visual. México: Paidós.

Alnaim, M. M., & Noaime, E. (2023). Evaluating public spaces in Hail, Saudi Arabia: A reflection on cultural changes and user perceptions. *Alexandria Engineering Journal, 71*, 51–72.

Alsaffar, N. H., & Alobaydi, D. (2023). Studying street configurations and land-uses in the downtown of Baghdad. In *AIP Conference Proceedings* (Vol. 2651, No. 1, p. 020064). AIP Publishing LLC.

Bao, L., Kusadokoro, M., Chitose, A., & Chen, C. (2023). Development of socially sustainable transport research: A bibliometric and visualization analysis. *Travel Behaviour and Society, 30*, 60–73.

Bibri, S. E., & Krogstie, J. (2020). Smart sustainable cities of the future: The untapped potential of big data analytics and context-aware computing for advancing sustainability. Springer.

Calthorpe, P. (2011). Urbanism in the age of climate change. Island Press.

Carmona Ochoa, G. (2015). Ciudad imaginaria y sociedad virtual. Las redes sociales virtuales como medio para el análisis de los imaginarios urbanos. Universidad de Guadalajara – Universidad Autónoma de Coahuila, México.

Carmona Ochoa, G. (2015). La netnografía como técnica de investigación social. *Papeles del CEIC, 2015/3(133)*, 1–15.

Carmona Ochoa, G. (2017). Imaginarios de la ciudad y la discapacidad en Internet. En A. Narváez et al. (Eds.), *Vivir la Discapacidad.*

Chen, C. W. (2023). Can smart cities bring happiness to promote sustainable development? Contexts and clues of subjective well-being and urban livability. *Developments in the Built Environment, 13*, 100108.

Discetti, R., & Anderson, V. (2023). The value of netnography for research in HRD. *Human Resource Development Review, 22*(1), 59–83.

Elfversson, E., Höglund, K., Sellström, A. M., & Pellerin, C. (2023). Contesting the growing city? Forms of urban growth and consequences for communal violence. *Political Geography, 100*, 102810.

Faggi, A., & Breuste, J. (2023). Relationship Between Form and Structure of Green Spaces and Uses: The Case of Parks in Buenos Aires City, Argentina. In *Making Green Cities: Concepts, Challenges and Practice* (pp. 445–459). Cham: Springer International Publishing.

Fallmann, J., Kumar, V., Barra, M., & Tost, H. (2023). The Impact of Imperviousness on Boundary Layer Mixing, Air Pollution and Cloud Formation over Urban Areas. In *Air Pollution Modeling and its Application XXVIII* (pp. 21–26). Cham: Springer International Publishing.

Flores Verduzco, M. (2021). *La Alameda Zaragoza. Colección Edificios Emblemáticos de Saltillo.* Gobierno Municipal de Saltillo. Instituto Municipal de Cultura de Saltillo. México.

Galal, L. A. (2023). The Influence of Sustainable Urban Mobility on the Urban Morphology of Copenhagen City. *MSA Engineering Journal, 2*(1), 112–129.

García-Palomares, J. C., Gutiérrez, J., & Mínguez, C. (2019). Urban planning and the smart city: Projects and principles to improve the quality of urban life. In *Information and Communication Technologies for Human Development and Social Inclusion*. IGI Global.

Gehl, J. (2010). *Cities for people.* Island Press.

Jacobs, J. (1961). *The Death and Life of Great American Cities.* Random House.

Jayasinghe, P. A., Derrible, S., & Kattan, L. (2023). Interdependencies between Urban Transport, Water, and Solid Waste Infrastructure Systems. *Infrastructures, 8*(4), 76.

Khavarian-Garmsir, A. R., Sharifi, A., Hajian Hossein Abadi, M., & Moradi, Z. (2023). From Garden City to 15-Minute City: A Historical Perspective and Critical Assessment. *Land, 12*(2), 512.

Kozinets, R. V. (2020). *Netnography: The Essential Guide to Qualitative Social Media Research*. SAGE Publications Ltd.
Levy, J. M. (2017). *Contemporary Urban Planning*. Routledge.
Lindón, A. (2007). La ciudad y la vida urbana a través de los imaginarios urbanos. *Revista EURE, XXXIII*(99), Santiago de Chile.
Liu, K., Xu, X., Zhang, R., Kong, L., Wang, W., & Deng, W. (2023). Impact of urban form on building energy consumption and solar energy potential: A case study of residential blocks in Jianhu, China. *Energy and Buildings, 280*, 112727.
Lojko, A., Yang, E., Holmes, J., Whitcomb, M., Dacic, N., Zhang, Y., Bassis, J., & Rood, R. (2023). Washtenaw County Green Infrastructure Development Analysis. Technical Report, University of Michigan, https://dx.doi.org/10.7302/7058.
Madanipour, A. (2011). *Urban Design, Space, and Society*. Palgrave Macmillan.
Moreno Basurto, L. Z. (2002). Educación, comunicación e imaginarios sociales. *Razón y Palabra, Número 25*. Estado de México.
Moudon, A. V. (2019). Urban morphology as an emerging interdisciplinary field. *Urban Morphology, 23*(1), 5–14.
Narváez, A., & Carmona, G. (2022). Etnografía y netnografía en la investigación sobre lo imaginario. Doi: 10.2307/j.ctv33mg9qb.9.
Piselli, C., Colladon, A. F., Segneri, L., & Pisello, A. L. (2022). Evaluating and improving social awareness of energy communities through semantic network analysis of online news. *Renewable and Sustainable Energy Reviews, 167*, 112792.
Recio, C. (2017). Espacios geográficos, urbanos, públicos y de tránsito de Saltillo. Siglos XVI al XX (pp.182–183). México: Escuela de Ciencias Sociales, Universidad Autónoma de Coahuila.
Rice, J. L., Cohen, D. A., Long, J., & Jurjevich, J. R. (2020). Contradictions of the climate-friendly city: new perspectives on eco-gentrification and housing justice. *International Journal of Urban and Regional Research, 44*(1), 145–165.
Rosa, S. D. (2021). Urban design for sustainability: A study on the interplay between urban form, social practices and environmental performance. *Sustainability, 13*(5), 2685.
Saunders, J., Budhdeo, S., Fadden-Hopper, K., Nienaber, A. M., Spundflasch, S., Surace, M., Nussio, F., Pirra, M., Navarro, A., Girdvainiene, I., Drambarean, T., & Ioannis, K. (2023). Integrating Transport Programmes for Sustainable Reduction in Urban Road Congestion—Best Practice Examples from Local Authorities Working with SUITS. In *Capacity Building in Local Authorities for Sustainable Transport Planning* (pp. 137–179). Singapore: Springer Nature Singapore.
Stepansky, K., Delbert, T., & Bucey, J. C. (2023). Therapeutic Impact of Engagement in Green Spaces.
Vanhanen, J. (2023). Counteracting inefficient urban development – an analysis of the negative effects of urban sprawl.
Villarreal, S., Carmona, G., & Torres, P. (2022). Escenarios turísticos a partir de sus rutas: imaginarios urbanos del cibernauta a cerca de la zona centro-sur del Estado de Sinaloa. In G. Carmona Ochoa (Coord), *Urbanismo y Arquitectura. 11 Aproximaciones* (1st edn., pp. 201–242). UAdeC. Labyrinthos.
Wang, P., Yang, Y., Ji, C., & Huang, L. (2023). Positivity and difference of influence of built environment around urban park on building energy consumption. *Sustainable Cities and Society, 89*, 104321.

Whitehand, J. W. R., & Gu, K. (2019). Urban morphology and conservation: A research agenda. *Urban Morphology, 23*(2), 143–156.

Ye, Y., Yue, W., & Li, Y. (2020). The relationship between urban morphology and sustainability: A bibliometric review. *Environment, Development and Sustainability, 22*(6), 4739–4759.

Zaldívar Bribiesca, L. (2002). Introducción. Los imaginarios sociales y la comunicación. *Razón y Palabra, Número 25*. Estado de México.

Zhao, P., Lu, Y., & Zhang, Y. (2021). Understanding the role of perception in urban sustainability: A systematic review and research agenda. *Cities, 110*, 103027.

CHAPTER 4

Sustainable Nanoscale Polymer-Based Edible Coatings

ABIGAIL A. BELMARES-CERDA,[1] JOSÉ SANDOVAL-CORTÉS,[1] RUTH E. BELMARES-CERDA,[1] KARINA REYES-ACOSTA,[1] MIGUEL Á. AGUILAR-GONZÁLEZ,[2] OLGA B. ALVAREZ-PÉREZ,[3] and CRISTÓBAL N. AGUILAR[1]

[1]*Departamento de Investigación en Alimentos, Facultad de Ciencias Químicas, Universidad Autónoma de Coahuila, Saltillo, Mexico*

[2]*Centro de Investigación y de Estudios Avanzados del Instituto Politécnico Nacional (CINVESTAV-IPN), Ramos Arizpe, Coahuila, Mexico*

[3]*Greencorp Biorganiks SA de CV, Blvd. Luis Donaldo Colosio, San Patricio, C.P., Saltillo, Coahuila, Mexico*

ABSTRACT

The production of minimally processed foods has strikingly increased due to the high demand for products that are 'ready to eat, cook, or use.' New technologies in food packaging are improving as a result of changing consumer demands or industrial production trends. These changes act as the driving force for the development of new packaging concepts that extend shelf-life while maintaining and monitoring food safety and quality. Innovations in food packaging have been limited mainly to small changes in packaging materials, which could be defined as passive packaging systems to meet marketing demands in the short term. New concepts of smart and intelligent packaging are expected to play an increasingly important role by offering innovative solutions to extend shelf-life or maintain, improve, or monitor food quality and safety. The application and usage of smart and intelligent

Sustainable Agriculture and Global Environmental Health: Toward the Efficient Management of Resources.
Mónica L. Chávez González, Cristóbal Noé Aguilar, Juan Carlos Contreras-Esquivel, and A. K. Haghi (Eds.)

packaging can extend the shelf-life of food or improve its organoleptic properties while preventing food losses. This chapter presents definitions, concepts, advances, challenges, and trends of sustainable nanoscale polymer-based edible coatings, including examples of some applications on food packaging materials.

4.1 INTRODUCTION

The use of safe edible layers and coatings has been implemented in the food industry as an alternative strategy for the preservation and improvement of nutritional, functional, and sensory properties over a long shelf life.

In recent years, their development has grown greatly due to the benefits they present compared to artificial coatings. One of the most important benefits is that edible films and coatings can be ingested; therefore, biodegradability, biocompatibility, as well as safety, are essential requirements (Kouhi et al., 2020).

Researchers and the food industry have looked for alternatives regarding the problems derived from the use of traditional non-degradable polymers. Natural edible polymers have several advantages, such as high availability, being renewable, ecological, non-toxic, and above all, being biodegradable (Ali & Ahmed, 2018).

The terms edible layers and coatings are usually used as a single expression; however, the difference between them is that to obtain a coating, the food is immersed in a liquid mixture that contains the biopolymers. On the other hand, the layers are systems that are formed in molds such as sheets and then applied to the product and/or food (Dhumal & Sarkar, 2018).

The application of edible layers and coatings is diverse, from the use in meat products, cereals, bakery, and confectionery to fruits and vegetables (Vukić et al., 2017).

Nowadays, the use of edible coatings in food processing has been accepted satisfactorily, thanks to its benefits on product quality and the environment. Some of these benefits are the preservation of the organoleptic properties, reduction in the loss of humidity, and the formation of a barrier for the exchange of gases between the internal atmosphere of the fruit and the external atmosphere (López et al., 2016). In this document, definitions, concepts, advances, challenges, and trends of edible coatings based on nanostructured biopolymers will be presented, including examples of some applications in food packaging materials (Table 4.1).

TABLE 4.1 Biopolymers and Nanocomposites Used in Edible Coatings

Biopolymers	Nanocomposite	Application	Evaluated Effects	References
Xanthan gum	Zinc oxide nanoparticles (ZnO)	Tomato and apple	Weight loss and antibacterial activity.	Joshy et al. (2020).
Alginate and chitosan	Zinc oxide nanoparticles (ZnO)	Guava	Weight loss, determination of texture and color, soluble solids, and titratable acidity.	Arroyo et al. (2020)
NA	Cellulose nanofibers	Pineapple	Weight loss and microbiological analysis.	Ghosh et al. (2021)
Chitosan and gallotanin	Methylcellulose and polyethylene glycol nanoparticles.	Ricotta cheese	Antimicrobial and antifungal activity, antioxidant activity, mechanical and barrier properties.	Kritchenkov et al. (2021)
NA	Chitosan nanoparticles	Bell pepper	Antioxidant activity (DPPH and ABTS), antibacterial activity, weight loss.	Hu et al. (2020)
NA	Zein nanofibers	Apple	Antioxidant activity (DPPH), weight loss.	Leena et al. (2020)
NA	Silver nanoparticles	Bell pepper	Antibacterial activity, texture and color analysis, total dissolved solids, moisture content.	Saravanakumar et al. (2020)
Chitosan	Silver nanoparticles	Pork meat	Antimicrobial activity, antioxidant activity activity, cytotoxicity evaluation.	Wu et al. (2019)
Chitosan	Magnetic cellulose nanofibers	Freshly cut pineapple	Thermal stability, mechanical properties, toxicity tests.	Ghosh et al. (2019)
Polylactic acid	Titanium dioxide nanoparticles Silver nanoparticles	Mango	Antimicrobial activity, weight loss, and firmness.	Chi et al. (2019)

NA: Not apply.

4.2 COMPOSITION OF EDIBLE COVERS

Currently, the interest in more natural foods that offer safety and are healthy, both for the consumer and the environment, has been increasing. So, food packaging plays an important role in the food industry. Edible layers and containers have been the focus of attention as they offer advantages such as being biodegradable and non-toxic, compared to containers based on synthetic polymers. In general, the layers as well as the edible coatings are formulated based on polysaccharides, proteins, and/or lipids of natural origin (Chen et al., 2020).

Next, there are some characteristics and/or properties of materials commonly used for the formulation and manufacture of edible films and coatings that are described.

4.2.1 POLYSACCHARIDES

Polysaccharides have various physical attributes such as good solubility, good flow behavior (a quality that allows the material to flow and fill a certain cavity), gelling potential and/or surface and interfacial characteristics that depend on the composition of monosaccharides, their types of bonds, chain forms, and level of polymerization (Saberi & Goldin, 2018).

On the other hand, coatings based on polysaccharides can present low barriers against humidity, but selective permeability to O_2 and CO_2.

Among the polysaccharides most used as materials for coatings are: cellulose, chitin, and chitosan, as they are renewable, biocompatible, non-toxic, degradable, and inexpensive (Sun et al., 2021). In addition, they contribute to preserving color, reducing weight loss, improving antimicrobial activity, and extending the quality of the fruit.

4.2.1.1 CELLULOSE

Cellulose is a polysaccharide formed by glucose monomers (Mishra et al., 2018) and is an important component of all-natural fibers such as cotton, pine, linen, and jute, among others. It is also the most abundant natural biopolymer in nature. Cellulose is characterized by being a fibrous, resistant, and insoluble biopolymer in water. Its function is to maintain the structure of plant cell walls (Sabapathi & Siddaramaiah, 2015). Among the important characteristics that cellulose offers are that it is renewable, biodegradable, and does not present toxicity (Nechyporchuk et al., 2016).

4.2.1.2 STARCH

Starch is a natural polymer formed by carbohydrates and can be found in wheat, rice, potatoes, and corn (Babu et al., 2013); however, the main source is potatoes (Shengjun Wu, 2019).

Specifically, starch is made up of two microstructures: amylose, whose molecules are linear, allowing its behavior to be similar to that of traditional synthetic polymers, and amylopectin, whose molecules are branched and therefore have a higher molecular weight compared to amylose (Li et al., 2011).

Starch is a promising edible polymer for food packaging as it has advantages such as being environmentally friendly, odorless, abundant, and affordable compared to other natural polymers (Shih & Zhao, 2021). It also contributes to delaying the ripening of fruits after harvest due to its low oxygen permeability. However, as a disadvantage, it presents insufficient mechanical properties due to its hydrophilic character (Pellá et al., 2020).

In addition to the food industry, and thanks to the aforementioned advantages, it is considered a candidate material in the chemical, textile, and pharmaceutical industries, among others. Starch granules behave differently under conditions such as heat, and temperature, and, in some cases, they present swelling, gelatinization, and retrogradation, and this influences their application as a polymer (Yun & Genyi, 2021). For the different areas of application, the starch is extracted from the plant and is called "native starch," or chemical modifications are carried out to achieve specific properties and is called "modified starch" (Najafi, Baghaie, & Ashori, 2016).

4.2.1.3 CHITOSAN

Chitosan is a natural polysaccharide derived from chitin and is the main component of the cell wall of fungi and exoskeletons of crustaceans and insects (Yanat & Schroen, 2021). Due to its natural origin, it has important characteristics such as biodegradability and non-toxicity (Wisejena et al., 2015); therefore, it is considered a GRAS material (generally recognized as safe), according to the United States Food and Drug Administration (USFDA) (Zahra, 2019).

Due to its biological properties such as antitumor, antifungal, antimicrobial, and antioxidant activity among others, it is considered an important bioactive compound (Zahra, 2019). In addition to the aforementioned properties, chitosan has other properties such as high solubility and high reactivity with various functional groups; this is due to the free amino groups that are part of its structure (Rocha et al., 2017).

Due to its biological characteristics, chitosan offers multiple applications, whether it is used alone or in conjunction with other components of a polymeric nature such as starch, gelatin, and alginate. Among the applications of chitosan are the food, pharmaceutical, and water treatment industries among others (Kong et al., 2010); it also acts as a clarifying agent for beverages, such as fruit juice, wines, beer, and tea (Rocha et al., 2017).

4.2.1.4 ALGINATE

Alginate is a linear polyuronic acid hydrocolloid (Día, 1998) that is obtained from acid or alkali treatments from brown algae and/or soil bacteria (Babu et al., 2013). The main sources are brown algae, generally of the Laminaria, Lessonia, and Macrocystis genera (Rhein-Knudsen & Meyer, 2021).

Attached to the cell wall of algae, alginate is in the form of insoluble calcium or magnesium alginate, which involves extraction methods that transform insoluble alginate to its soluble form (Rhein-Knudsen & Meyer, 2021).

Among the advantages that alginate has, it is worth mentioning that it is accessible, easily extracted, and inexpensive, in addition to its good rheological properties, which makes it one of the most widely used biopolymeric materials in the food industry (Borazjani et al., 2017).

4.2.1.5 PECTIN

Pectin, whose name means "frozen" or "curdled," is found in its natural form as a heterogeneous polysaccharide of high molecular weight, composed of three domains: linear homogalacturonans (HG), branched rhamnogalacturonans-I (RG-I), and branched rhamnogalacturonanes-II (RG-II). The RG-I domain is highlighted as the region with the highest bioactivity. However, in its natural form, pectin exhibits poor bioavailability and bioactivity in vivo (Zhi et al., 2017).

Pectin is commercially produced from the peel of citrus fruits as a by-product of the extraction of lime, lemon, orange, or apple pomace juices.

Pectin has multiple uses in the fields of medicine, the chemical industry, and the food industry. Specifically, it is used as a biomaterial for the production of edible coatings due to its biocompatibility (Chen et al., 2020) and its gelling properties. This is why it is widely used in the food industry (Nieto, 2009). It is important to mention that its significance in the food industry lies in its ability to create gels in the presence of Ca^{2+} ions at low pH. In addition to ions, the degree of methoxylation also influences the formation of the gel, as pectin can be found in two forms: low methoxy pectin and high methoxy pectin (Thakur et al., 2009).

4.2.2 PROTEINS

Proteins are important in nutraceutical aspects. They are also used to produce various bioproducts such as fibers, adhesives, pharmaceutical products, and edible films and coatings (Muhammad et al., 2018). The high demand for the replacement of petroleum-based polymeric materials with renewable and ecological materials has generated great interest in the research and development of films and coatings based on natural polymers such as proteins. These can be of plant origin such as zein, soy, wheat, and animal gluten such as casein, whey, collagen, and egg albumin (Dangaran et al., 2009).

Proteins are characterized by their amino acid content, and it is their sequence that defines their function and; therefore, their uses and applications. Proteins are generally considered suitable polymers for making biodegradable films and coatings because of their ability to retain moisture, gas, and aromatic flavors, although they have disadvantages such as poor mechanical strength and high water permeability (Ahmed, 2019). One of the most attractive applications of proteins is their emulsifying capacity. Being an amphiphilic compound, that is, having both hydrophilic and lipophilic properties, it can achieve better homogenization between the oil phase and the aqueous phase of the polymeric matrix, and in this way, achieve greater stability of the emulsion (Burgos-Díaz et al., 2016).

4.2.2.1 PROTEINS OF ANIMAL'S ORIGIN

Animal products are the main source of protein in the diet of consumers worldwide (Mingle et al, 2021). Although proteins of animal origin are preferred for their quality and nutritional value, there are studies on the damage they cause to health (Ashaye et al., 2011).

4.2.2.1.1 Milk Proteins: Casein

Casein is the main protein component of bovine milk, presenting itself in the form of casein micelles, specifically as spherical aggregates (Tang et al., 2019). Casein micelles are colloidal systems formed by four different casein units (Grewal et al., 2021): αs1, αs2, β-casein, and κ-casein present in 40%, 10%, 35%, and 15% respectively (Castellanos et al., 2021).

Casein represents 80% of milk proteins, so its use in the food industry has been increasing thanks to its properties, as well as caseinate derivatives that are currently used in milk-derived products and those that are not derivatives (Gorji et al., 2014).

Milk proteins provide nutritional value to food; in addition, they act as functional and bioactive ingredients. Edible casein-based films offer advantages such as good water vapor permeability, good oxygen barrier, improve the strength and integrity of the coating, and allow them to adhere better to food products (Joanna Tkaczewska, 2020).

4.2.2.1.2 Plasma Proteins: Albumin

Serum albumin is the most abundant protein in the bloodstream. It performs various functions such as the transport of endogenous and exogenous molecules, as well as helping to maintain osmotic pressure and pH. Albumin synthesis is carried out in the liver, as well as most of the plasma proteins, with a 19-day half-life (Divsalar et al., 2012). In the area of health, serum albumin is very important because it has advantageous characteristics such as being biocompatible, biodegradable, non-immunogenic, and it is also not toxic (Aguilera-Garrido et al., 2021). Due to its emulsifying capacity, serum albumin is widely used in the food industry. This capacity is due to the hydrophobic and hydrophilic groups it possesses (Lam & Nickerson, 2013).

4.2.2.2 PLANT-BASED PROTEINS

Vegetable proteins are made up of different fractions: glutenin, globulin, albumin, and prolamin, from higher to lower amounts respectively. The use of proteins of plant origin in the food area has grown considerably compared to proteins of animal origin, as they are less allergenic. For this reason, the use of proteins of vegetable origin has shown a greater impact in recent years, not only in the development of new products but also in the diet of consumers (Nesterenko et al., 2013).

4.2.2.2.1 Soy Protein

Soy protein is a natural polymer used to develop textile fibers and molded parts for decades. In the food area, soy protein has received great interest in food preservation, specifically in the formation of edible films as it provides flexibility, softness, and transparency in contrast to other types of proteins of vegetable origin, in addition to presenting good oxygen permeability (Cho et al., 2007).

Soy protein has emulsifying and gelling properties, as well as good solubility. These are generally influenced by pH, heat treatment, and the presence of salts among other ingredients such as oils or carbohydrates (Nesterenko et al., 2013).

4.2.3 WAXES, LIPIDS, AND RESINS

Lipids and resins are not considered natural polymers; however, they are edible and biodegradable, allowing the formation of edible films and coatings. Most of these materials are solid at room temperature, and they act as a barrier against water. Thanks to this property, films and/or coatings formulated based on lipids are resistant to water and have low surface energy, due to their hydrophobicity. As an example, we have beeswax, candelilla wax, carnauba wax, cocoa butter, acetylated monoglycerides, fatty acids, and resins (Noemí, 2010).

4.2.3.1 WAXES

Waxes are a mixture of various molecules (such as hydrocarbons, fatty acids, and sterols among others) of a nonpolar nature. Waxes are classified as natural and synthetic, with natural waxes being a defense for plants against water loss, therefore, they are present on their surface (Rocha et al., 2013). The natural sources of wax are vegetables (such as carnauba, candelilla, and sugar cane wax) and animals (such as beeswax and lanolin) (Galus & Kadzinska, 2015).

4.2.3.1.1 Plant-Based Waxes

1. **Candelilla Wax:** It is a material that is extracted from candelilla (*Euphorbia antisyphilitica*), which in its raw form is brown, and once refined, it turns yellow (Alvarado et al., 2013). Due to its hydrophobic nature, it is an alternative as a barrier against humidity (Kowalczyk & Biendl, 2016). Thanks to this, within the diverse uses of candelilla wax, it is applied in the food industry mainly and the cosmetic industry as an ingredient for lip balms and deodorant bars (Rojas-Molina et al., 2011). In edible coatings for fruits and vegetables, it contributes to maintaining adequate levels of O_2 and CO_2, which reduces the development of microorganisms (Zamudio et al., 2016).
2. **Carnauba Wax:** It is extracted from the dust found on the leaves of the carnauba palm tree (*Copernicia prunifera*) and is mainly composed of free fatty acids, alcohols, and long-chain esters (Dantas et al., 2013). Carnauba wax has high hardness and low solubility (Encyclopedic Dictionary of Polymers, 2007). Its main application is as a material for edible coatings on fruits to provide shine and delay moisture loss (Mee & Sea, 2018).

4.2.3.1.2 ANIMAL'S ORIGIN WAXES

1. **Bee Wax:** It is the result of the secretion of the wax-producing glands of bees of the genus *Apis mellifera* and *Apis cerana* (Saralaya et al., 2021).

   Its application as a component in edible coatings is important as it improves the barrier against moisture and mechanical properties, as well as enhances the appearance of fresh fruits (Ochoa et al., 2017).
2. **Lanolin:** It is derived from wool fat, which is secreted by the sebaceous glands of sheep and acts as a natural lubricant that protects sheep's wool against environmental factors such as sun, rain, and air. Lanolin is made up of a mixture of high molecular-weight esters, polyesters, and fatty acids (Schlossman & McCarthy, 1978). By having a high content of sterols, lanolin has the ability to absorb water, preventing moisture from penetrating the surface of products and thus preventing damage to surfaces (Broddy & Sudhakar, 2018).

4.3 NANOMATERIALS AND THEIR APPLICATION

Emerging technologies offer an alternative in the food packaging area by providing packaging materials that will guarantee food products with a long shelf life, keeping them safe, and thus guaranteeing their quality (López et al., 2016).

Nanotechnology is based on the application of materials at the atomic and molecular level (1–100 nanometers) (Manickam et al., 2017) and integrates various disciplines such as physics, chemistry, biotechnology, and engineering (Bajpai et al., 2018). The approximate impact of nanotechnology in 2020 was $3 trillion on the global economy (Duncan, 2011).

Unlike conventional materials, nanomaterials have a greater surface area per unit mass, which guarantees greater accessibility to a given mass of nanomaterials (Clement, Iseli, Karote, & Cremer, Rajagopalan, 2012).

A nanomaterial is defined as a material that has a size of 1–100 nm and can be obtained by chemical synthesis and vapor deposition, mainly (Park et al., 2020). These nanomaterials can be formulated from polymers of natural or synthetic origin (Karna & Geun-Pyo, 2021). In this context, nanocomposites result from the combination of materials with a nanometric scale structure with another existing material or biomolecule, resulting in a larger particle size at 100 nm (Bajpai et al., 2018).

According to their chemical and physical properties, shape, and size, nanoparticles (nanomaterials) can be classified into different groups as shown in Figure 4.1 (Naseem & Durrani, 2021).

Nanomaterials provide a variety of functionalities due to their physicochemical properties such as particle size and surface area (Allan et al., 2021). For example, in the area of nanomedicine, they are an alternative for the diagnosis and treatment of neurodegenerative diseases such as Alzheimer's, and they also have a high therapeutic potential by facilitating the storage and targeted administration of drugs (Xu et al., 2021) (Figure 4.1).

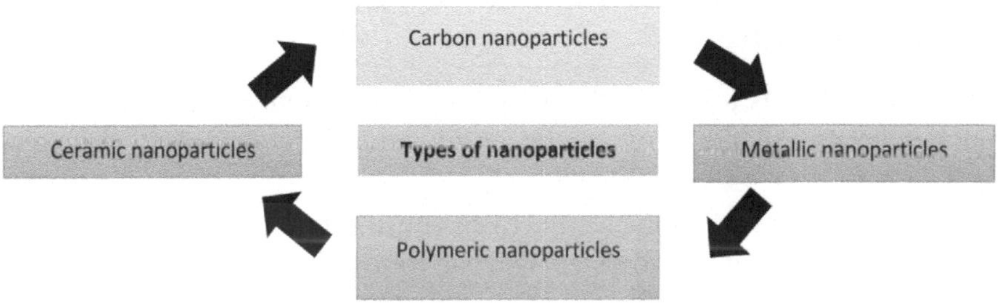

FIGURE 4.1 Some types of nanoparticles.

4.3.1 *NANOMATERIAL MANUFACTURING METHODS*

The effects obtained in nanoscale materials, such as magnetic, optical, thermal, and electrical properties, have been a focus of attention in the area of research in different fields, such as biomedicine (Xiong et al., 2021). These properties are reflected as nanomaterials have a small size (nanometric) and, therefore, a larger specific surface area.

Due to the growing interest in the application of nanomaterials, various synthesis methods have been developed. These are based on two general principles: 1) "top-down" where a large material is reduced using physical and chemical procedures, and 2) "bottom-up" where atomic-sized materials are joined to form larger materials such as films or nanostructured materials (Rodríguez-Félix et al., 2021).

These nanomaterials can be obtained by different methods, such as chemical synthesis; however, their cost is high and they can contain toxic chemical compounds in their structure that represent a biological hazard for the consumer and the environment (Behravan et al., 2019). Another synthesis method is low-pressure vapor deposition. In contrast to chemical synthesis, it is an ecological and cost-effective alternative (Jin et al., 2021) (Figure 4.2).

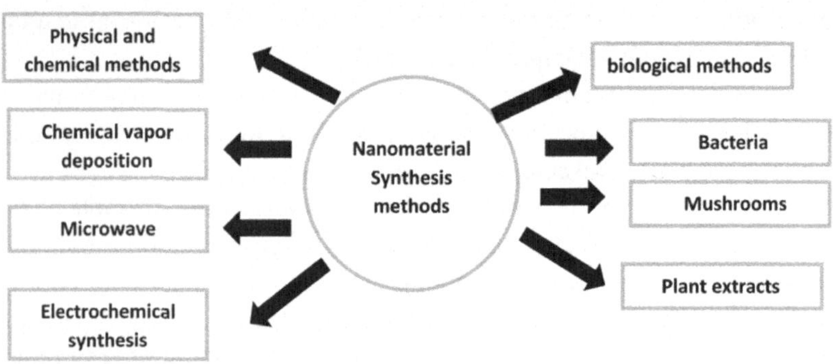

FIGURE 4.2 Nanomaterial synthesis methods.

4.4 NANO-EMULSIONS

A nanoemulsion is a system that consists of two immiscible liquid phases and can be formulated in two ways: oil in water and water in oil (Divsalar et al., 2012).

Nanoemulsions are characterized by having a submicroscopic size (1 μm), they behave like a colloidal system and can transport various molecules such as essential oils and other antimicrobial compounds (Hamidreza, Asghar, & Hasti, 2021).

These nanoemulsions are used in different fields of application, such as the pharmaceutical, cosmetic, and food industries. It is important to mention that, in the food industry, nanoemulsions are used in the encapsulation, protection, and storage of food (Borthakur et al., 2016). Nanoemulsions are prepared by various methods, such as dispersion and condensation. Due to their particle size, the main application of nanoemulsions is the elaboration of nanoparticles, through the mini-emulsion polymerization process (Solans et al., 2005).

Some nano emulsions benefits are:

- Non-toxic;
- They function as carrier agents;
- They act as transporters of bioactive substances;
- They improve stability and physical resistance;
- They have better transparency.

4.5 TYPES OF NANOMATERIALS

As mentioned above, nanomaterials are those materials that are on a scale of 1–100 nm. Within this group, there are carbon nanotubes, graphene, metallic nanoparticles, and metallic oxides, among others. Some examples are briefly described below.

4.5.1 CARBON NANOMATERIALS

Carbon nanomaterials have attracted attention thanks to their excellent qualities such as high thermal stability, good electrical conductivity, and high contact surface, in addition to the fact that carbon is completely inert (Majidi et al., 2020).

Carbon nanomaterials (CNM) and their derivatives (carbon nanotubes, graphite, graphene/graphene oxide, and fullerenes) have unique characteristics and applications. CNMs generally provide electrical conductivity, mechanical strength, thermal conductivity, photoluminescence, transparency, and construction durability, as well as excellent surface and antimicrobial activities against pathogens (Azizi-Lalabadi et al., 2020). Thanks to the properties they offer, within this group of nanomaterials, carbon nanotubes have various applications as additives for polymers and catalysts, lithium battery anodes, and gas discharge tubes in telecommunications networks, among others (Zaporotskova et al., 2016).

One of the synthesis methods of these materials is based on thermal radiation, an example is microwave heating (Schwenke & Hoeppener, 2015).

4.5.2 METALLIC NANOPARTICLES

Metallic nanoparticles can be obtained by electrochemical methods and by microbial synthesis, the latter being of great importance due to their profitability, simplicity, and environmental safety, among other advantages (Elahian et al., 2017). Generally, this group of nanomaterials is manufactured by hydrothermal/solvothermal processes at a temperature ranging from 100 to 200°C (Xiong et al., 2021). Within this group of nanomaterials are silver, titanium, and zinc nanoparticles, which, together with other biopolymers applied in films, can improve the gas barrier and antimicrobial activity (Chi et al., 2021).

4.5.3 METAL OXIDE NANOPARTICLES

According to Diez Pascual, the metal oxide nanoparticles incorporated into biopolymers considerably enhance the antibacterial properties. This is due to certain properties such as size, shape, and chemical composition. Metallic nanoparticles are used as materials for food packaging as they reduce contamination by pathogens (Saravanakumar et al., 2020). However, future research is necessary since the information is limited, this to the recent application and development of nanocomposites. Among the most common metal oxide nanoparticles are zinc oxide, titanium oxide, copper oxide, and magnesium oxide, among others (Ana María Diez Pascual, 2019) (Figure 4.3).

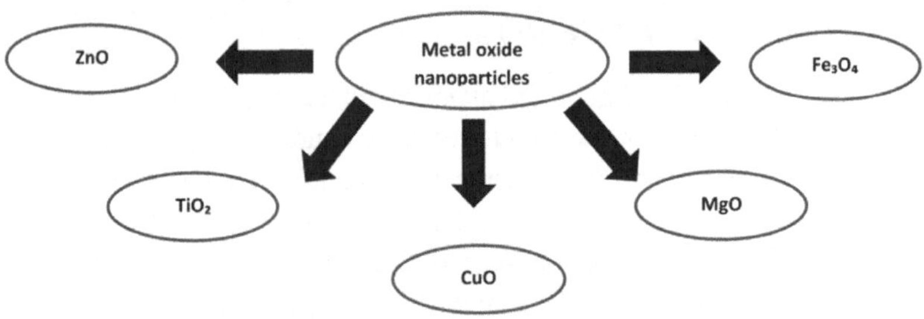

FIGURE 4.3 Types of metal oxide nanoparticles.

4.5.4 NATURAL POLYMERS

Currently, hydrocarbon-based plastics are the best-known and most-used materials for food packaging due to their rapid production, profitability, and excellent mechanical properties. However, most of these synthetic plastics present a problem as they are not biodegradable, which can cause environmental issues. On the other hand, natural polymers offer a promising alternative as packaging materials (Chawla et al., 2021).

Natural polymers or biopolymers can be obtained from polysaccharides, proteins, lipids, and even produced by microorganisms through fermentation. One of the advantages offered by these materials is that they can be edible and have antioxidant and antimicrobial properties (Behjat & Mina, 2020).

Edible coatings based on natural polymers act by forming a semi-permeable barrier, reducing water loss, and thus prolonging the useful life and quality of the food (Moreira et al., 2020).

In recent years, interest in natural bio-based polymers has grown considerably because they are naturally occurring. Some examples are starch, polylactic acid, and polyhydroxyalkanoates, among others (Babu et al., 2013).

However, despite their nature and advantages such as being ecological and providing protection, safety, and quality to food-based products, biopolymers have low mechanical and barrier properties in contrast to hydrocarbon-based materials, which limits their application in food packaging. So, the use of nanoscale compounds is a useful alternative to optimize the properties of biopolymers and/or thermoplastics of synthetic origin (Mihindukulasuriya & Lim, 2014).

A solution to the aforementioned disadvantages is the use of nanoscale polymers. Polymeric nanocomposites result from the application of an inert filler (on a scale of 1–100 nm) throughout the polymer. Among the most commonly used fillers are cellulose nanoparticles, chitin or chitosan nanoparticles, and starch nanocrystals, to name a few (Duncan, 2011). Figure 4.4 shows some natural and bio-based polymers generally used in the food industry.

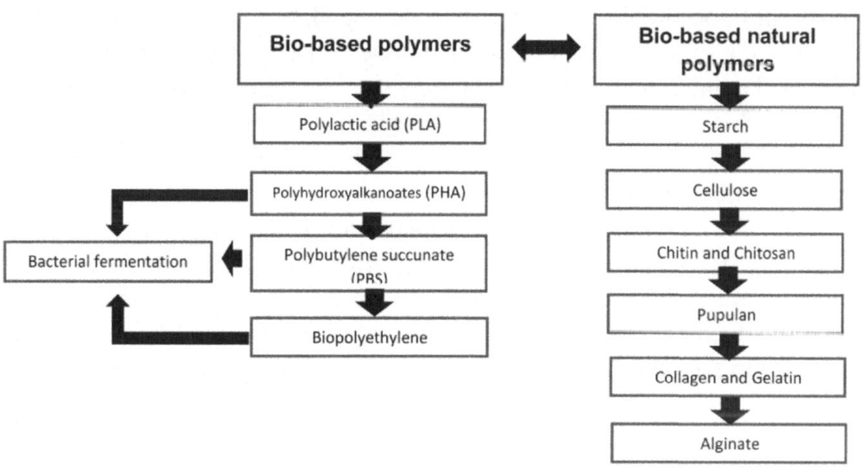

FIGURE 4.4 Bio-based polymers.

4.6 SAFETY OF NANOMATERIALS IN FOOD

The concern for the safety and innocuousness of compounds and additives in food is increasing. Therefore, consumers are leaning towards the use of materials of natural origin that are not toxic.

In the food industry, a container, in addition to fulfilling the function of containing and protecting the product, aims to extend its useful life, guaranteeing the preservation of its organoleptic characteristics and nutritional value. Migration is a phenomenon that occurs between the components of the container and the food and is a relevant factor for risk assessment in terms of toxicity and environmental impact (Videira-Quintela et al., 2021).

Nanomaterials, due to their submicroscopic size, migrate faster than conventional additives. Therefore, it is essential that all materials used in the food and medicine area be evaluated in detail to rule out any risk to consumer health and the environment (Xavier et al., 2021).

The use of nanomaterials in different fields of industry is increasing. In turn, the possibility of harmful unwanted effects (Bergamaschi et al., 2015) is also increasing. Due to the previously mentioned advantages that nanotechnology offers, novel products are developed quickly compared to regulatory frameworks.

Current methods for evaluating food safety and security, such as cell culture, microbiological tests, and chemical tests, require long periods of time and several stages for their analysis. They are also expensive and require professional experts and special tools (Zahra & Seid, 2020).

To standardize and facilitate the risk assessment of nanomaterials, the European Food Safety Authority (EFSA), together with contributions from experts and external institutions that assess the risks of nanomaterials, published a guide that covers the aspects that should be considered for food safety, such as physicochemical characterization, exposure assessment, and hazard identification using in vitro and in vivo methods, as well as toxicological studies (Hardy et al., 2018).

Assessing the physicochemical properties of the nanomaterials of interest is relevant. EFSA recommends applying at least one screening and confirmatory method. The screening methods are based on light scattering and particle movement, for example, DLS or the Brunauer-Emmett-Teller (BET) method. On the other hand, the confirmatory methods are based on electron microscopy (EM), the latter can be scanning or transmission (Wohlleben et al., 2017; Peters et al., 2014).

The evaluation of nanomaterials from their characterization (evaluation of physicochemical properties) to their *in vivo* functionality is important. *In vitro*, genotoxicity evaluation of nanomaterials is relevant to rule out genetic mutations and/or chromosomal aberrations. Toxicity tests are also important; these include the evaluation of cell viability, induction of oxidative stress (ROS formation), and induction of inflammation (Xavier et al., 2021).

KEYWORDS

- **cell viability**
- **edible coating**
- **genetic mutations**
- **nanomaterial synthesis methods**
- **nanomaterials**
- **natural polymers**
- **oxidative stress**

REFERENCES

Aguilera-Garrido, A., del Castillo-Santaella, T., Yang, Y., Galisteo-González, F., Gálvez-Ruiz, M. J., Molina-Bolívar, J. A., Holgado-Terriza, J. A., Cabrerizo-Vilchez, M. A., & Maldonado-Valderrama, J. (2021). Applications of serum albumins in delivery systems: Differences in interfacial behaviour and interacting abilities with polysaccharides. *Advances in Colloid and Interface Science, 290*, 102365. https://doi.org/10.1016/j.cis.2021.102365.

Ali, A., & Ahmed, S. (2018). Desarrollo de hidrogeles a partir de polímeros comestibles. En T. Gutiérrez (Ed.), *Polímeros Para Aplicaciones Alimentarias*. Springer. https://doi.org/10.1007/978-3-319-94625-2_21.

Allan, J., Belz, S., Hoeveler, A., Hugas, M., Okuda, H., Patri, A., Rauscher, H., Silva, P., Slikker, W., Sokull-Kluettgen, B., Tong, W., & Anklam, E. (2021). Regulatory landscape of nanotechnology and nanoplastics from a global perspective. *Regulatory Toxicology and Pharmacology, 122*, 104885. https://doi.org/10.1016/j.yrtph.2021.104885.

Alvarado, C. J., Galindo, S. A., Bermúdez, L. B., Berumen, C. P., Orta, C. Á., & Garza, J. A. V. (2013). Cera de Candelilla y sus aplicaciones. *Avances en Química, 8*(2), 105–110. ISSN: 1856–5301.

Arif, M., Chia, L. S., & Pauls, K. P. (2018). Bioproductos a base de proteínas. En G. Chen, R. Weselake, & S. Singer (Eds.), *Plant Bioproducts*. Springer. https://doi.org/10.1007/978-1-4939-8616-3_9.

Arroyo, B. J., Bezerra, A. C., Oliveira, L. L., Arroyo, S. J., Melo, E. A. de, & Santos, A. M. P. (2020). Antimicrobial active edible coating of alginate and chitosan add ZnO nanoparticles applied in guavas (*Psidium guajava* L.). *Food Chemistry, 309*, 125566. https://doi.org/10.1016/j.foodchem.2019.125566.

Ashaye, J., Gaziano, L., & Djoussé, L. (2011). Red meat consumption and risk of heart failure in male physicians. *Nutrition, Metabolism and Cardiovascular Diseases, 21*(12), 941–946. https://doi.org/10.1016/j.numecd.2010.03.009.

Azizi-Lalabadi, M., Hashemi, H., Feng, J., & Jafari, S. M. (2020). Carbon nanomaterials against pathogens; the antimicrobial activity of carbon nanotubes, graphene/graphene oxide, fullerenes, and their nanocomposites. *Advances in Colloid and Interface Science, 284*, 102250. https://doi.org/10.1016/j.cis.2020.102250.

Babu, R. P., O'Connor, K., & Seeram, R. (2013). Current progress on bio-based polymers and their future trends. *Progress in Biomaterials, 2*(8). https://doi.org/10.1186/2194-0517-2-8.

Bajpai, V. K., Kamle, M., Shukla, S., Mahato, D. K., Chandra, P., Hwang, S. K., Kumar, P., Huh, Y. S., & Han, Y.-K. (2018). Prospects of using nanotechnology for food preservation, safety, and security. *Journal of Food and Drug Analysis, 26*(4), 1201–1214. https://doi.org/10.1016/j.jfda.2018.06.011.

Barbosa-Canovas G. (eds) Food Engineering Interfaces. Serie de Ingeniería de Alimentos. Springer, Nueva York, NY. https://doi.org/10.1007/978-1-4419-7475-4_27.

Behravan, M., Panahi, A. H., Naghizadeh, A., Mahdavi, R., & Mirzapour, A. (2019). Facile green synthesis of silver nanoparticles using *Berberis vulgaris* leaf and root aqueous extract and its antibacterial activity. *International Journal of Biological Macromolecules, 124*, 148–154. https://doi.org/10.1016/j.ijbiomac.2018.11.101.

Bergamaschi, E., Poland, C., Guseva Canu, I., & Prina-Mello, A. (2015). The role of biological monitoring in nano-safety. *Nano Today, 10*(3), 274–277. https://doi.org/10.1016/j.nantod.2015.02.001.

Borazjani, N. J., Tabarsa, M., You, S., & Rezaei, M. (2017). Effects of extraction methods on molecular characteristics, antioxidant properties and immunomodulation of alginates from *Sargassum angustifolium*. *International Journal of Biological Macromolecules, 101*, 703–711. https://doi.org/10.1016/j.ijbiomac.2017.03.128.

Borthakur, P., Boruah, P. K., Sharma, B., & Das, M. R. (2016). Nanoemulsion: preparation and its application in food industry. In A. M. Grumezescu (Ed.), *Nanotechnology in the Agri-Food Industry, Emulsions* (pp. 153–191). Academic Press. https://doi.org/10.1016/B978-0-12-804306-6.00005-2.

Broddy, S., & Sudhakar, K. V. (2017). Corrosion Inhibition: Investigation of Lanolin Coating on the Corrosion Characteristics of Low Carbon Steel in Simulated Sea Water. *Protection of Metals and Physical Chemistry of Surfaces, 53*, 1125–1130. https://doi.org/10.1134/S2070205117060077.

Burgos-Díaz, C., Wandersleben, T., Marqués, A. M., & Rubilar, M. (2016). Multilayer emulsions stabilized by vegetable proteins and polysaccharides. *Current Opinion in Colloid & Interface Science, 25*, 51–57. https://doi.org/10.1016/j.cocis.2016.06.014.

Cera de carnauba. (2007). En: Gooch JW (eds) Diccionario enciclopédico de polímeros. Springer, Nueva York, NY. https://doi.org/10.1007/978-0-387-30160-0_1943.

Chen, Y., Xu, L., Wang, Y., Chen, Z., Zhang, M., & Chen, H. (2020). Characterization and functional properties of a pectin/tara gum based edible film with ellagitannins from the unripe fruits of *Rubus chingii* Hu. *Food Chemistry, 325*, 126964. https://doi.org/10.1016/j.foodchem.2020.126964.

Chi, H., Song, S., Luo, M., Zhang, C., Li, W., Li, L., & Qin, Y. (2019). Effect of PLA nanocomposite films containing bergamot essential oil, TiO_2 nanoparticles, and Ag nanoparticles on shelf life of mangoes. *Scientia Horticulturae, 249*, 192–198. https://doi.org/10.1016/j.scienta.2019.01.059.

Cho, S. Y., Park, J. W., Batt, H. P., & Thomas, R. L. (2007). Edible films made from membrane processed soy protein concentrates. *LWT – Food Science and Technology, 40*(3), 418–423. https://doi.org/10.1016/j.lwt.2006.02.003.

Clement, K., Iseli, A., Karote, D., Cremer, J., & Rajagopalan, S. (2012). Materiales nanoestructurados: aplicaciones industriales. In J. Kent (Ed.), *Manual de Química Industrial y Biotecnología*. Springer. https://doi.org/10.1007/978-1-4614-4259-2_9.

Dantas, S., Magalhães, A. N., Matos, T. A., Gouveia, W. O., & Lopes, G. S. (2013). Characterization of Carnauba Wax Inorganic Content. *Journal of the American Oil Chemists' Society, 90*, 1475–1483. https://doi.org/10.1007/s11746-013-2300-6.

Dhumal, C. V., & Sarkar, P. (2018). Películas y recubrimientos comestibles compuestos de biopolímeros de grado alimenticio. *Journal of Food Science and Technology, 55*, 4369–4383. https://doi.org/10.1007/s13197-018-3402-9.

Día, D. F. (1998). Alginatos. In D. L. Kaplan (Ed.), *Biopolymers from Renewable Resources. Sistemas Macromoleculares: Enfoque de Materiales* (pp. 5–22). Springer. https://doi.org/10.1007/978-3-662-03680-8_5.

Diez-Pascual, A. M. (2019). Nanocomposites antibacterianos basados en polímeros termoendurecibles derivados de aceites vegetales y nanopartículas de óxidos metálicos. *Polímeros, 11*, 1790. https://doi.org/10.3390/polym11111790.

Divsalar, A., Saboury, A. A., Nabiuni, M., Zare, Z., Kefayati, M. E., & Seyedarabi, A. (2012). Effect analysis of a newly designed nanoemulsion targeting human serum albumin for drug delivery. *Colloids and Surfaces B: Biointerfaces*, *98*, 80–84. https://doi.org/10.1016/j.colsurfb.2012.04.036.

Duncan, T. V. (2011). Applications of nanotechnology in food packaging and food safety: Barrier materials, antimicrobials and sensors. *Journal of Colloid and Interface Science, 363*(1), 1–24. https://doi.org/10.1016/j.jcis.2011.07.017.

Elahian, F., Reiisi, S., Shahidi, A., & Mirzaei, S. A. (2017). High-throughput bioaccumulation, biotransformation, and production of silver and selenium nanoparticles using genetically engineered *Pichia pastoris*. *Nanomedicine: Nanotechnology, Biology and Medicine, 13*(3), 853–861. https://doi.org/10.1016/j.nano.2016.10.009.

Flores-López, M. L., Cerqueira, M. A., de Rodríguez, D. J., & Vicente, A. A. (2016). Perspectivas sobre la utilización de recubrimientos comestibles y recubrimientos nanolaminados para la extensión del almacenamiento poscosecha de frutas y hortalizas. *Food Engineering Reviews, 8*, 292–305. https://doi.org/10.1007/s12393-015-9135-x.

Galus, S., & Kadzinska, J. (2015). Aplicaciones alimentarias de películas y recubrimientos comestibles a base de emulsión. *Tendencias en Ciencia y Tecnología de los Alimentos, 45*(2), 273–283. doi: 10.1016/j.tifs.2015.07.011.

George, J., Sabapathi, S. N., & Siddaramaiah. (2015). Nanocompuestos a base de polímeros solubles en agua que contienen nanocristales de celulosa. In V. Thakur & M. Thakur (Eds.), *Nanocomposites de Polímeros Ecológicos. Materiales Estructurados Avanzados* (Vol. 75). Springer. https://doi.org/10.1007/978-81-322-2470-9_9.

Ghorbani Gorji, S., Ghorbani Gorji, E., Mohammadifar, M. A., & Zargaraan, A. (2014). Complexation of sodium caseinate with gum tragacanth: Effect of various species and rheology of coacervates. *International Journal of Biological Macromolecules, 67*, 503–511. https://doi.org/10.1016/j.ijbiomac.2014.02.037.

Ghosh, T., Nakano, K., Mulchandani, N., & Katiyar, V. (2021). Curcumin loaded iron functionalized biopolymeric nanofibre reinforced edible nanocoatings for improved shelf life of cut pineapples. *Food Packaging and Shelf Life, 28*, 100658. https://doi.org/10.1016/j.fpsl.2021.100658.

Ghosh, T., Teramoto, Y., & Katiyar, V. (2019). Influence of Nontoxic Magnetic Cellulose Nanofibers on Chitosan Based Edible Nanocoating: A Candidate for Improved Mechanical, Thermal, Optical, and Texture Properties. *Journal of Agricultural and Food Chemistry, 67*(15), 4289–4299. https://doi.org/10.1021/acs.jafc.8b05905.

Grewal, M. K., Vasiljevic, T., & Huppertz, T. (2021). Influence of calcium and magnesium on the secondary structure in solutions of individual caseins and binary casein mixtures. *International Dairy Journal, 112*, 104879. https://doi.org/10.1016/j.idairyj.2020.104879.

Hammam, A. R. A. (2019). Tecnología, aplicaciones y características de películas y recubrimientos comestibles: una revisión. *SN Applied Sciences, 1*, 632. https://doi.org/10.1007/s42452-019-0660-8.

Hardy, A., Benford, D., Halldorsson, T., Jeger, M. J., Knutsen, H. K., More, S., & Mortensen, A. (2018). Guidance on risk assessment of the application of nanoscience and nanotechnologies in the food and feed chain: Part 1, human and animal health. *EFSA Journal, 16*(7), 5327. https://doi.org/10.2903/j.efsa.2018.5327.

Hu, X., Saravanakumar, K., Sathiyaseelan, A., & Wang, M.-H. (2020). Chitosan nanoparticles as edible surface coating agent to preserve the fresh-cut bell pepper (*Capsicum annuum* L. var. grossum (L.) Sendt). *International Journal of Biological Macromolecules, 165*(A), 948–957. https://doi.org/10.1016/j.ijbiomac.2020.09.176.

Jin, H., Sun, Y., Sun, Z., Yang, M., & Gui, R. (2021). Zero-dimensional sulfur nanomaterials: Synthesis, modifications and applications. *Coordination Chemistry Reviews, 438*, 213913. https://doi.org/10.1016/j.ccr.2021.213913.

Joshy, K. S., Jose, J., Li, T., Thomas, M., Shankregowda, A. M., Sreekumaran, S., Kalarikkal, N., & Thomas, S. (2020). Application of novel zinc oxide reinforced xanthan gum hybrid system for edible coatings. *International Journal of Biological Macromolecules, 151*, 806–813. https://doi.org/10.1016/j.ijbiomac.2020.02.085.

Kazemeini, H., Azizian, A., & Adib, H. (2021). Inhibition of *Listeria monocytogenes* growth in turkey fillets by alginate edible coating with *Trachyspermum ammi* essential oil nano-emulsion. *International Journal of Food Microbiology, 344*, 109104. https://doi.org/10.1016/j.ijfoodmicro.2021.109104.

Kong, M., Chen, X. G., Xing, K., & Park, H. J. (2010). Antimicrobial properties of chitosan and mode of action: A state of the art review. *International Journal of Food Microbiology, 144*(1), 51–63. https://doi.org/10.1016/j.ijfoodmicro.2010.09.012.

Kouhi, M., Prabhakaran, M. P., & Ramakrishna, S. (2020). Edible polymers: An insight into its application in food, biomedicine and cosmetics. *Trends in Food Science & Technology, 103*, 248–263. https://doi.org/10.1016/j.tifs.2020.05.025.

Kowalczyk, D., & Biendl, M. (2016). Physicochemical and antioxidant properties of biopolymer/candelilla wax emulsion films containing hop extract – A comparative study. *Food Hydrocolloids, 60*, 384–392. https://doi.org/10.1016/j.foodhyd.2016.04.010.

Kritchenkov, A. S., Egorov, A. R., Volkova, O. V., Artemjev, A. A., Kurliuk, A. V., Le, T. A., Truong, H. H., Le-Nhat-Thuy, G., Thi, T. V. T., Tuyen, N. V., & Khrustalev, V. N. (2021). Novel biopolymer-based nanocomposite food coatings that exhibit active and smart properties due to a single type of nanoparticles. *Food Chemistry, 343*, 128676. https://doi.org/10.1016/j.foodchem.2020.128676.

Kumar, S., Mukherjee, A., & Dutta, J. (2020). Chitosan based nanocomposite films and coatings: Emerging antimicrobial food packaging alternatives. *Trends in Food Science & Technology, 97*, 196–209. https://doi.org/10.1016/j.tifs.2020.01.002.

Lam, R. S. H., & Nickerson, M. T. (2013). Food proteins: A review on their emulsifying properties using a structure–function approach. *Food Chemistry, 141*(2), 975–984. https://doi.org/10.1016/j.foodchem.2013.04.038.

Lara-Castellanos, M. J., Azuara, E., Jimenez-Fernandez, V. M., Luna-Solano, G., & Jimenez, M. (2021). Effect of casein replacement by modified casein on physicochemical, textural, sensorial properties and microbiological stability of fresh cheese. *International Dairy Journal, 112*, 104864. https://doi.org/10.1016/j.idairyj.2020.104864.

Leena, M. M., Yoha, K. S., Moses, J. A., & Anandharamakrishnan, C. (2020). Edible coating with resveratrol loaded electrospun zein nanofibers with enhanced bioaccessibility. *Food Bioscience, 36*, 100669. https://doi.org/10.1016/j.fbio.2020.100669.

Li, M., Liu, P., Zou, W., Yu, L., Xie, F., Pu, H., Liu, H., & Chen, L. (2011). Extrusion processing and characterization of edible starch films with different amylose contents. *Journal of Food Engineering, 106*(1), 95–101. https://doi.org/10.1016/j.jfoodeng.2011.04.021.

Mahmoudi Najafi, S. H., Baghaie, M., & Ashori, A. (2016). Preparation and characterization of acetylated starch nanoparticles as drug carrier: Ciprofloxacin as a model. *International Journal of Biological Macromolecules, 87*, 48–54. https://doi.org/10.1016/j.ijbiomac.2016.02.030.

Majidi, H. J., Mirzaee, A., Jafari, S. M., Amiri, M., Shahrousvana, M., & Babaei, A. (2020). Fabrication and characterization of graphene oxide-chitosan-zinc oxide ternary nano-hybrids for the corrosion inhibition of mild steel. *International Journal of Biological Macromolecules, 148*, 1190–1200. https://doi.org/10.1016/j.ijbiomac.2019.11.060.

Manickam, V., Velusamy, R. K., Lochana, R., Amiti, Rajendran, B., & Ramasamy, T. (2017). Genotoxicidad de los nanomateriales en los alimentos. In S. Ranjan, N. Dasgupta, & E. Lichtfouse (Eds.), *Nanocience in Food and Agriculture 4. Sustainable Agriculture Reviews* (Vol. 24). Springer. https://doi.org/10.1007/978-3-319-53112-0_4.

Mihindukulasuriya, S. D. F., & Lim, L.-T. (2014). Nanotechnology development in food packaging: A review. *Trends in Food Science & Technology, 40*(2), 149–167. https://doi.org/10.1016/j.tifs.2014.09.009.

Mingle, C. L., Darko, G., Asare-Donkor, N. K., Borquaye, L. S., & Woode, E. (2021). Patterns in protein consumption in Ghanaian cities. *Scientific African, 11*, e00684. https://doi.org/10.1016/j.sciaf.2020.e00684.

Mishra, R. K., Ha, S. K., Verma, K., & Tiwari, S. K. (2018). Recent progress in selected bio-nanomaterials and their engineering applications: An overview. *Journal of Science: Advanced Materials and Devices, 3*(3), 263–288. https://doi.org/10.1016/j.jsamd.2018.05.003.

Mohammadi, Z., & Jafari, S. M. (2020). Detection of food spoilage and adulteration by novel nanomaterial-based sensors. *Advances in Colloid and Interface Science, 286*, 102297. https://doi.org/10.1016/j.cis.2020.102297.

Moreira, B. R., Pereira-Júnior, M. A., Fernandes, K. F., & Batista, K. A. (2020). An ecofriendly edible coating using cashew gum polysaccharide and polyvinyl alcohol. *Food Bioscience, 37*, 100722. https://doi.org/10.1016/j.fbio.2020.100722.

Naseem, T., & Durrani, T. (2021). The role of some important metal oxide nanoparticles for wastewater and antibacterial applications: A review. *Environmental Chemistry and Ecotoxicology, 3*, 59–75. https://doi.org/10.1016/j.enceco.2020.12.001.

Nechyporchuk, O., Belgacem, M. N., & Bras, J. (2016). Production of cellulose nanofibrils: A review of recent advances. *Industrial Crops and Products, 93*, 2–25. https://doi.org/10.1016/j.indcrop.2016.02.016.

Nesterenko, A., Alric, I., Silvestre, F., & Durrieu, V. (2013). Vegetable proteins in microencapsulation: A review of recent interventions and their effectiveness. *Industrial Crops and Products, 42*, 469–479. https://doi.org/10.1016/j.indcrop.2012.06.035.

Nieto, M. B. (2009). Estructura y función de películas y recubrimientos comestibles a base de goma de polisacárido. In K. Huber & M. Embuscado (Eds.), *Películas y Recubrimientos Comestibles Para Aplicaciones Alimentarias*. Springer. https://doi.org/10.1007/978-0-387-92824-1_3.

Ochoa, T. A., Almendárez, B. E. G., Reyes, A. A., Pastrana, D. M. R., López, G. F. G., Belloso, O. M., & González, C. R. (2017). Design and Characterization of Corn Starch Edible Films Including Beeswax and Natural Antimicrobials. *Food Bioprocess Technol, 10*(1), 103–114. https://doi.org/10.1007/s11947-016-1800-4.

Oregel-Zamudio, E., Aguilar, C. N., Oyoque-Salcedo, G., Angoa-Pérez, M. V., & Mena-Violante, H. G. (2016). Caracterización fisicoquímica de películas comestibles a base de cera de candelilla. *Revista Iberoamericana de Tecnología Postcosecha, 17*(1), 1–7.

Park, S., Jeon, Y., Han, T., Kim, S., Gwon, Y., & Kim, J. (2020). Nanoscale manufacturing as an enabling strategy for the design of smart food packaging systems. *Food Packaging and Shelf Life, 26*, 100570. https://doi.org/10.1016/j.fpsl.2020.100570.

Pellá, M. C. G., Silva, O. A., Pellá, M. G., Beneton, A. G., Caetano, J., Simões, M. R., & Dragunski, D. C. (2020). Effect of gelatin and casein additions on starch edible biodegradable films for fruit surface coating. *Food Chemistry, 309*, 125764. https://doi.org/10.1016/j.foodchem.2019.125764.

Rhein-Knudsen, N., & Meyer, A. S. (2021). Chemistry, gelation, and enzymatic modification of seaweed food hydrocolloids. *Trends in Food Science & Technology, 109*, 608–621. https://doi.org/10.1016/j.tifs.2021.01.052.

Rocha, J. C. B., Lopes, J. D., Mascarenhas, M. C. N., Arellano, D. B., Guerreiro, L. M. R., & Cunha, R. L. D. (2013). Thermal and rheological properties of organogels formed by sugarcane or candelilla wax in soybean oil. *Food Research International, 50*(1), 318–323. https://doi.org/10.1016/j.foodres.2012.10.043.

Rocha, M. A. M., Coimbra, M. A., & Nunes, C. (2017). Applications of chitosan and their derivatives in beverages: a critical review. *Current Opinion in Food Science, 15*, 61–69. https://doi.org/10.1016/j.cofs.2017.06.008.

Rodríguez-Félix, F., López-Cota, A. G., Moreno-Vásquez, M. J., Graciano-Verdugo, A. Z., Quintero-Reyes, I. E., Del-Toro-Sanchez, C. L., & Tapia-Hernandez, J. A. (2021). Sustainable-green synthesis of silver nanoparticles using safflower (*Carthamus tinctorius* L.) waste extract and its antibacterial activity. *Heliyon, 7*(4), e06923. https://doi.org/10.1016/j.heliyon.2021.e06923.

Rojas Molina, R., Saucedo Pompa, S., De León Zapata, M. A., Jasso Cantú, D., & Aguilar, C. N. (2011). Pasado, presente y futuro de la candelilla. *Revista mexicana de ciencias forestales, 2*(6), 7–18. http://www.scielo.org.mx/scielo.php?script=sci_arttext&pid=S2007-11322011000400003#.YWA-9b9tRNA.twitter (accessed on 28 May 2024).

Saberi, B., & Golding, J. B. (2018). Aplicación poscosecha de recubrimientos comestibles a base de biopolímeros para mejorar la calidad de los productos hortícolas frescos. In T. Gutiérrez (Ed.), *Polímeros Para Aplicaciones Alimentarias* (p. 9). Springer. https://doi.org/10.1007/978-3-319-94625-2_9.

Saralaya, S., Jayanth, B. S., Thomas, N. S., Sunil, S. M., & Thomas, N. S. (2021). Bee wax and honey—a primer for OMFS. *Oral Maxillofacial Surgery, 25*, 1–6. https://doi.org/10.1007/s10006-020-00893-0.

Saravanakumar, K., Hu, X., Chelliah, R., Oh, D. H., Kathiresan, K., & Wang, M. H. (2020). Biogenic silver nanoparticles-polyvinylpyrrolidone based glycerosomes coating to expand the shelf life of fresh-cut bell pepper (*Capsicum annuum* L. var. *grossum* (*L.*) Sendt). *Postharvest Biology and Technology, 160*, 111039. https://doi.org/10.1016/j.postharvbio.2019.111039.

Schlossman, M. L., & McCarthy, J. P. (1978). Lanolin and its derivatives. *Journal of the American Oil Chemists' Society, 55*, 447–450. https://doi.org/10.1007/BF02911911.

Shariatinia, Z. (2019). Pharmaceutical applications of chitosan. *Advances in Colloid and Interface Science, 263*, 131–194. https://doi.org/10.1016/j.cis.2018.11.008.

Shih, Y. T., & Zhao, Y. (2021). Development, characterization and validation of starch based biocomposite films reinforced by cellulose nanofiber as edible muffin liner. *Food Packaging and Shelf Life, 28*, 100655. https://doi.org/10.1016/j.fpsl.2021.100655.

Solans, C., Izquierdo, P., Nolla, J., Azemar, N., & Garcia-Celma, M. J. (2005). Nano-emulsions. *Current Opinion in Colloid & Interface Science, 10*(3–4), 102–110. https://doi.org/10.1016/j.cocis.2005.06.004.

Sun, X., Wu, Q., Picha, D. H., Ferguson, M. H., Ndukwe, I. E., & Azadi, P. (2021). Comparative performance of bio-based coatings formulated with cellulose, chitin, and chitosan nanomaterials suitable for fruit preservation. *Carbohydrate Polymers, 259*, 117764. https://doi.org/10.1016/j.carbpol.2021.117764.

Tajeddin, B., & Arabkhedri, M. (2020). Polymers and food packaging. In M. A. A. AlMaadeed, D. Ponnamma, & M. A. Carignano (Eds.), *Polymer Science and Innovative Applications* (pp. 525–543). Elsevier. https://doi.org/10.1016/B978-0-12-816808-0.00016-0.

Tang, M., Zhu, Y., Li, D., Adhikari, B., & Wang, L. (2019). Rheological, thermal and microstructural properties of casein/κ-carrageenan mixed systems. *LWT, 113*, 108296. https://doi.org/10.1016/j.lwt.2019.108296.

Thakur, B. R., Singh, R. K., Handa, A. K., & Rao, M. A. (2009). Chemistry and uses of pectin-A review. *Critical Reviews in Food Science and Nutrition, 37*(1), 47–73. https://doi.org/10.1080/10408399709527767.

Tkaczewska, J. (2020). Peptides and protein hydrolysates as food preservatives and bioactive components of edible films and coatings – A review. *Trends in Food Science & Technology, 106*, 298–311. https://doi.org/10.1016/j.tifs.2020.10.022.

Videira-Quintela, D., Martin, O., & Montalvo, G. (2021). Recent advances in polymer-metallic composites for food packaging applications. *Trends in Food Science & Technology, 109*, 230–244. https://doi.org/10.1016/j.tifs.2021.01.020.

Vukić, M., Grujić, S., & Odzaković, B. (2017). Aplicación de películas y recubrimientos comestibles en la producción de alimentos. En: Pellicer E. et al. (eds) Avances en aplicaciones de biomateriales industriales. Springer, Cham. https://doi.org/10.1007/978-3-319-62767-0_7.

Wang, Y., & Zhang, G. (2021). The preparation of modified nano-starch and its application in food industry. *Food Research International, 140*, 110009. https://doi.org/10.1016/j.foodres.2020.110009.

Wijesena, R. N., Tissera, N., Kannangara, Y. Y., Amaratunga, G. A. J., & de Silva, K. M. N. (2015). A method for top down preparation of chitosan nanoparticles and nanofibers. *Carbohydrate Polymers, 117*, 731–738. https://doi.org/10.1016/j.carbpol.2014.10.055.

Won, M. Y., & Min, S. C. (2018). Coating Satsuma mandarin using grapefruit seed extract-incorporated carnauba wax for its preservation. *Food Science and Biotechnology, 27*, 1649–1658. https://doi.org/10.1007/s10068-018-0327-z.

Wu, S. (2019). Extending shelf-life of fresh-cut potato with cactus Opuntia dillenii polysaccharide-based edible coatings. *International Journal of Biological Macromolecules, 130*, 640–644. https://doi.org/10.1016/j.ijbiomac.2019.03.022.

Wu, Z., Zhou, W., Pang, C., Deng, W., Xu, C., & Wang, X. (2019). Multifunctional chitosan-based coating with liposomes containing laurel essential oils and nanosilver for pork preservation. *Food Chemistry, 295*, 16–25. https://doi.org/10.1016/j.foodchem.2019.05.114.

Xavier, M., Parente, I. A., Rodrigues, P. M., Cerqueira, M. A., Pastrana, L., & Gonçalves, C. (2021). Safety and fate of nanomaterials in food: The role of in vitro tests. *Trends in Food Science & Technology, 109*, 593–607. https://doi.org/10.1016/j.tifs.2021.01.050.

Xiong, G., Jia, J., Zhao, L., Liu, X., Zhang, X., Liu, H., & Zhou, W. (2021). Non-thermal radiation heating synthesis of nanomaterials. *Science Bulletin, 66*(4), 386–406. https://doi.org/10.1016/j.scib.2020.08.037.

Xu, Y., Zhao, M., Zhou, D., Zheng, T., & Zhang, H. (2021). The application of multifunctional nanomaterials in Alzheimer's disease: A potential theranostics strategy. *Biomedicine & Pharmacotherapy, 137*, 111360. https://doi.org/10.1016/j.biopha.2021.111360.

Yanat, M., & Schroën, K. (2021). Preparation methods and applications of chitosan nanoparticles; with an outlook toward reinforcement of biodegradable packaging. *Reactive and Functional Polymers, 161*, 104849. https://doi.org/10.1016/j.reactfunctpolym.2021.104849.

Zaporotskova, I. V., Boroznina, N. P., Parkhomenko, Y. N., & Kozhitov, L. V. (2016). Carbon nanotubes: Sensor properties. A review. *Modern Electronic Materials, 2*(4), 95–105. https://doi.org/10.1016/j.moem.2017.02.002.

Zaritzky, N. (2010). Recubrimientos comestibles para mejorar la calidad y seguridad de los alimentos. In J. Aguilera, R. Simpson, J. Welti-Chanes, D. Bermudez-Aguirre, & G. Barbosa-Canovas (Eds.), *Food Engineering Interfaces* (Food Engineering Series). Springer. https://doi.org/10.1007/978-1-4419-7475-4_27.

Zhi, Z., Chen, J., Li, S., Wang, W., Huang, R., Liu, D., Ding, T., Linhardt, R. J., Chen, S., & Ye, X. (2017). Fast preparation of RG-I enriched ultra-low molecular weight pectin by an ultrasound accelerated Fenton process. *Scientific Reports, 7*, 541. https://doi.org/10.1038/s41598-017-00572-3.

CHAPTER 5

Removal of Dyes from Water: Techniques and Materials

L. F. CANO SALAZAR,[1] V. J. CRUZ DELGADO,[2] T. E. FLORES GUÍA,[1] and G. SORIA ARGÜELLO[3]

[1]Departamento de Materiales Avanzados, Facultad de Ciencias Químicas, Universidad Autónoma de Coahuila, Saltillo, Coahuila, Mexico

[2]Departamento de Procesos de Transformación, Centro de Investigación en Química Aplicada, Saltillo, Coahuila, Mexico

[3]Investigador por México CONACYT, Departamento de Química Macromolecular y Nanomateriales, Centro de Investigación en Química Aplicada, Saltillo, Coahuila, Mexico

ABSTRACT

The increase in the population worldwide, as well as anthropogenic activities, among which is industrialization, are the main causes of various types of environmental pollution. One of the most worrying sources, due to all the repercussions it can have, is water contamination. There are several causes of contamination of this resource, one of the main ones is the textile industry, which consumes large quantities of the vital liquid, and its effluents contain different types of pollutants, which are usually released without prior treatment. Some of the contaminants present are detergents, inorganic salts, heavy metals, and dyes, among others; the latter being a complex problem since they are chemically, photo catalytically, and biologically stable, as well as being highly persistent in nature; which generates an ecological imbalance and severe health problems. For the removal of dyes in water effluents, various methods have been tested, among which can be mentioned

Sustainable Agriculture and Global Environmental Health: Toward the Efficient Management of Resources.
Mónica L. Chávez González, Cristóbal Noé Aguilar, Juan Carlos Contreras-Esquivel, and A. K. Haghi (Eds.)

techniques of a chemical, physicochemical, and biological nature, as well as a combination of these. For the aforementioned, the purpose of this chapter is to briefly describe some of the techniques used to remove dyes from water, such as electrocoagulation, ozonation, plasma, biological processes, photodegradation, oxidation, and adsorption, as well as the agents, materials, or compounds used in recent years.

5.1 INTRODUCTION

The increase in the use of sustainable techniques for the treatment of contaminated water from different industrial processes is increasing. Among these industries, textiles stand out for the most severe water pollution, due to the high consumption of water associated with each of the stages of the process as a whole and the discharge of dyes, detergents, salts, etc., in the effluent. The discharge of these effluents to water bodies affects their biochemical oxygen demand (BOD), chemical oxygen demand (COD), total dissolved solids (TDS), total suspended solids (TSS), and pH, resulting in particular importance whenever these tributaries are usually discharged into bodies of water without any prior treatment, causing cross-contamination and preventing it from being used for human consumption and/or agricultural activities (Islam et al., 2022). Some of the dyes used can be considered as recalcitrant contaminants because they have a very stable chemical structure, and can resist attack by various microorganisms or any degradation mechanism, whether biological or chemical. On the other hand, various studies have revealed that irrigation with textile wastewater can cause the soil to accumulate mutagenic and genotoxic agents, with detrimental effects on crops and devastating effects on humans who consume the crops (Khan et al., 2022).

In the work of Boretti and Rosa, data and projections of the demand, sources, quality, and other ecological changes related to water are shown for the year 2050. The objective to highlight is the interrelationship between the population, economic growth, demand, resources, and pollution of water, which ultimately drive water scarcity, and the relevance of these aspects in a local, rather than a global, perspective (Boretti & Rosa, 2019). For this reason, there is a strong demand for technologies that allow the removal of these pollutants from effluents and thus make it possible to recycle the significant volume of water that is consumed during industrial processes. In this work, the main processes involved in the removal of pigments present in water from the textile industry are presented, emphasizing the most promising techniques and materials. Likewise, the possibility of combining different processes is considered in order to provide clean and innocuous water suitable for human consumption.

Some of the techniques that will be addressed are considered advanced oxidation processes (AOP), where the main oxidizing agent is a radical species called hydroxyl (OH•) with a high oxidizing capacity and very short reaction times. Due to the high reactivity of these species, it is possible to eliminate both organic and inorganic compounds, thus achieving a reduction in COD, TOC, and toxicity in treated wastewater. Because the generation of these radicals can come from different sources, there are different associated processes. It is also considered the biological treatment of wastewater carried out through the use of different microorganisms, among which bacteria and/or consortia of microorganisms stand out. Finally, the adsorption process will be addressed, which consists of the concentration of molecules or particles on the surface of a solid. The adsorbent can be any solid material that has activated sites, to which the adsorbate can bind. Derived from the above, there is a wide variety of materials used as adsorbents designed according to the characteristics of the pollutant to be removed.

Each one of the aforementioned processes allows the elimination of pollutants (oils and fats, heavy metals, colloids, organic molecules, pigments/dyes, etc.) in suspension, dissolved or emulsified present in wastewater, from different industrial activities such as electroplating, food, paper, leather, steel, textiles, laundries, and urban wastewater. Taking into consideration that universal access to water increases at an accelerated rate, as well as the population and the productive processes that make use of it, the main techniques and materials involved in the treatment of wastewater are presented below, placing emphasis on water contaminated with dyes derived from the textile industry.

5.2 ELECTROCOAGULATION

Electrocoagulation (EC) is an advanced, versatile, easy-to-operate, eco-friendly, and low-footprint process for the treatment of a wide variety of contaminants present in water, such as the removal of total suspended solids (TSS), total organic carbon (TOC), chemical oxygen demand (COD), biochemical oxygen demand (BOD), turbidity, heavy metals, emulsified oils, microorganisms, and dyes, among others. It is considered a combination of techniques such as electrochemistry, coagulation, and flotation. Since its introduction more than a century ago, EC has been used to treat effluents from different industries, such as textiles, meat and dairy processors, electroplating, agrochemicals, pharmaceuticals, and wastewater (Boinpally et al., 2023).

5.2.1 PRINCIPLE

The process occurs in an electrolytic solution, allowing the transfer of ions between electrodes. In the EC cell, positive ions travel to the cathode and are reduced; simultaneously, negative ions travel to the anode and are oxidized. The EC unit consists of an electrolytic cell with a cathode and anode connected externally to a power source and immersed in an electrolytic solution. When current is passed through the electrolytic cell, an anode oxidizes and the metal dissociates into di- or trivalent metal ions, releasing an equal number of electrons. Although the electrodes are made of the same material, the dissolution occurs exclusively at the anode (Das et al., 2022).

The water reduction process at the cathode results in the formation of hydrogen gas and hydroxyl ions. The metal introduced into the anode dissociates into a metal ion (Mn^+). The metal ions formed at the cathode combine with the hydroxyl group to form the metal hydroxide, the pH of the solution having an influence on whether soluble or insoluble monomeric or polymeric metal hydroxides are formed. The metal hydroxides M(OH)n are very good pollutant adsorbents and have a high tendency to form bonds with them and form flocs since they have a large surface area. The gases H_2 and O_2 are essential in the removal of flocs that do not settle by gravity. The H_2 gas that is evolved lifts the flocs to the surface of the solution, this process is called electroflotation and these flocs can be removed by filtration. The oxygen produced at the anode forms hydrogen peroxide, which is an intermediate and aids in the oxidation of toxic and non-toxic species. The heavier flocs that form tend to sink to the bottom and form sludge, which can be efficiently removed in several ways (Boinpally et al., 2023; Naje et al., 2017).

5.2.2 FACTORS AFFECTING EC

Throughout the reviewed literature, various parameters/factors have been found that significantly affect the performance of the EC cell. Next, the main parameters are listed and grouped according to the component of the EC cell that is involved (Boinpally et al., 2023; Naje et al., 2017; Taheri, 2022).

5.2.2.1 ELECTRODES

5.2.2.2.1 Separation Distance

It is an essential parameter since it is necessary to maintain an optimal space between the electrodes that allows for maximizing the efficiency of contaminant removal. The efficiency of contaminant removal increases as the distance between the electrodes goes from the minimum to the optimum distance.

5.2.2.2.2 Equipment

Al and Fe electrodes are widely preferred among all other electrodes in the EC process due to their high contaminant removal efficiency, cost-effectiveness, and easy availability.

5.2.2.2.3 Configuration

The electrode configuration mode not only affects the contaminant removal efficiency but also the operating cost and energy consumption of the EC process. The most commonly used electrode arrangements are bipolar electrodes in series connection (BP-S) monopolar electrodes in series connection (MP-S) and parallel (MP-P). Depending on the electrical polarity, each electrode in a monopolar configuration mode functions as either the cathode or anode in the electrochemical cell.

5.2.2.2.4 Design

High porosity fixed bed reactors/electrodes, rotating cylinders, and vertical and horizontal plates have been used successfully and have gained considerable attention in recent years for the removal of contaminants from wastewater.

5.2.2.2 SOLUTION

5.2.2.2.1 pH

It is a significant factor affecting the electrochemical process. The pH of the solution fluctuates with time throughout the electrocoagulation process, and it is observed that not all contaminants are removed from the solution at the same pH; rather, the final pH of the effluent has a significant impact on the efficiency of the solution removal of contaminants.

5.2.2.2.2 Temperature

It is another factor that influences EC and is independent of other parameters. It is known that as the temperature increases, the conductivity of the water increases, which in turn increases the removal efficiency of the process, also increases the turbidity removal, and reduces the current consumption by simplifying the EC process.

5.2.2.2.3 Agitation

It is an essential parameter in the EC procedures because it improves the kinetic processes of mass transfer by increasing the mobility of the ions in the solution. With increasing agitation speed, the particle collision speed increases, as well as the release of metal ions, hydroxyl groups, and the formation of flocs.

5.2.2.2.4 Electrolysis Time

Contaminant removal efficiency is also a function of electrolysis time. Contaminant removal efficiency increases with increasing electrolysis time. But beyond the optimal electrolysis time, the contaminant removal efficiency becomes constant and does not increase with increasing time.

5.2.2.3 ENERGY

5.2.2.3.1 Source Type

In EC processes, electrical current is applied to dose the coagulant in several different ways: direct current (DC), where the current flows in only one direction, alternating current (AC) with a sinusoidal current, or polarity reversal (PR), where the polarity of the electrodes alternates with time. The effect of the current mode (DC, AC, or PR) on contaminant removal depends on several variables, including the electrode material, the reversal range of polarity or AC frequency, and the type of contaminant.

5.2.2.3.2 Current Density

It is defined as the current per unit area, which defines the number of metal ions released from the electrode. Current density is a critical parameter in EC as it influences bubble generation rate, coagulant dose rate, floc size, and evolution, all of which have an impact on EC efficiency.

5.2.3 MAIN DYES REMOVED

In Table 5.1, recent studies on the removal of different dyes associated with the textile industry are listed, emphasizing the different parameters involved, the electrode material used, and as a point of comparison, the color reduction expressed as removal efficiency.

TABLE 5.1 Main Colorants Removed from Water by Electrocoagulation

Dye	Parameters	Electrodes	Removal Efficiency (%)	References
Reactive Red 231 (azo dye)	I = 4 A, t = 20 min, C = 2 g/L	Al–Al	96.5	Lach et al. (2022)
Brilliant green (cationic dye)	I = 10 mA/cm^2, pH = 8.45, t = 27.8 min	Fe–Fe	96.1	Mariah & Pak (2020)
Yellow 10 gw (disperse dye)	I = 60 A/m^2, t = 5 min, pH = 7.5	Al–Al	99.5	Kalivel et al. (2021)
	I = 60 A/m^2, t = 15 min, pH = 9	Cu–Cu	97.4	
Red 1 (disperse dye)	I = 30 A/m^2, t = 15 min, C = 0.5 g/L, pH = 4	Fe–Fe	96	Ardhan et al. (2022)
Reactive Blue 21 (disperse dye)	I = 30 A/m^2, t = 15 min, C = 0.5 g/L, pH = 8.4	Fe–Fe	95	
Reactive black 5 (azo dye)	I = 8.33 A/m^2, t = 12.5 min, pH = 6.5	Al–Al	99.9	Keyikoglu et al. (2019)
	I = 4.17 A/m^2, t = 10 min, pH = 6.32	F-F	98.0	
Methyl orange (azo dye)	I = 10 mA/cm^2, t = 50 min, pH = 6, C = 100 mg/L	Fe–Fe	98	Wu et al. (2021)
Methyl orange (azo dye)	I = 0.11 mA/cm^2, t = 60 min, pH = 4, C = 100 mg/L	Fe–Fe	91.1	Akter & Islam (2022)
Reactive blue 2 (azo dye)	I = 18.6 mA/cm^2, t = 6 min, pH = 3, C = 5 mg/L	Fe–Fe	95	Yaqub et al. (2023)

TABLE 5.1 *(Continued)*

Dye	Parameters	Electrodes	Removal Efficiency (%)	References
Acid brown 14	I = 41.7 A/m^2, t = 9 min, pH = 7, C = 300 mg/L	Al–Al	59	Taheri (2022)
Acid orange 7	I = 229 A/m^2, t = 12 min, pH = 5.5, C = 150 mg/L	Al–Al	81	
Acid red 18	I = 229 A/m^2, t = 12 min, pH = 5.5, C = 150 mg/L	Al–Al	3.4	
Blue indigo	I = 47 V, pH = 7.5, C = 60 mg/L	Fe–Fe	93.9	Hendaoui et al. (2021)
Reactive red 35 (anionic dye)	I = 43.4 A/m^2, t = 60 min, pH = 8, C = 70 mg/L	Al–Al	83.1	Moneer et al. (2021)
Disperse yellow 56 (anionic dye)	I = 43.4 A/m$_2$, t = 60 min, pH = 8, C = 50 mg/L	Al–Al	91.2	
Royal blue tiafix	I = 6 A, t = 5 min, pH = 3, C = 0.14 mg/L	Fe–Fe	95	Rodrigues et al. (2020)
Brilliant green	I = 6 mA/m^2, t = 30 min, pH = 6, C = 660 mg/L	Fe–Fe	100	Márquez et al. (2022)
Reactive yellow 17	I = 52.1 mA/m^2, t = 60 min, pH = 6, C = 60 mg/L	SS–Fe	99.8	Moneer et al. (2022)
Industrial dye mixture	I = 15 mA/m^2, t = 23 min, pH = 5.5	Al–Al	97	Khorram & Fallah (2018)

Note: I: current intensity or current; t: time; C: dye concentration.

5.3 OZONATION

Due to the chemical stability and complexity of many dyes, biological oxidation is not sufficient to promote their degradation. Therefore, oxidation treatments are necessary, where one of the most important oxidative processes is ozonation. Ozonation is a versatile, effective, and environmentally friendly chemical treatment that has been widely studied in the removal of water contaminants, including pathogenic microorganisms, organic compounds, and industrial wastes, including dyes (Bal & Thakur, 2021; Nghi et al., 2018; Sarayu et al., 2007). Ozone (O_3) is an unstable gas with a high electrochemical potential (2.08 V); therefore, it is highly oxidizing and capable of reacting with organic compounds through direct or indirect chemical reactions, degrading dyes present in water (Sarayu et al., 2007; Shajeelammal et al., 2022).

5.3.1 *OZONATION PROCESS*

In general, the process of removing dyes from water is carried out in an ozone reactor, where an ozone generator converts oxygen into ozone, which is injected into the water using gas diffusers that promote its interaction with the contaminated water. The dyes present in the water are absorbed on the surface of the ozone bubbles; then oxidation reactions take place that degrade the dyes, and finally, the by-products or residues of the process are eliminated (Chávez et al., 2023; Malik et al., 2020; Shajeelammal et al., 2022). The most important key parameters of the process are the flow and concentration of ozone, volume or flow rate of the liquid, pH, temperature, conductivity, and dye concentration in the water. Other key factors in the efficiency of the process are the reactor morphology, the ozone diffusion method, and the residence time of ozone in the water, as all of these determine the interaction of ozone with contaminants (Epelle et al., 2023; Mehrjouei et al., 2015; Pang et al., 2023).

The oxidation of organic compounds can occur due to the contact of molecular ozone or hydroxyl radicals. Molecular ozone is selective and preferentially degrades chromophore groups. With a higher gas flow and/or concentration, the amount of ozone available to interact with water contaminants increases, and therefore the oxidation of dyes increases, achieving greater efficiency in the degradation of compounds. An increase in the concentration of dyes in the water decreases the removal of dyes and increases the demand for ozone, which is attributed to the increase and oxidation of secondary products produced in the degradation of dyes (Hu et al., 2019; Mehrjouei et al., 2015; Soares et al., 2006a).

One of the key factors in the efficiency of the process is the pH of the solution. Direct oxidation reactions are more selective and predominate in acidic solutions, while radical oxidation reactions are less selective and predominate under basic conditions. Several studies have shown that increasing the pH favors mineralization and decreases decolorization, which is due to the formation of hydroxyl radicals, which have a greater oxidation capacity than molecular ozone (Sarayu et al., 2007; Shajeelammal et al., 2022).

It has been observed that increasing the temperature in the ozonation process decreases the degradation of the dyes due to the decrease in the solubility of ozone in the solution, thus reducing the ozone available for oxidation reactions (Mehrjouei et al., 2015). On the other hand, it has been found that higher solution conductivity (solutions prepared by adding different concentrations of sodium chloride) reduces the degradation efficiency of dyes but favors mineralization. This is attributed to the presence of sodium chloride which promotes the decomposition of ozone into hydroxyl radicals (Mehrjouei et al., 2015; Qiu et al., 2023; Tehrani-Bagha et al., 2010).

5.3.2 OZONATION PROCESS MECHANISMS

The ozonation process mainly occurs through two mechanisms: direct reactions of dissolved ozone molecules and indirect reactions with radical species (OH y HO_2), which are formed by the decomposition of ozone in water. In general, reactions between hydroxyl ions and ozone promote the formation of a superoxide anion radical O_2 and a hydroperoxide radical HO_2. The ozone and hydroxyl radicals generated in aqueous solution can open aromatic rings and oxidize inorganic and organic compounds to higher oxidation states (Guo et al., 2012a).

The interaction of decomposed ozone with OH involves a series of intermediate steps that produce OH radicals and additional oxygen. The reaction occurs according to equation (1) (Foster et al., 2012).

$$3O_3 + OH^- + H^+ \rightarrow 2OH + 4O_2 \tag{1}$$

The ozone interacts with hydrogen peroxide to ultimately produce OH. In this case, an intermediate step involving the formation of peroxide is required, with the hydroperoxide radical (HOO) acting as an intermediary, where two ozone molecules react with peroxide to produce two molecules of OH (Eqns. (2)–(4)) (Foster et al., 2012).

$$H_2O_2 \rightarrow HO_2^- + H_3O^+ \tag{2}$$

$$HO_2^- + O_3 \rightarrow HO_2 + O_3^- \tag{3}$$

$$2O_3 + H_2O_2 \rightarrow 2OH + 3O_2 \tag{4}$$

In general, ozonation has proven to be an efficient process in degrading most commercial dyes, with removal efficiencies above 80%, except for dispersed and sulfur dyes where values lower than 50% have been achieved (Dias et al., 2019; Soares et al., 2006a). It has been observed that ozonation can break the chemical bonds of azo dyes and convert the high molecular weight compounds of dyes into shorter organic acid molecules, leading to color removal and improved biodegradability. Therefore, ozonation is a very useful process in the removal or partial oxidation of dyes, which can improve the biodegradability of these compounds, especially in the textile industry.

5.3.3 PROCESS ADVANTAGES

Ozonation has a high efficiency in the removal of dyes. It does not produce halogenated compounds compared to processes that use chlorine. It does not generate residues in the water. It is a fast and easy-to-implement process. The treatment can eliminate other contaminants such as pathogenic microorganisms and organic compounds.

5.3.4 PROCESS DISADVANTAGES

Ozonation can result in an incomplete degradation of some inorganic molecules and the generation of toxic intermediate compounds. In many treatments, it is necessary to include complementary processes, which increases the investment cost and energy consumption. Additionally, there is little control over degradation reactions.

5.4 PLASMA

Plasma is a technique that has been widely used in many environmental remediation processes. In recent years, this technique has attracted a lot of interest in the removal of contaminants in water. Plasma is the fourth state of matter, a (partially) ionized gas that contains particles with electric charge (ions, electrons, and neutral particles). It is a highly reactive medium that interacts through collisions, responds to electric and magnetic fields, and has a neutral global electric charge (Bilińska & Gmurek, 2021; Pacheco et al., 2015; Pipliya et al., 2023; Vaiano et al., 2022).

According to the electronic temperature, plasma is classified as non-equilibrium thermal plasma (cold plasma) and equilibrium thermal plasma (thermal plasma). In cold plasmas, the electron temperature (T_e) is much higher than the ion temperature (T_i) and neutral species temperature (T_n), which remain at temperatures close to ambient temperature. In contrast, both T_e and T_i are maintained at a high temperature in thermal plasmas, resulting in a high global temperature for this type of plasma (Kumar et al., 2021; Lieberman & Lichtenberg, 2005).

Thermal plasma has a high degree of ionization, high power consumption, low selectivity in reactions, and a high overall temperature, making it widely used in the gasification and vitrification of contaminants (Pacheco et al., 2015; Salazar-Torres et al., 2015). On the other hand, cold plasmas are discharges that maintain a low degree of ionization, high selectivity in reactions, low consumption of electrical energy, and a low global temperature. These characteristics make this type of plasma an ideal candidate for the treatment of water contaminated with dyes (Kumar et al., 2021; Pacheco et al., 2015).

5.4.1 PLASMA PROCESS MECHANISMS

The interaction of water with the plasma discharge produces dissociation, ionization, and vibrational and rotational excitation of water molecules, which leads to the formation of species such as HO^* and H^* (equations 1–4). Because of this, treatments for the removal of dye contaminants in water are considered advanced oxidation processes (AOP) (Kumar et al., 2021; Liu et al., 2022). According to the conditions of each process, there can be many chemical reactions that produce different reactive species. Plasma discharges can break the structure of molecules due to the generation of reactive species in the ionization process, some of these species can be ($HO^{\bullet}, H^{\bullet}O^{\bullet}, HO_2^{\bullet}, NO_2^{\bullet}, O^{\bullet}, O_3,$ and H_2O_2). Additionally, other effects such as UV radiation, shockwaves, and high local temperatures also occur, contributing to the degradation or elimination of contaminants (Liu et al., 2022; Vaiano et al., 2022).

The dissociation process is shown in Equation 1:

$$H_2O + e \rightarrow OH^{\bullet} + H^{\bullet} + e \tag{1}$$

The ionization process occurs according to the reactions in equations 2 and 3:

$$H_2O + e \rightarrow 2e + H_2O^{+} \tag{2}$$

$$H_2O^{+} + H_2O \rightarrow OH^{\bullet} + H_3O^{+} \tag{3}$$

The vibration/excitation process is given by Equation 4:

$$H_2O + e \rightarrow H_2O^{*} + e \tag{4}$$

Plasma-generated reactive species have high oxidation potentials and can react and remove many stable organic compounds at low temperatures, and atmospheric pressure and without using any additional chemicals. For example, the plasma-generated oxidizing agents $HO^{\bullet}, O^{\bullet}, O_3$, and H_2O_2, have oxidation potentials of 2.86 V, 2.42 V, 2.07 V, and 1.78 V, respectively (Kumar et al., 2021). In addition to the contributions of oxidizing species to contaminant removal, the presence of reducing species in plasma provides the potential to remove contaminants from water through reductive degradation mechanisms (Gao et al., 2022; Iervolino et al., 2020; Zeghioud et al., 2020).

5.4.1.1 UV RADIATION EFFECT

Particle collisions in an excited state produce active species that subsequently change to a relaxed state with lower energy, resulting in the emission of ultraviolet light. When an organic compound is irradiated with UV light, it absorbs the radiation and proceeds through an excitation-relaxation process, which can decompose the original molecule into new molecules (Lee et al., 2023).

5.4.1.2 TEMPERATURE EFFECT

In cold plasmas, the electronic temperature induces the production of localized regions of high temperature. In these regions, water dissociation and contaminant pyrolysis are induced. This phenomenon can occur mostly in discharges with a higher degree of ionization, which generate higher temperatures during their operation (Jiang et al., 2014; Rahimpour et al., 2019).

5.4.1.3 SHOCKWAVES EFFECT

In some plasma discharges, the interaction of plasma with the liquid medium causes intense pressure changes that produce powerful acoustic waves, better known as shockwaves. This phenomenon can occur due to the immersion of high electrical energy directly into the liquid or by bubbles imploding in the process. The resulting shockwaves can indirectly induce chemical and pyrolytic reactions in the liquid through electrohydraulic cavitation (Jiang et al., 2014; Vaiano et al., 2022).

5.4.2 PLASMA DISCHARGE TYPES

There are different types of plasma discharge that can be used in the treatment of contaminated water, such as corona discharge, dielectric barrier discharge (DBD), and atmospheric pressure plasma jet. Depending on the characteristics of each discharge, water treatment can be performed by applying the plasma directly to the liquid surface, inside the liquid, or using bubble diffusers in the liquid (Guo et al., 2022; Kumar et al., 2021; Pipliya et al., 2023; Rahimpour et al., 2019; Roy et al., 2023).

5.4.2.1 DIELECTRIC BARRIER DISCHARGES

Dielectric barrier discharges are luminous discharges that are characterized by having at least one dielectric material between their electrodes. They require a gas flow and in most cases, the use of a noble gas is necessary. Due to their morphology, this type of discharge mainly interacts indirectly or in the post-discharge phase with the aqueous medium through diffusers, outside the reaction medium (Meropoulis & Aggelopoulos, 2023; Rahimpour et al., 2019; Roy et al., 2023; Yang et al., 2022).

5.4.2.2 CORONA DISCHARGES

Corona discharges are filamentary discharges and therefore the treatment area is reduced. These discharges can be carried out through ambient air without the need for additional working gases. They can impact directly on the liquid or through the liquid by immersing one of its electrodes (Guo et al., 2022; Nippatlapalli et al., 2022).

5.4.2.3 ATMOSPHERIC PRESSURE PLASMA JET DISCHARGES

Atmospheric pressure plasma jet discharges produce a plasma flame, which is generally confined to the size of the reactor nozzle, which is small. They require a constant flow of working gas and can directly affect the surface or be immersed inside the liquid. The electrodes of these reactors are exposed to the discharge and can erode over time (Kumar et al., 2021; Topolovec et al., 2022).

5.4.3 EFFECT OF PLASMA OPERATING PARAMETERS

The plasma is a technique that involves many physical and chemical processes; therefore, the type and properties of the plasma depend on the generation conditions used. Some of these parameters are the type of electrical excitation, the applied electrical power, the working gas, the reactor morphology, the operating pressure, the gas flow, and others (Guo et al., 2022; Kumar et al., 2021; Nam et al., 2022).

Plasma has proven to be an innovative and effective technique in the removal of dyes present in water. In most treatments, efficiencies of degradation higher than 90% have been achieved, positioning it as a good alternative to conventional treatments with low degradation percentages (Jiang et al., 2014; Liu et al., 2022; Rahimpour et al., 2019). To achieve higher efficiencies, plasma treatments have also been used in combination with other processes, such as heterogeneous catalysis, where superior efficiencies have been achieved in the decomposition of contaminants. These improvements are attributed to the formation of hydroxyl radicals due to photocatalytic reactions in the presence of ultraviolet radiation generated by the plasma (Bilińska & Gmurek, 2021; Nam et al., 2022; Vaiano et al., 2022).

5.4.4 PROCESS ADVANTAGES

Plasma promotes the abundant formation of highly reactive species without the addition of chemical agents. Plasma treatments do not produce contaminant by-products and during processing create less toxic transformation intermediates that are biodegradable. They can eliminate other contaminating compounds present in water, and pathogenic microorganisms, and are environmentally friendly processes.

5.4.5 PROCESS DISADVANTAGES

Process disadvantages include high investment costs for equipment, some systems requiring high energy consumption, some discharges needing noble gases such as argon or helium to operate, scaling up to an industrial level being complex, and some equipment having low treatment rates or working in batch mode.

5.5 BIOLOGICAL PROCESSES

In recent years, attention has turned towards the disposal of dyes, due to the fact that in several parts of the world, the textile industry represents one of the most important economic activities. One of the continents that has traditionally dedicated itself to this activity is Asia, since India and China represent 50% of the global textile market in both production and exportation (Ghaly et al., 2013; Ito et al., 2016; Lotito et al., 2014; Rybczyńska-Tkaczyk et al., 2020).

The problem with the use of dyes in different industries lies in the fact that the wastewater from these processes reaches bodies of water, causing the waterways to become contaminated and reducing the penetration of light, thereby inhibiting the natural process of photosynthesis. This phenomenon causes a significant reduction in the oxygen content of the water, affecting the ecosystem that is related to these bodies of water. The chemical nature of dyes makes them complex compounds that accumulate in both water and living organisms with high toxicity, mutagenicity, and high resistance to degradation, making dyes a great biological risk (Alhassani et al., 2007; Jassim et al., 2023).

For these reasons, and as mentioned in previous sections, it is necessary to seek suitable technologies or methods to reduce or eliminate dyes from bodies of water, and one of these techniques is biological processes or treatments.

Biological treatments, unlike some physicochemical or photochemical processes, can be considered a low-cost and effective alternative for the treatment of wastewater from the textile industry (Siddique et al., 2017). Among these biological processes, aerobic and anaerobic biodegradation can be highlighted. Aerobic biodegradation consists of the breakdown of the compound, in this case, dyes, by the presence of microorganisms and oxygen in the environment, meanwhile, anaerobic transformation is the process in which a compound is degraded by the action of microorganisms in the absence of oxygen.

Basically, biological processes for the degradation of pollutants are known as biotransformation or biomodification, where processes occur within microorganisms after the adsorption of certain compounds or indirectly during the process exudates are generated that bind to the chemical or pollutant to change its properties (Speight, 2018).

For instance, in 2019, Al Farraj et al. proposed the biodegradation of Solvent Green 3 (SG3), which is a widely used dye in the textile and pharmaceutical industries, through the biological action of *Hortaea sp.* (halophilic bacteria). Although the main objective of this research was the degradation and detoxification of the dye, the effect on chemical oxygen demand, metabolites produced during the SG3 decolorization process, enzyme activity, and bacterial growth were also studied comprehensively. Their main findings were that *Hortaea* sp. can completely degrade SG3, meaning it transforms it into non-toxic compounds. First, it degrades it into 1,4-diaminoanthracene-9,10-dione and p-toluidine, and then into phthalic acid and 4-hydroxybenzoic acid, respectively. Finally, a way to validate that the by-products of this biotransformation are not toxic was to test the effects of SG3 and the generated compounds on two plants (*Sorghum vulgare* and *Triticum aestivum*). In general, both plants achieved better germination percentages, plumule, and radical lengths when exposed to the by-products of SG3 degradation than when in contact with SG3 itself (Al-Farraj et al., 2019).

Another study published in 2018 by Sheng et al. reported the results of the aerobic degradation study of two water-soluble azo dyes, Sudan I and Acid Orange 7 (AO7). Untreated azo dyes can cause serious environmental problems and hazardous effects on human health (Ali, 2010). To biodegrade the aforementioned dyes, aerobic conditions were used in parallel mixed-flow reactors. After a 40-day process, the chemical oxygen demand (COD) in the reactor containing Sudan I dye reached a percentage of 85 (close to that of the glucose reactor, which was used as a reference), while in the case of the AO7 reactor, the COD was less than 60%. These data suggest that the high efficiency of Sudan I dye removal is due to a process of biosorption, and not biodegradation and mineralization. In the case of the reactor containing AO7, the presence of NH_3-N ions, SO_4^{2-} ions, and components like tryptophan were found, confirming that AO7 can be biodegraded into compounds derived from benzene and inorganic ions. From this study, it can be concluded that biosorption or biodegradation phenomena are different for two azo dyes with different solubility in water, but it can be said that they are similar for dyes that have a similar chemical structure (Sheng et al., 2018).

On the other hand, the fungus *Cylindrocephalum aurelium* RY06 (*C. aurelium* RY06) was used to study the decolorization and biotransformation of the azo dye Mordant Orange-1 (MO-1). In this research, various culture variables such as agitation rate, pH, carbon, and nitrogen sources, as well as surfactant were studied. The authors concluded that the maximum decolorization of MO-1 reached a percentage of 86% when using a combination of variables that includes glucose as the carbon source, ammonium nitrate as the nitrogen source, a working pH of 4, agitation of 100 rpm, and Tween 80 as surfactant. This work lays the foundation for the use of *C. aurelium* RY06 for the decolorization of dyes (Mostafa et al., 2019).

Rybczyńska-Tkaczyk et al. in 2020, reported the use of *Bjerkandera adusta* CCBAS 930 (*B. adusta* CCBAS 930) cultures as potential biotransformers of anthraquinone dyes. They also evaluated the phytotoxicity, ecotoxicity, and cytotoxicity of the dyes after being treated with B. adusta CCBAS 930. The anthraquinone dyes used were Alizarin Blue Black B (ABBB) and Acid Blue 129 (AB129). The noteworthy results of this research are that the strain of *B. adusta* CCBAS 930 exhibited decolorization activity with ABBB and AB129 dyes at an initial concentration of 0.01%. The secondary products generated after treatment with the strain were toxic to plants (*Lepidium sativum L.*), bacteria (*Vibrio fischeri*), and the BJ skin fibroblast cell line despite the degree of decolorization. Therefore, further research is necessary to delve deeper into the subject (Rybczyńska-Tkaczyk et al., 2020).

As mentioned previously, in various industrial applications such as wool, cotton, nylon, and leather dyeing, among others, synthetic dyes are used which have an impact on the environment and health (Alhassani et al., 2007). For these reasons, in 2023, the removal of synthetic dyes using *Klebsiella* spp. (Gram-negative bacteria) was proposed, where the research group used different concentrations and incubation conditions of *Klebsiella* for the decolorization of some colors separately and together. The azo-type dyes used were Evans blue (EB), bright green (BG), and rose Bengal (RB). After an exhaustive study, they found that this type of bacteria has great potential for the removal of BG and EB. However, the best results were for the removal of RB, achieving removal percentages of 95–81%, respectively, with initial dye concentrations in the range of 0.01–1 mg/L (Jassim et al., 2023).

Regarding the concern for environmental issues related to water, different mathematical models and technologies are currently being integrated to help find the appropriate conditions for the bioremediation of soils and waters contaminated with dyes from the textile industry and other sources. These approaches, known as omics-based, propose several steps to carry out bioremediation in the current era, which consist of generating a deeper understanding of bioremediation pathways, functional characterization and identification of proteins and genes, using mathematical models to evaluate toxicity data, acquiring, analyzing, processing, transmitting, and standardizing protocols, identifying new biomarkers, and integrating data for their application with multiomics approaches (Patil et al., 2023).

As can be seen, it is of vital importance to restore good quality to the waters that have been discharged with effluents from the textile industry (which is one of the main polluters due to dyes), all with the aim of promoting environmental remediation, as well as human health. In this case, biological processes can be of great support to achieve this objective. Likewise, it is necessary to continue with research and data collection work to achieve this milestone.

5.6 PHOTODEGRADATION

Photodegradation is an alternative technique for industrial water treatment, with its main advantages being its low cost, the absence of polluting by-products, and the fact that it is an environmentally friendly process. A photocatalytic reaction implies heterogeneous catalysis where a catalyst (semiconductors, nanomaterials, nanocomposites, etc.) absorbs energy from light, producing hole/electron pairs and the catalyst in contact with the molecule to be degraded. This process can occur in an aqueous medium or a gas (Ayodhya & Veerabhadram, 2018; Chiu et al., 2019).

The degradation process involves various reactions in photoactive materials and can be observed in homogeneous and/or heterogeneous systems. The most commonly used process is heterogeneous, as unlike the homogeneous process (which uses photosensitive molecules solubilized in the medium, which can hardly be eliminated), the reaction takes place in the presence of other compounds that act as photocatalysts and can be easily separated from the reaction medium in later stages. This process generally utilizes sunlight in the presence of semiconductor material to accelerate the degradation and destruction of potentially toxic molecules. It begins with the adsorption of the dye molecules on the catalyst surface to subsequently degrade them (Ayodhya & Veerabhadram, 2018).

For a material to be considered a photocatalyst, it must have an appropriate bandgap, good photostability, high carrier mobility, high surface area, and efficient light absorption (Challagulla et al., 2017; Rafiq et al., 2021). Efficient photodegradation strongly depends on the type of material used (amount and efficiency) and on the operational parameters of the process, such as the concentration of the dye, the pH of the medium, the reaction temperature, and the intensity of irradiation, to mention a few (Ayodhya & Veerabhadram, 2018; Rafiq et al., 2021).

The initial concentration of the dye strongly affects the reaction, since its absorption on the surface of the photocatalyst can influence the absorption of photons and the subsequent generation of active species, and a high concentration can decrease the amount of light that penetrates the reaction solution which may result in a decrease in the efficiency of photodegradation (Vaiano et al., 2021). On the other hand, increasing the light intensity linearly favors the speed of photodegradation, however, once the active sites are completed, the degradation speed presents a maximum value regardless of how much the light intensity is increased (Lyubimenko, 2019). In addition to the parameters, temperature plays a crucial role in photodegradation. Although a low temperature favors the adsorption of the dye, it can decrease the number of active sites. On the contrary, an increase in temperature decreases the adsorption of the reactants (since due to the high kinetic energy the molecules can escape from the catalyst surface), but the recombination of charge carriers increases (Groeneveld et al., 2023). For this reason, the optimal temperature for the photodegradation of dyes is between 25 and 80°C (Shams-Ghahfarokhi & Nezamzadeh-Ejhieh, 2015).

In Figure 5.1, a general scheme of the mechanism of photo-degradation of organic dyes is presented. As a first stage, the irradiation of the material with incident photons is presented. If the energy of the incident light is equivalent to the energy of the semiconductor band gap, the process of excitation of electrons begins, which will allow them to pass from the valence band (generating holes in it) to the conduction band. In the second stage, there is a recombination between the electrons and the holes generated due to the low energy stability, generating thermal energy or presenting the beginning of the third stage, which begins with the formation of reactive radical species (superoxide radicals and hydroxyl) to diffuse from the surface of the photocatalyst and react with the adsorbed molecules promoting redox reactions. Finally, the degradation of the

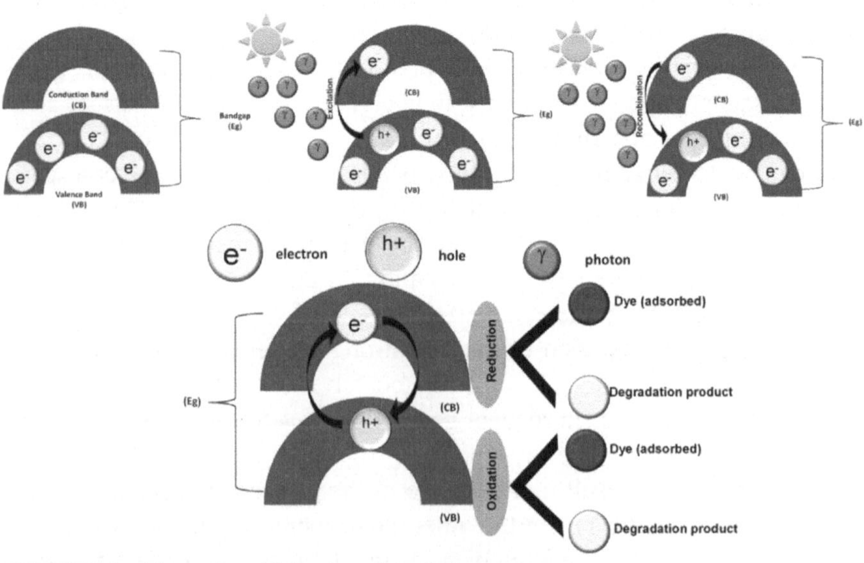

FIGURE 5.1 Mechanism of the photodegradation process.

pollutant is carried out through reduction and oxidation reactions caused by the movement of electrons between the valence and conduction bands (Ayodhya & Veerabhadram, 2018; Pirhashemi et al., 2018).

There is a wide variety of materials used for the photodegradation of dyes in aqueous media, the most widely used being metal oxides. These materials can be classified into first-, second-, and third-generation photocatalysts, the main difference between them being the separation between the valence band and the conduction band. The first-generation photocatalysts are of a single component and require UV light to start the photodegradation process since they have high bandgap and interfacial charge recombination. Despite this, oxides of zinc, nickel, tin, zirconium, and manganese have been used in the degradation of anionic and cationic dyes and different types of chromophores (Saeed et al., 2022). However, despite the good results that these semiconductor oxides initially presented by themselves, it was necessary to modify them to satisfy the requirements of large redox potentials and improve the range of light absorption, giving rise to second-generation photocatalysts (Groeneveld et al., 2023). Modifications that have been studied include downsizing at the nanometer scale, the addition of a co-catalyst to search a heterojunction, sensitization with dyes, and doping with metals and nonmetals (Kocijan et al., 2021; Maruthupandy et al., 2020; Saeed et al., 2022; Tomar et al., 2020). Third-generation photocatalysts, on the other hand, consist of the immobilization of the photocatalyst on various inert solid supports or substrates, and unlike the second-generation ones, they can be removed without the need for post-separation treatment. Third generation photocatalysts include boron nitrides, carbon nitrides, metal oxides, metallic nanocrystals, and many others deposited on various organic and inorganic substrates (Saeed et al., 2022; Sarkar et al., 2020).

5.7 OXIDATION

Many of the dyes used in the textile industry have extremely stable molecular structures, so removing them from effluents is complicated. Biological processes are inefficient for their removal, and traditional physical methods such as coagulation and adsorption are not enough since they only transfer the contaminant to another medium, causing secondary contamination (Sonu et al., 2021; Tang & An, 1995). For this reason, chemical oxidation is a method that has gained great interest due to its high efficiency and ease of operation. Advanced oxidation processes (AOPs) are techniques that use the formation of hydroxyl radicals to oxidize organic industrial waste in effluents. This type of oxidation can be considered heterogeneous ozonation (Venkatesh et al., 2017), photocatalytic oxidation (Sonu et al., 2021), atmospheric air cold plasma (Sarangapani et al., 2016), integrating biological methods (Ledakowicz & Pázdzior, 2021), and electrochemical oxidation (Cañizares et al., 2006), or homogeneous oxidation (Fenton oxidation; Morshed et al., 2020).

The ozonation is considered a heterogeneous oxidation process. Ozone, due to its extremely high redox potential, has been used to decompose dyes to simpler species that can follow a natural biological treatment; however, its use is limited due to its relatively high cost to implement on a large scale (Chaturvedi & Jaiswal, 2022; Venkatesh et al., 2017; Yong et al., 2005). Ozone reacts in an aqueous solution with some organic compounds (mainly unsaturated and aromatic molecules) through a direct reaction with molecular ozone or through a radical reaction involving the hydroxyl radical produced by the decomposition of ozone in water. Oxidation occurs when these hydroxyl radicals instantly react with organic compounds. These radicals are the decomposition products of ozone in aqueous solutions and in turn accelerate the decomposition of ozone, generating a chain mechanism with initiation, propagation, and termination, giving rise to ozonolysis (Soares et al., 2006b; Venkatesh et al., 2017).

However, in some processes, ozonation is not sufficient to completely degrade certain inorganic compounds and generate secondary toxic products. In order to obtain more efficient processes, ozonation is used in combination with other treatment techniques such as adsorption (Khamparia & Jaspal, 2017). Some other strategies to reduce these drawbacks include the use of hydrogen peroxide and/or UV radiation to generate free radicals that increase the oxidation potential. Other processes that have demonstrated high efficiency are catalytic ozonation processes, where a greater decomposition of O_3 into free radicals is desired, which can promote reactions with higher oxidation potential (Dias et al., 2019; Polat et al., 2015). Metal oxides such as MgO, Co_3O_4, ZnO, TiO_2, and Al_2O_3 are some of the catalysts that have facilitated the conversion of organic pollutants into less toxic or inert forms, reducing their risk of environmental and public health contamination (Li et al., 2023).

Ozonation, unlike other advanced oxidation process techniques, can eliminate turbidity and work at basic pH and low temperatures with good degradation efficiency, with the advantage that it is easily implemented, and the effects on subsequent processes are minimal due to the rapid decomposition of residual ozone into oxygen. However, ozone generation requires high energy and in some cases, only partial mineralization of the dyes is achieved, so it can only be used as a pretreatment (Venkatesh et al., 2017). The main factors that influence ozonolysis are the ozone dose, the initial concentration of the dye, the initial pH, the reaction time, and the temperature of the solution (Kasiri et al., 2013; Otieno et al., 2019).

Catalytic ozonation, on the other hand, is more effective in terms of cost-benefit, especially when heterogeneous catalysts such as sulfides, transition oxides, and semiconductors are used, to name a few. This technique has been used to improve efficiency in ozonation processes by combining ozone with hydrogen peroxide, UV radiation, metal ions, or heterogeneous catalysts (Faria et al., 2009). Ozonation, unlike other advanced oxidation process techniques, can eliminate turbidity and work at basic pH and low temperatures with good degradation efficiency, with the advantage that it is easily implemented, and the effects on subsequent processes are minimal due to the rapid decomposition of residual ozone into oxygen. However, ozone generation requires high energy and in some cases, only partial mineralization of the dyes is achieved, so it can only be used as a pretreatment (Venkatesh et al., 2017). The main factors that influence ozonolysis are the ozone dose, the initial concentration of the dye, the initial pH, the reaction time, and the temperature of the solution (Kasiri et al., 2013; Otieno et al., 2019).

Catalytic ozonation, on the other hand, is more effective in terms of cost-benefit, especially when heterogeneous catalysts such as sulfides, transition oxides, and semiconductors are used, to name a few. This technique has been used to improve efficiency in ozonation processes by combining ozone with hydrogen peroxide, UV radiation, metal ions, or heterogeneous catalysts (Faria et al., 2009).

Another of the most used techniques is catalytic oxidation, which is based on the production of active radicals during a reaction in the presence of a catalyst under relatively mild conditions. Unlike photocatalysis, which generally uses metal oxides and semiconductors, this type of technique uses hydrogen peroxide, copper ions, graphene oxide, and sulfate radicals, to name a few. The latter has attracted attention in recent years since they can be activated by heat, ultrasound, and transition metal catalysts. The advantage of this technique is that the radicals generated have a higher oxidation potential than that of hydroxyl radicals and can react in a wide pH range. In catalytic oxidation, many of the processes take place on the catalyst surface, so parameters such as surface area, porosity, stability, and generation of oxidant radicals are essential to define the overall reaction performance (Javaid & Qazi, 2019).

For their part, electrochemical oxidation processes are based on the in-situ generation of oxidizing species that can mineralize organic pollutants into CO_2, water, and inorganic ions. This can be achieved by using anodes that have high oxygen potential, corrosion stability, or oxidizing materials such as chloride, hypochlorite, and ozone. In the anode, two processes occur; if it is direct electrolysis, oxidation occurs on the surface of the electrode; on the other hand, if it is indirect electrolysis, mediators are needed to improve conductivity and generate oxidation products on the anodic surface (Morsi et al., 2011). Anodes with a high overpotential for oxygen evolution (such as oxides of lead, tin, and boron-doped diamond) favor the electrochemical oxidation of organic compounds, converting them into CO_2. In addition, these materials have been found to have good mechanical and chemical resistance under acid conditions, and high current density (da Costa Soares et al., 2020; Martínez-Huitle & Panizza, 2018; Singh et al., 2016).

The Fenton process is a homogeneous oxidation method in which a mixture of hydrogen peroxide and iron salts is added directly to the wastewater to degrade it through the appearance of hydroxyl radicals from the catalytic decomposition of hydrogen peroxide. This process generates redox reactions that allow the ferrous ions consumed during the Fenton reaction to be regenerated through the additional reaction of ferric ions with hydrogen peroxide. The main disadvantage of this homogeneous oxidation process is that it uses single-use reagents, and the regeneration of ferrous ions is slow, which causes a decrease in degradation efficiency and generates a large amount of waste (Morshed et al., 2020; Qian et al., 2020). For this reason, this process has been used accompanied by catalysts seeking to reduce iron sludge resulting from its oxidation (Morshed et al., 2019; Yue et al., 2014).

The techniques mentioned here have presented good results in the removal of dyes in both model and wastewater; however, their efficiency does not only depend on the material with which they work or the synthesis method used. It is necessary to optimize the operating parameters to obtain the best results.

5.8 ADSORPTION

As mentioned throughout this document, water is an essential element for the world, both for ecosystems and for human beings, and the amount of fresh water is limited. It is predicted that by the year 2025, half of the world's population will live in areas with water scarcity (Dotto & McKay, 2020; World Health Organization, 2022).

Due to the presence of a wide variety of contaminants in water, it is necessary to search for or design materials, technologies, or processes that can help combat this problem. So far, there are various procedures that have been used for the decontamination of this resource. This section focuses on the process of adsorption, which is based on the use of a fluid (in this case water) that contains a contaminant (adsorbate) and a solid phase known as an adsorbent. Simply put, the process involves transferring the contaminant(s) present in the liquid phase to the solid phase, which is the adsorbent, in order to purify the water (Bonilla Petricolet et al., 2019; Dotto & McKay, 2020). This method has become very popular due to its cost-effectiveness, high quality of treated effluents, ease of implementation and operation, and the possibility of the adsorbent being reused (Cano-Salazar et al., 2019; Dotto & McKay, 2020).

The effectiveness of the adsorption process mainly depends on the characteristics of the solid phase or adsorbent. This refers to the physical and chemical characteristics of the material, including pore size and volume, point of zero charge, particle size, chemical composition, and specific surface area, among others. Over the decades, various types of adsorbents have been tested, including agricultural and industrial byproducts, carbonaceous materials, clays, sludges, marine materials, minerals, and polymers, to name a few (De Gisi et al., 2016; Senthil Kumar et al., 2019). Table 5.2 summarizes some adsorbent materials that have been used for the removal of dyes from water, as well as the main characteristics and findings during these procedures.

As can be observed, efforts to contribute to improving the quality of water contaminated by dyes are of great interest to the scientific community as it has a direct impact on environmental well-being and public health. Adsorption is an affordable process for this purpose, however, there are still many areas of opportunity in the search for adsorbents that are increasingly efficient to achieve this goal.

5.9 CONCLUSION

Currently, the need to improve the quality of water contaminated by various types of dyes is a global concern, which has led researchers to implement or develop different chemical, physical, physicochemical, biological, and electrochemical techniques to counteract the damage caused by these pollutants.

TABLE 5.2 Adsorbent Materials Used for the Removal of Dyes

Adsorbent	Dye	Characteristics	References
Soursop waste (peel, seeds, pulp fiber) / sugarcane bagasse.	Methylene blue (MB)	The adsorbents were dried, pulverized, and sieved to use particle sizes smaller than 0.60 mm. The experiments were carried out in batch mode. Three blocks of experiments were conducted: the first varied the amount of adsorbent, the second the agitation speed, and the third the initial concentration of MB. The relevant results show that MB removal with sugarcane bagasse can reach up to 90.4% using 0.5 g of adsorbent, 50 mg/L, and an agitation speed of 110 rpm, while for soursop residues, the maximum removal capacity of the dye is 91.6% with an initial MB concentration of 100 mg/L and 110 rpm of agitation. The data fit the Langmuir model, and it is concluded that the adsorption is a multi-molecular process based on physisorption.	Meili et al. (2019)
Magnetite intercalated bentonite	Congo red (CR)	Bentonite clay was modified with magnetite nanoparticles (Fe_3O_4). The resulting composite showed higher efficiency in removing dye compared to acid-activated bentonite. The optimal removal conditions were a contact time of 105 minutes, 0.6 g of the composite, an initial concentration of 10 mg/L, and a pH of 4, achieving up to 94.9% adsorption of CR. The experimental data was fitted to the Langmuir model and a pseudo-second-order kinetic model.	Belachew & Bekele (2020)
Water treatment residues (WTR)	Indigo carmine (IC)	WTRs are non-hazardous materials that are generated during water treatment to produce potable water. Two particle sizes were used, <2 mm (bWTR) and <100 nm (nWTR). The nWTR showed great effectiveness in removing IC, achieving over 90% dye adsorption using 0.1 g of the material and 40 mL of solutions ranging from 10–500 mg/L at pH 5. Kinetic studies were fitted to a first-order model, while equilibrium data were fitted to the Langmuir model, from which a q_{max} of 172.4 mg/g was obtained for nWTR, compared to bWTR, which yielded a q_{max} of 30.86 mg/g.	El-Kammah et al. (2022)
Nanoparticles of hydroxyapatite (HAp) modified with polyethylene glycol (PEG).	Indigo carmine (IC)	The synthesis of HAp nanoparticles with PEG was carried out by the chemical precipitation method and was characterized by FTIR, XRD, EDS, and FESEM. It was found that the synthesized compound can effectively adsorb IC from water samples. The optimal values to achieve maximum adsorption capacity in this system were established at 25°C, neutral pH, 0.06 g of adsorbent, initial concentration of 700 mg/L, and contact time of 1 h. The kinetics fit the pseudo-second-order model. The q_{max} obtained from the Sips model was 129.3 mg/g.	Zolgharnein et al. (2023)

TABLE 5.2 *(Continued)*

Adsorbent	Dye	Characteristics	References
Zeolite	Malachite green (MG)	The zeolite used in this study was chemically activated and used for the removal of MG. The experimental data were fitted to the Langmuir model, and the maximum adsorption capacity was established at 83.33 mg/g at 25°C. Positive values of $\Delta G°$, ΔH, and ΔS indicate that the adsorption process is not spontaneous, is endothermic, and that randomness increases at the solid-solution interface during adsorption. The data were fitted to a pseudo-second-order kinetic model, suggesting that the process is chemisorption.	Abbas & Trari (2023)
Anionic (DAC) and cationic (c-DAC) dialdehyde nanocelluloses	Alizarin red S (ARS), brilliant cresyl blue (BCB), acid green 25 (AG25), Congo red (CR), Bismarck brown Y (BBY), and acid brown M (ABM)	DAC demonstrated greater affinity towards cationic dyes (BBY and BCB) and c-DAC showed higher adsorption capacity for anionic dyes (CR, AG25, ABM, and ARS). For CR, various variables were studied (contact time, pH, ionic strength, and initial dye concentration), and both materials were able to remove CR in a highly alkaline and high ionic strength medium. The q_{max} values were 129.6 mg/g (DAC) and 540.3 mg/g (c-DAC), which are very close to those of other nanocellulose-based materials. DAC can be reused for up to 5 cycles and can be adjusted for a column process.	Huang et al. (2023)
Nanoporous hydrogel	Methyl orange (MO)	The hydrogel was synthesized by the gas-blowing method using poly (3-sulfopropyl acrylate-co-acrylic acid-co-acrylamide) grafted onto salep. The hydrogel has excellent swelling capacity, meaning it is super absorbent. The adsorption of MO was tested with different contact times and initial concentrations. The q_{max} was established at 400 mg/g using 0.06 g of hydrogel. The kinetic process fits a pseudo-second order.	Bardajee et al. (2023)
Coral-like TiO_2/organosilane hybrid particles	Methyl orange (MO)	The kinetic studies indicated that the adsorption process between the material and MO followed a pseudo-second-order and fitted to the Langmuir model, which indicates that the type of adsorption is chemical. The maximum adsorption capacity of MO was 369.3 mg/g at a pH of 3.3.	Tu et al. (2023)
Graphene sand composite (GSC)	Congo red (CR)	The optimal parameters for this process were an initial concentration of 20 mg/L of CR, 1 g of adsorbent, and a contact time of 150 min. The process fits a pseudo-second-order kinetic model and the Freundlich isotherm. The maximum adsorption capacity is 5.53 mg/g. The negative values of the changes in Gibbs free energy, enthalpy, and entropy indicate the feasibility of the adsorption process, as well as its spontaneity and exothermic nature. The material can be reused for up to 4 cycles.	Abdullah Sani et al. (2023)

In this section, seven different techniques that are widely used for decolorization, degradation, or removal of dyes in water were addressed. Through the sections that make up this work, it can be appreciated that the techniques discussed here have significant advantages as well as some disadvantages. For example, electrocoagulation is a very versatile and easy-to-operate technique; however, the procedure has multiple steps that can increase the cost of treating contaminated water. The same applies to ozonation, which, although very effective in eliminating dyes, can generate toxic products. Plasma technology is costly, and biological processes are highly efficient, low-cost, and generally easy to operate; nevertheless, like the adsorption process, the waste they can produce is an alternate problem to the primary one. In contrast, photodegradation does not generate waste; nevertheless, the search for appropriate materials for this method can be complicated.

Despite the advantages and disadvantages that each of these techniques may have, the need to remedy the environmental damage caused by the discharge of dyes into bodies of water remains latent. Efforts should be directed at optimizing processes, searching for materials (waste, inorganic, organic compounds), and combining techniques to eliminate these pollutants from the environment.

ACKNOWLEDGMENTS

The authors acknowledge the support of Consejo Nacional de Ciencia y Tecnología (CONACyT) for financial support (grant FORDECYT-PRONACES/6660/2020) and through grant 321156 (LANIAUTO).

KEYWORDS

- **adsorption**
- **dyes**
- **ecology**
- **health problems**
- **materials**
- **oxidation**
- **photodegradation**
- **plasma operating parameters**
- **water pollution**

REFERENCES

Abbas, M., & Trari, M. (2023). Contribution of zeolite to remove malachite green in aqueous solution by adsorption processes: Kinetics, isotherms and thermodynamic studies. *Textile Research Journal*. Advance online publication. *93*(15–16), https://doi.org/10.1177/00405175231165336.

Abdullah Sani, N. S., Ang, W. L., Mohammad, A. W., Nouri, A., & Mahmoudi, E. (2023). Sustainable synthesis of graphene sand composite from waste cooking oil for dye removal. *Scientific Reports, 13*(1), Article 1. https://doi.org/10.1038/s41598-023-27477-8.

Akter, S., & Islam, Md. S. (2022). Effect of additional Fe^{2+} salt on electrocoagulation process for the degradation of methyl orange dye: An optimization and kinetic study. *Heliyon, 8*(8), e10176. https://doi.org/10.1016/j.heliyon.2022.e10176.

Al Farraj, D. A., Elshikh, M. S., Al Khulaifi, M. M., Hadibarata, T., Yuniarto, A., & Syafiuddin, A. (2019). Biotransformation and detoxification of Antraquione dye Green 3 using halophilic *Hortaea sp. International Biodeterioration and Biodegradation, 140*, 72–77. https://doi.org/10.1016/j.ibiod.2019.03.011.

Alhassani, H. A., Rauf, M. A., & Ashraf, S. S. (2007). Efficient microbial degradation of Toluidine Blue dye by *Brevibacillus sp. Dyes and Pigments, 75*(2), 395–400. https://doi.org/10.1016/j.dyepig.2006.06.019.

Ali, H. (2010). Biodegradation of synthetic dyes – A review. *Water, Air, and Soil Pollution, 213*(1–4), 251–273. https://doi.org/10.1007/s11270-010-0382-4.

Ardhan, N., Tongpadungrod, P., & Phalakornkule, C. (2022). Effects of auxiliary chemicals and dye solubility on chemical oxygen demand reduction of dyes by electrocoagulation with Fe electrode. *Materials Today: Proceedings, 52*, 2529–2533. https://doi.org/10.1016/j.matpr.2021.10.446.

Ayodhya, D., & Veerabhadram, G. (2018). A review on recent advances in photodegradation of dyes using doped and heterojunction based semiconductor metal sulfide nanostructures for environmental protection. *Materials Today Energy, 9*, 83–113. https://doi.org/10.1016/j.mtener.2018.05.007.

Bal, G., & Thakur, A. (2021). Distinct approaches of removal of dyes from wastewater: A review. *Materials Today: Proceedings, 50*, 1575–1579. https://doi.org/10.1016/j.matpr.2021.09.119.

Bardajee, G. R., Mahmoodian, H., Boraghi, S. A., Elmizadeh, H., Ziarani, N. B., Rezanejad, Z., Tyagi, I., Gaur, R., & Javadian, H. (2023). Nanoporous hydrogel absorbent based on salep: Swelling behavior and methyl orange adsorption capacity. *Environmental Research, 225*. Advance online publication. https://doi.org/10.1016/j.envres.2023.115571.

Belachew, N., & Bekele, G. (2020). Synergy of magnetite intercalated bentonite for enhanced adsorption of Congo red dye. *Silicon, 12*(3), 603–612. https://doi.org/10.1007/s12633-019-00152-2.

Bilińska, L., & Gmurek, M. (2021). Novel trends in AOPs for textile wastewater treatment. Enhanced dye by-products removal by catalytic and synergistic actions. *Water Resources and Industry, 26*. https://doi.org/10.1016/j.wri.2021.100160.

Boinpally, S., Kolla, A., Kainthola, J., Kodali, R., & Vemuri, J. (2023). A state-of-the-art review of the electrocoagulation technology for wastewater treatment. *Water Cycle, 4*, 26–36. https://doi.org/10.1016/j.watcyc.2023.01.001.

Bonilla Petricolet, A., Mendoza Castillo, D. I., Luiz Dotto, G., & Duran Valle, C. J. (2019). Adsorption in Water Treatment. In *Chemistry Molecular Sciences and Chemical Engineering* (Vol. 1, pp. 1–21). Cent Conf. https://doi.org/10.1016/b978-0-12-409547-2.14390-2.

Boretti, A., & Rosa, L. (2019). Reassessing the projections of the World Water Development Report. *Npj Clean Water, 2*(1), 15. https://doi.org/10.1038/s41545-019-0039-9.

Cañizares, P., Gadri, A., Lobato, J., Nasr, B., Paz, R., Rodrigo, M. A., & Saez, C. (2006). Electrochemical oxidation of azoic dyes with conductive-diamond anodes. *Industrial and Engineering Chemistry Research, 45*(10), 3468–3473. https://doi.org/10.1021/ie051427n.

Cano-Salazar, L. F., Martínez-Luévanos, A., Claudio-Rizo, J. A., Carrillo-Pedroza, F. R., Montemayor, S. M., & Rangel-Mendez, J. R. (2019). Synthesis, structural characterization and Cu(II) adsorption behavior of manganite (γ-MnOOH) nanorods. *RSC Advances, 10*(1), 179–186. https://doi.org/10.1039/c9ra09652c.

Challagulla, S., Nagarjuna, R., Ganesan, R., & Roy, S. (2017). TiO_2 synthesized by various routes and its role on environmental remediation and alternate energy production. *Nano-Structures and Nano-Objects, 12*, 147–156. https://doi.org/10.1016/j.nanoso.2017.10.002.

Chaturvedi, A., & Jaiswal, R. P. (2022). Optimization for minimizing the cost of ozonation of highly concentrated textile dyeing wastewater in a bubble column reactor. *Environmental Science and Pollution Research, 29*(58), 88018–88026. https://doi.org/10.1007/s11356-022-21800-y.

Chávez, A. M., Beltrán, F. J., López, J., Javier Rivas, F., & Álvarez, P. M. (2023). On the importance of reactions in the proximity of the gas–water interface: Application to direct ozone reactions of antibiotics in water. *Chemical Engineering Journal, 458*. https://doi.org/10.1016/j.cej.2023.141408.

Chiu, Y. H., Chang, T. F. M., Chen, C. Y., Sone, M., & Hsu, Y. J. (2019). Mechanistic insights into photodegradation of organic dyes using heterostructure photocatalysts. *Catalysts, 9*(5). https://doi.org/10.3390/catal9050430.

da Costa Soares, I. C., da Silva, Á. R. L., de Moura Santos, E. C. M., dos Santos, E. V., da Silva, D. R., & Martínez-Huitle, C. A. (2020). Understanding the electrochemical oxidation of dyes on platinum and boron-doped diamond electrode surfaces: experimental and computational study. *Journal of Solid State Electrochemistry, 24*(11–12), 3245–3256. https://doi.org/10.1007/s10008-020-04813-w.

Das, P. P., Sharma, M., & Purkait, M. K. (2022). Recent progress on electrocoagulation process for wastewater treatment: A review. *Separation and Purification Technology, 292*, 121058. https://doi.org/10.1016/j.seppur.2022.121058.

De Gisi, S., Lofrano, G., Grassi, M., & Notarnicola, M. (2016). Characteristics and adsorption capacities of low-cost sorbents for wastewater treatment: A review. In *Sustainable Materials and Technologies (Vol. 9, pp. 10–40).* Elsevier B.V. https://doi.org/10.1016/j.susmat.2016.06.002.

Dias, N. C., Bassin, J. P., Sant'Anna, G. L., & Dezotti, M. (2019). Ozonation of the dye Reactive Red 239 and biodegradation of ozonation products in a moving-bed biofilm reactor: Revealing reaction products and degradation pathways. *International Biodeterioration and Biodegradation, 144*. https://doi.org/10.1016/j.ibiod.2019.104742.

Dotto, G. L., & McKay, G. (2020). Current scenario and challenges in adsorption for water treatment. *Journal of Environmental Chemical Engineering, 8*(4). https://doi.org/10.1016/j.jece.2020.103988.

El-Kammah, M., Elkhatib, E., Gouveia, S., Cameselle, C., & Aboukila, E. (2022). Cost-effective ecofriendly nanoparticles for rapid and efficient indigo carmine dye removal from wastewater: Adsorption equilibrium, kinetics and mechanism. *Environmental Technology and Innovation, 28*. https://doi.org/10.1016/j.eti.2022.102595.

Epelle, E. I., Macfarlane, A., Cusack, M., Burns, A., Okolie, J. A., Mackay, W., Rateb, M., & Yaseen, M. (2023). Ozone application in different industries: A review of recent developments. In *Chemical Engineering Journal (Vol. 454)*. Elsevier B.V. https://doi.org/10.1016/j.cej.2022.140188.

Faria, P. C. C., Órfão, J. J. M., & Pereira, M. F. R. (2009). Activated carbon and ceria catalysts applied to the catalytic ozonation of dyes and textile effluents. *Applied Catalysis B: Environmental, 88*(3–4), 341–350. https://doi.org/10.1016/j.apcatb.2008.11.002.

Foster, J., Sommers, B. S., Gucker, S. N., Blankson, I. M., & Adamovsky, G. (2012). Perspectives on the interaction of plasmas with liquid water for water purification. In *IEEE Transactions on Plasma Science (Vol. 40, No. 5 Part 1, pp. 1311–1323)*. https://doi.org/10.1109/TPS.2011.2180028.

Gao, X., Huang, K., Zhang, A., Wang, C., Sun, Z., & Liu, Y. (2022). Simultaneous degradation of glucocorticoids and sterilization using bubbling corona discharge plasma based systems: A promising terminal water treatment facility for hospital wastewater. *Chemical Engineering Journal, 430*. https://doi.org/10.1016/j.cej.2021.132845.

Ghaly, A., Ananthashankar, R., Alhattab, M., & Ramakrishnan, V. (2013). Production, Characterization and Treatment of Textile Effluents: A Critical Review. *Journal of Chemical Engineering & Process Technology, 05*(01). https://doi.org/10.4172/2157-7048.1000182.

Ghuge, S. P., & Saroha, A. K. (2018). Catalytic ozonation of dye industry effluent using mesoporous bimetallic Ru-Cu/SBA-15 catalyst. *Process Safety and Environmental Protection, 118*, 125–132. https://doi.org/10.1016/j.psep.2018.06.033.

Groeneveld, I., Kanelli, M., Ariese, F., & van Bommel, M. R. (2023). Parameters that affect the photodegradation of dyes and pigments in solution and on substrate – An overview. In *Dyes and Pigments (Vol. 210). Elsevier* Ltd. https://doi.org/10.1016/j.dyepig.2022.110999.

Guo, H., Wang, Y., Liao, L., Li, Z., Pan, S., Puyang, C., Su, Y., Zhang, Y., Wang, T., Ren, J., & Li, J. (2022). Review on remediation of organic-contaminated soil by discharge plasma: Plasma types, impact factors, plasma-assisted catalysis, and indexes for remediation. In *Chemical Engineering Journal* (Vol. 436). Elsevier B.V. https://doi.org/10.1016/j.cej.2022.135239.

Guo, Y., Yang, L., & Wang, X. (2012). The Application and Reaction Mechanism of Catalytic Ozonation in Water Treatment. *Journal of Environmental & Analytical Toxicology, 02*(06). https://doi.org/10.4172/2161-0525.1000150.

Hendaoui, K., Trabelsi-Ayadi, M., & Ayari, F. (2021). Optimization and mechanisms analysis of indigo dye removal using continuous electrocoagulation. *Chinese Journal of Chemical Engineering, 29*, 242–252. https://doi.org/10.1016/j.cjche.2020.07.065.

Hu, E., Shang, S., & Chiu, A. K. L. (2019). Removal of reactive dyes in textile effluents by catalytic ozonation pursuing on-site effluent recycling. *Molecules (Basel, Switzerland), 24*(15). https://doi.org/10.3390/molecules24152755.

Hu, E., Wu, X., Shang, S., Tao, X. M., Jiang, S. X., & Gan, L. (2016). Catalytic ozonation of simulated textile dyeing wastewater using mesoporous carbon aerogel supported copper oxide catalyst. *Journal of Cleaner Production, 112*, 4710–4718. https://doi.org/10.1016/j.jclepro.2015.06.127.

Huang, X., Hadi, P., Joshi, R., Alhamzani, A. G., & Hsiao, B. S. (2023). A Comparative Study of Mechanism and Performance of Anionic and Cationic Dialdehyde Nanocelluloses for

Dye Adsorption and Separation. *ACS Omega, 8*(9), 8634–8649. https://doi.org/10.1021/acsomega.2c07839.

Iervolino, G., Vaiano, V., & Palma, V. (2020). Enhanced azo dye removal in aqueous solution by H_2O_2 assisted non-thermal plasma technology. *Environmental Technology and Innovation, 19*. https://doi.org/10.1016/j.eti.2020.100969.

Islam, T., Repon, R., Islam, T., Sarwar, Z., & Rahman, M. M. (2022). Impact of textile dyes on health and ecosystem: a review of structure, causes, and potential solutions. *Environmental Science and Pollution Research, 30*(4), 9207–9242. https://doi.org/10.1007/s11356-022-24398-3.

Ito, T., Adachi, Y., Yamanashi, Y., & Shimada, Y. (2016). Long-term natural remediation process in textile dye-polluted river sediment driven by bacterial community changes. *Water Research, 100*, 458–465. https://doi.org/10.1016/j.watres.2016.05.050.

Jassim, H. J., Sayeb, A., & Al-Azzawi, N. (2023). Removal of synthetic dyes from water samples contaminated with dyes using Klebsiella spp. *Journal of Survey in Fisheries Sciences, 10*(3S), 2163–2170.

Javaid, R., & Qazi, U. Y. (2019). Catalytic oxidation process for the degradation of synthetic dyes: An overview. In *International Journal of Environmental Research and Public Health* (Vol. 16, Issue 11). MDPI AG. https://doi.org/10.3390/ijerph16112066.

Jiang, B., Zheng, J., Qiu, S., Wu, M., Zhang, Q., Yan, Z., & Xue, Q. (2014). Review on electrical discharge plasma technology for wastewater remediation. In *Chemical Engineering Journal* (Vol. 236, pp. 348–368). https://doi.org/10.1016/j.cej.2013.09.090.

Kalivel, P., Jagadeesh, T., Kavitha, S., Padmanabhan, D., Palanichamy, J., & Murphy, A. (2021). Comparative study on removal of yellow 10 gw dye from aqueous solution using Al, Cu electrodes in electrocoagulation. *Materials Today: Proceedings, 47*, 807–813. https://doi.org/10.1016/j.matpr.2020.10.561.

Kasiri, M. B., Modirshahla, N., & Mansouri, H. (2013). Decolorization of organic dye solution by ozonation; Optimization with response surface methodology. *International Journal of Industrial Chemistry, 4*(1), 3. https://doi.org/10.1186/2228-5547-4-3.

Keyikoglu, R., Can, O. T., Aygun, A., & Tek, A. (2019). Comparison of the effects of various supporting electrolytes on the treatment of a dye solution by electrocoagulation process. *Colloid and Interface Science Communications, 33*, 100210. https://doi.org/10.1016/j.colcom.2019.100210.

Khamparia, S., & Jaspal, D. K. (2017). Adsorption in combination with ozonation for the treatment of textile waste water: a critical review. In *Frontiers of Environmental Science and Engineering* (Vol. 11, Issue 1). Higher Education Press. https://doi.org/10.1007/s11783-017-0899-5.

Khan, S., Zeyad, M. T., & Malik, A. (2022). Genotoxicity assessment of textile waste contaminated soil and characterization of textile dye degradation by a novel indigenous bacterium Ochrobactrum intermedium BS39. *Chemosphere, 299*, 134082. https://doi.org/10.1016/j.chemosphere.2022.134082.

Khorram, A. G., & Fallah, N. (2018). Treatment of textile dyeing factory wastewater by electrocoagulation with low sludge settling time: Optimization of operating parameters by RSM. *Journal of Environmental Chemical Engineering, 6*(1), 635–642. https://doi.org/10.1016/j.jece.2017.12.054.

Kocijan, M., Ćurković, L., Ljubas, D., Mužina, K., Bačić, I., Radošević, T., Podlogar, M., Bdikin, I., Otero-Irurueta, G., Hortigüela, M. J., & Gonçalves, G. (2021). Graphene-based TiO_2 nanocomposite for photocatalytic degradation of dyes in aqueous solution under solar-like radiation. *Applied Sciences (Switzerland), 11*(9). https://doi.org/10.3390/app11093966.

Kumar, A., Škoro, N., Gernjak, W., & Puač, N. (2021). Cold atmospheric plasma technology for removal of organic micropollutants from wastewater—a review. *European Physical Journal D, 75*(11). https://doi.org/10.1140/epjd/s10053-021-00283-5.

Lach, C. E., Pauli, C. S., Coan, A. S., Simionatto, E. L., & Koslowski, L. A. D. (2022). Investigating the process of electrocoagulation in the removal of azo dye from synthetic textile effluents and the effects of acute toxicity on Daphnia magna test organisms. *Journal of Water Process Engineering, 45*, 102485. https://doi.org/10.1016/j.jwpe.2021.102485.

Ledakowicz, S., & Pázdzior, K. (2021). Recent achievements in dyes removal focused on advanced oxidation processes integrated with biological methods. *Molecules, 26*(4). https://doi.org/10.3390/molecules26040870.

Lee, H., Park, S., Young Park, J., Kim, J., & Choe, W. (2023). Ultraviolet Light-Driven gaining of hydroxyl and nitrogen oxide radicals in Plasma–Treated water. *Chemical Engineering Journal, 458*. https://doi.org/10.1016/j.cej.2023.141425.

Li, M., Fu, L., Deng, L., Hu, Y., Yuan, Y., & Wu, C. (2023). A tailored and rapid approach for ozonation catalyst design. *Environmental Science and Ecotechnology, 15*. https://doi.org/10.1016/j.ese.2023.100244.

Lieberman, M. A., & Lichtenberg, A. J. (2005). *Principles of Plasma Discharges and Materials Processing*. John Wiley & Sons, Inc. https://doi.org/10.1002/0471724254.

Liu, Y., Song, J. W., Bao, J., Shen, X. J., Li, C. L., Wang, X., & Shao, L. X. (2022). Optimized Removal of Azo Dyes from Simulated Wastewater through Advanced Plasma Technique with Novel Reactor. *Water (Switzerland), 14*(19). https://doi.org/10.3390/w14193152.

Lotito, A. M., De Sanctis, M., Di Iaconi, C., & Bergna, G. (2014). Textile wastewater treatment: Aerobic granular sludge vs activated sludge systems. *Water Research, 54*, 337–346. https://doi.org/10.1016/j.watres.2014.01.055.

Malik, S. N., Ghosh, P. C., Vaidya, A. N., & Mudliar, S. N. (2020). Hybrid ozonation process for industrial wastewater treatment: Principles and applications: A review. In *Journal of Water Process Engineering* (Vol. 35). Elsevier Ltd. https://doi.org/10.1016/j.jwpe.2020.101193.

Manivel, A., Lee, G. J., Chen, C. Y., Chen, J. H., Ma, S. H., Horng, T. L., & Wu, J. J. (2015). Synthesis of MoO_3 nanoparticles for azo dye degradation by catalytic ozonation. *Materials Research Bulletin, 62*, 184–191. https://doi.org/10.1016/j.materresbull.2014.11.016.

Mariah, G. K., & Pak, K. S. (2020). Removal of brilliant green dye from aqueous solution by electrocoagulation using response surface methodology. *Materials Today: Proceedings, 20*, 488–492. https://doi.org/10.1016/j.matpr.2019.09.175.

Márquez, A. A., Coreño, O., & Nava, J. L. (2022). Removal of brilliant green tannery dye by electrocoagulation. *Journal of Electroanalytical Chemistry, 911*, 116223. https://doi.org/10.1016/j.jelechem.2022.116223.

Martínez-Huitle, C. A., & Panizza, M. (2018). Electrochemical oxidation of organic pollutants for wastewater treatment. In *Current Opinion in Electrochemistry* (Vol. 11, pp. 62–71). Elsevier B.V. https://doi.org/10.1016/j.coelec.2018.07.010.

Maruthupandy, M., Qin, P., Muneeswaran, T., Rajivgandhi, G., Quero, F., & Song, J. M. (2020). Graphene-zinc oxide nanocomposites (G-ZnO NCs): Synthesis, characterization and their photocatalytic degradation of dye molecules. *Materials Science and Engineering B: Solid-State Materials for Advanced Technology, 254*. https://doi.org/10.1016/j.mseb.2020.114516.

Mehrjouei, M., Müller, S., & Möller, D. (2015). A review on photocatalytic ozonation used for the treatment of water and wastewater. In *Chemical Engineering Journal* (Vol. 263, pp. 209–219). Elsevier. https://doi.org/10.1016/j.cej.2014.10.112.

Meili, L., Lins, P. V. S., Costa, M. T., Almeida, R. L., Abud, A. K. S., Soletti, J. I., Dotto, G. L., Tanabe, E. H., Sellaoui, L., Carvalho, S. H. V., & Erto, A. (2019). Adsorption of methylene blue on agroindustrial wastes: Experimental investigation and phenomenological modelling. *Progress in Biophysics and Molecular Biology, 141*, 60–71. https://doi.org/10.1016/j.pbiomolbio.2018.07.011.

Meropoulis, S., & Aggelopoulos, C. A. (2023). Plasma microbubbles vs gas-liquid DBD energized by low-frequency high voltage nanopulses for pollutants degradation in water: Destruction mechanisms, composition of plasma-activated water and energy assessment. *Journal of Environmental Chemical Engineering, 11*(3), 109855. https://doi.org/10.1016/j.jece.2023.109855.

Moneer, A. A., El-Mallah, N. M., El-Sadaawy, M. M., Khedawy, M., & Ramadan, M. S. H. (2021). Kinetics, thermodynamics, isotherm modeling for removal of reactive Red 35 and disperse yellow 56 dyes using batch bi-polar aluminum electrocoagulation. *Alexandria Engineering Journal, 60*(4), 4139–4154. https://doi.org/10.1016/j.aej.2021.02.061.

Moneer, A. A., El-Mallah, N. M., Ramadan, M. S. H., & Shaker, A. M. (2022). The effect of a novel electrode on the removal of Reactive Yellow 17 dye by Electrocoagulation technique. *The Egyptian Journal of Aquatic Research, 48*(3), 191–197. https://doi.org/10.1016/j.ejar.2022.04.002.

Morshed, M. N., Bouazizi, N., Behary, N., Vieillard, J., Thoumire, O., Nierstrasz, V., & Azzouz, A. (2019). Iron-loaded amine/thiol functionalized polyester fibers with high catalytic activities: A comparative study. *Dalton Transactions, 48*(23), 8384–8399. https://doi.org/10.1039/c9dt00937j.

Morsi, M. S., Al-Sarawy, A. A., & El-Dein, W. A. S. (2011). Electrochemical degradation of some organic dyes by electrochemical oxidation on a Pb/PbO_2 electrode. *Desalination and Water Treatment, 26*(1–3), 301–308. https://doi.org/10.5004/dwt.2011.1926.

Mostafa, A. A. F., Elshikh, M. S., Al-Askar, A. A., Hadibarata, T., Yuniarto, A., & Syafiuddin, A. (2019). Decolorization and biotransformation pathway of textile dye by *Cylindrocephalum aurelium*. *Bioprocess and Biosystems Engineering, 42*(9), 1483–1494. https://doi.org/10.1007/s00449-019-02144-3.

Naje, A. S., Chelliapan, S., Zakaria, Z., Ajeel, M. A., & Alaba, P. A. (2017). A review of electrocoagulation technology for the treatment of textile wastewater. *Reviews in Chemical Engineering, 33*(3). https://doi.org/10.1515/revce-2016-0019.

Nam, S. N., Choong, C. E., Hoque, S., Farouk, T. I., Cho, J., Jang, M., Snyder, S. A., Meadows, M. E., & Yoon, Y. (2022). Catalytic non-thermal plasma treatment of endocrine disrupting compounds, pharmaceuticals, and personal care products in aqueous solution: A review. In *Chemosphere* (Vol. 290). Elsevier Ltd. https://doi.org/10.1016/j.chemosphere.2021.133395.

Nghi, N. H., Cuong, L. C., Dieu, T. V., Ngu, T., & Oanh, D. T. Y. (2018). Ozonation process and water disinfection. *Vietnam Journal of Chemistry, 56*(6), 717–720. https://doi.org/10.1002/vjch.201800076.

Nippatlapalli, N., Ramakrishnan, K., & Philip, L. (2022). Enhanced degradation of complex organic compounds in wastewater using different novel continuous flow non-thermal pulsed corona plasma discharge reactors. *Environmental Research, 203*. https://doi.org/10.1016/j.envres.2021.111807.

Otieno, B., Apollo, S., Kabuba, J., Naidoo, B., & Ochieng, A. (2019). Ozonolysis Post-Treatment of Anaerobically Digested Distillery Wastewater Effluent. *Ozone: Science and Engineering, 41*(6), 551–561. https://doi.org/10.1080/01919512.2019.1593818.

Pacheco, J., Soria, G., Pacheco, M., Valdivia, R., Ramos, F., Frías, H., Durán, M., & Hidalgo, M. (2015). Greenhouse gas treatment and H_2 production, by warm plasma reforming. *International Journal of Hydrogen Energy, 40*(48), 17165–17171. https://doi.org/10.1016/j.ijhydene.2015.08.062.

Pang, H., Zhang, J., Allinson, M., Gray, S., & Scales, P. J. (2023). A chemical credit framework to predict the removal performance of organic chemicals of concern from water through an ozonation process. *Water Research, 232*. https://doi.org/10.1016/j.watres.2023.119671.

Patil, S. M., Tamboli, A. S., Bhalkar, B. N., Surawase, S. N., Watharkar, A. D., & Jeon, B.-H. (2023). Systems biology and multiomics integration in biological treatment of textile wastewater. In *Current Developments in Bioengineering and Biotechnology* (pp. 711–742). Elsevier. https://doi.org/10.1016/b978-0-323-91235-8.00011-5.

Pipliya, S., Kumar, S., Babar, N., & Srivastav, P. P. (2023). Recent trends in non-thermal plasma and plasma activated water: Effect on quality attributes, mechanism of interaction and potential application in food & agriculture. In *Food Chemistry Advances* (Vol. 2). Elsevier Ltd. https://doi.org/10.1016/j.focha.2023.100249.

Pirhashemi, M., Habibi-Yangjeh, A., & Rahim Pouran, S. (2018). Review on the criteria anticipated for the fabrication of highly efficient ZnO-based visible-light-driven photocatalysts. In *Journal of Industrial and Engineering Chemistry* (Vol. 62, pp. 1–25). Korean Society of Industrial Engineering Chemistry. https://doi.org/10.1016/j.jiec.2018.01.012.

Polat, D., Balci, I., & Özbelge, T. A. (2015). Catalytic ozonation of an industrial textile wastewater in a heterogeneous continuous reactor. *Journal of Environmental Chemical Engineering, 3*(3), 1860–1871. https://doi.org/10.1016/j.jece.2015.04.020.

Qian, H., Hou, Q., Yu, G., Nie, Y., Bai, C., Bai, X., & Ju, M. (2020). Enhanced removal of dye from wastewater by Fenton process activated by core-shell $NiCo_2O_4$@FePc catalyst. *Journal of Cleaner Production, 273*. https://doi.org/10.1016/j.jclepro.2020.123028.

Qiu, J., Wang, J., Ren, M., Yang, X., Zhang, J., Zhang, X., Cao, H., & Xie, Y. (2023). Comprehensive effect of water matrix on catalytic ozonation of chloride contained saline wastewater. *Water Research, 234*. https://doi.org/10.1016/j.watres.2023.119827.

Rafiq, A., Ikram, M., Ali, S., Niaz, F., Khan, M., Khan, Q., & Maqbool, M. (2021). Photocatalytic degradation of dyes using semiconductor photocatalysts to clean industrial water pollution. In *Journal of Industrial and Engineering Chemistry* (Vol. 97, pp. 111–128). Korean Society of Industrial Engineering Chemistry. https://doi.org/10.1016/j.jiec.2021.02.017.

Rahimpour, M., Taghvaei, H., Zafarnak, S., Rahimpour, M. R., & Raeissi, S. (2019). Post-discharge DBD plasma treatment for degradation of organic dye in water: A comparison with different plasma operation methods. *Journal of Environmental Chemical Engineering, 7*(4). https://doi.org/10.1016/j.jece.2019.103220.

Rodrigues, A. R., Seki, C. C., Ramalho, L. S., Argondizo, A., & Silva, A. P. (2020). Electrocoagulation in a fixed bed reactor – Color removal in batch and continuous mode. *Separation and Purification Technology, 253*, 117481. https://doi.org/10.1016/j.seppur.2020.117481.

Roy, N. C., Maira, N., Pattyn, C., Remy, A., Delplancke, M. P., & Reniers, F. (2023). Mechanisms of reducing energy costs for nitrogen fixation using air-based atmospheric

DBD plasmas over water in contact with the electrode. *Chemical Engineering Journal, 461*. https://doi.org/10.1016/j.cej.2023.141844.

Rybczyńska-Tkaczyk, K., Korniłłowicz-Kowalska, T., Szychowski, K. A., & Gmiński, J. (2020). Biotransformation and toxicity effect of monoanthraquinone dyes during Bjerkandera adusta CCBAS 930 cultures. *Ecotoxicology and Environmental Safety, 191*. https://doi.org/10.1016/j.ecoenv.2020.110203.

Saeed, M., Muneer, M., Haq, A. ul, & Akram, N. (2022). Photocatalysis: an effective tool for photodegradation of dyes—a review. In *Environmental Science and Pollution Research* (Vol. 29, Issue 1, pp. 293–311). Springer Science and Business Media Deutschland GmbH. https://doi.org/10.1007/s11356-021-16389-7.

Salazar-Torres, J. A., Pacheco-Sotelo, J., Valdivia-Barrientos, R., Pacheco-Pacheco, M., Ramos-Flores, F., Soria-Arguello, G., & Ibañez-Olvera, M. (2015). Impulse Three Phase Power Supply Used for a Gliding Plasma Discharge. *Journal of Physics: Conference Series, 591*, 012062. https://doi.org/10.1088/1742-6596/591/1/012062.

Sarangapani, C., Misra, N. N., Milosavljevic, V., Bourke, P., O'Regan, F., & Cullen, P. J. (2016). Pesticide degradation in water using atmospheric air cold plasma. *Journal of Water Process Engineering, 9*, 225–232. https://doi.org/10.1016/j.jwpe.2016.01.003.

Sarayu, K., Swaminathan, K., & Sandhya, S. (2007). Assessment of degradation of eight commercial reactive azo dyes individually and in mixture in aqueous solution by ozonation. *Dyes and Pigments, 75*(2), 362–368. https://doi.org/10.1016/j.dyepig.2006.06.011.

Sarkar, S., Ponce, N. T., Banerjee, A., Bandopadhyay, R., Rajendran, S., & Lichtfouse, E. (2020). Green polymeric nanomaterials for the photocatalytic degradation of dyes: a review. In *Environmental Chemistry Letters* (Vol. 18, Issue 5, pp. 1569–1580). Springer. https://doi.org/10.1007/s10311-020-01021-w.

Senthil Kumar, P., Janet Joshiba, G., Femina, C. C., Varshini, P., Priyadharshini, S., Arun Karthick, M. S., & Jothirani, R. (2019). A critical review on recent developments in the low-cost adsorption of dyes from wastewater. *Desalination and Water Treatment, 172*, 395–416. https://doi.org/10.5004/dwt.2019.24613.

Shajeelammal, J., Mohammed, S., Prathish, K. P., Jeeva, A., Asok, A., & Shukla, S. (2022). Treatment of real time textile effluent containing azo reactive dyes via ozonation, modified pulsed low frequency ultrasound cavitation, and integrated reactor. *Journal of Hazardous Materials Advances, 7*, 100098. https://doi.org/10.1016/j.hazadv.2022.100098.

Shams-Ghahfarokhi, Z., & Nezamzadeh-Ejhieh, A. (2015). As-synthesized ZSM-5 zeolite as a suitable support for increasing the photoactivity of semiconductors in a typical photodegradation process. *Materials Science in Semiconductor Processing, 39*, 265–275. https://doi.org/10.1016/j.mssp.2015.05.022.

Sheng, S., Liu, B., Hou, X., Wu, B., Yao, F., Ding, X., & Huang, L. (2018). Aerobic biodegradation characteristic of different water-soluble Azo Dyes. *International Journal of Environmental Research and Public Health, 15*(1). https://doi.org/10.3390/ijerph15010035.

Siddique, K., Rizwan, M., Shahid, M. J., Ali, S., Ahmad, R., & Rizvi, H. (2017). Textile wastewater treatment options: A critical review. In *Enhancing Cleanup of Environmental Pollutants* (Vol. 2, pp. 183–207). Springer International Publishing. https://doi.org/10.1007/978-3-319-55423-5_6.

Singh, S., Lo, S. L., Srivastava, V. C., & Hiwarkar, A. D. (2016). Comparative study of electrochemical oxidation for dye degradation: Parametric optimization and mechanism identification. *Journal of Environmental Chemical Engineering, 4*(3), 2911–2921. https://doi.org/10.1016/j.jece.2016.05.036.

Soares, O. S. G. P., Órfão, J. J. M., Portela, D., Vieira, A., & Pereira, M. F. R. (2006a). Ozonation of textile effluents and dye solutions under continuous operation: Influence of operating parameters. *Journal of Hazardous Materials, 137*(3), 1664–1673. https://doi.org/10.1016/j.jhazmat.2006.05.006.

Soares, O. S. G. P., Órfão, J. J. M., Portela, D., Vieira, A., & Pereira, M. F. R. (2006b). Ozonation of textile effluents and dye solutions under continuous operation: Influence of operating parameters. *Journal of Hazardous Materials, 137*(3), 1664–1673. https://doi.org/10.1016/j.jhazmat.2006.05.006.

Sonu, Sharma, S., Dutta, V., Raizada, P., Hosseini-Bandegharaei, A., Thakur, V., Nguyen, V. H., Vanle, Q., & Singh, P. (2021). An overview of heterojunctioned $ZnFe_2O_4$ photocatalyst for enhanced oxidative water purification. *Journal of Environmental Chemical Engineering, 9*(5). https://doi.org/10.1016/j.jece.2021.105812.

Speight, J. G. (2018). Biological Transformations. In *Reaction Mechanisms in Environmental Engineering* (pp. 269–306). Elsevier. https://doi.org/10.1016/b978-0-12-804422-3.00008-0.

Taheri, M. (2022). Techno-economical aspects of electrocoagulation optimization in three acid azo dyes' removal comparison. *Cleaner Chemical Engineering, 2*, 100007. https://doi.org/10.1016/j.clce.2022.100007.

Tang,' W. Z., & An, H. (1995). UV/TiO, Photocatalytic Oxidation of Commercial Dyes in Aqueous Solutions. In *Chemosphere* (Vol. 31, Issue 9).

Tehrani-Bagha, A. R., Mahmoodi, N. M., & Menger, F. M. (2010). Degradation of a persistent organic dye from colored textile wastewater by ozonation. *Desalination, 260*(1–3), 34–38. https://doi.org/10.1016/j.desal.2010.05.004.

Tomar, R., Abdala, A. A., Chaudhary, R. G., & Singh, N. B. (2020). Photocatalytic degradation of dyes by nanomaterials. *Materials Today: Proceedings, 29*, 967–973. https://doi.org/10.1016/j.matpr.2020.04.144.

Topolovec, B., Škoro, N., Puač, N., & Petrovic, M. (2022). Pathways of organic micropollutants degradation in atmospheric pressure plasma processing – A review. In *Chemosphere* (Vol. 294). Elsevier Ltd. https://doi.org/10.1016/j.chemosphere.2022.133606.

Tu, M., Yu, J., Wang, J., Shi, X., Fu, Z., Hu, S., Zhong, M., & Fei, Z. (2023). Coral-like TiO_2/organosilane hybrid particles with rapid adsorption of methyl orange. *Separation and Purification Technology, 309*. https://doi.org/10.1016/j.seppur.2022.123000.

Vaiano, V., Chianese, L., Rizzo, L., & Iervolino, G. (2021). Visible light driven oxidation of arsenite to arsenate in aqueous solution using Cu-doped ZnO supported on polystyrene pellets. *Catalysis Today, 361*, 69–76. https://doi.org/10.1016/j.cattod.2020.01.011.

Vaiano, V., Miranda, L. N., Pepe, G., Basilicata, M. G., Campiglia, P., & Iervolino, G. (2022). Catalytic non-thermal plasma process for the degradation of organic pollutants in aqueous solution. *Journal of Environmental Chemical Engineering, 10*(3). https://doi.org/10.1016/j.jece.2022.107841.

Venkatesh, S., Venkatesh, K., & Quaff, A. R. (2017). Dye decomposition by combined ozonation and anaerobic treatment: Cost effective technology. *Journal of Applied Research and Technology, 15*(4), 340–345. https://doi.org/10.1016/j.jart.2017.02.006.

World Health Organization. (2022). Drinking-water. [Fact sheet]. https://www.who.int/en/news-room/fact-sheets/detail/drinking-water (accessed on 28 May 2024).

Wu, Z., Dong, J., Yao, Y., Yang, Y., & Wei, F. (2021). Continuous flowing electrocoagulation reactor for efficient removal of azo dyes: Kinetic and isotherm studies of adsorption. *Environmental Technology & Innovation, 22*, 101448. https://doi.org/10.1016/j.eti.2021.101448.

Yang, J., Zeng, D., Hassan, M., Ma, Z., Dong, L., Xie, Y., & He, Y. (2022). Efficient degradation of Bisphenol A by dielectric barrier discharge non-thermal plasma: Performance, degradation pathways and mechanistic consideration. *Chemosphere, 286*. https://doi.org/10.1016/j.chemosphere.2021.131627.

Yaqub, A., Raza, H., Ajab, H., Shah, S. H., Shad, A., & Bhatti, Z. A. (2023). Decolorization of reactive blue-2 dye in aqueous solution by electrocoagulation process using aluminum and steel electrodes. *Journal of Hazardous Materials Advances, 9*, 100248. https://doi.org/10.1016/j.hazadv.2023.100248.

Yong, K., Wu, J., & Andrews, S. (2005). Heterogeneous catalytic ozonation of aqueous reactive dye. *Ozone: Science and Engineering, 27*(4), 257–263. https://doi.org/10.1080/01919510591005888.

Yue, L., Wang, K., Guo, J., Yang, J., Luo, X., Lian, J., & Wang, L. (2014). Enhanced electrochemical oxidation of dye wastewater with Fe_2O_3 supported catalyst. *Journal of Industrial and Engineering Chemistry, 20*(2), 725–731. https://doi.org/10.1016/j.jiec.2013.06.001.

Zeghioud, H., Nguyen-Tri, P., Khezami, L., Amrane, A., & Assadi, A. A. (2020). Review on discharge Plasma for water treatment: mechanism, reactor geometries, active species and combined processes. *Journal of Water Process Engineering* (Vol. 38). Elsevier Ltd. https://doi.org/10.1016/j.jwpe.2020.101664.

Zolgharnein, J., Rajabalipour, F., & Dermanaki Farahani, S. (2023). Indigo Carmine Dye Adsorptive Removal by Polyethylene Glycol-Modified Hydroxyapatite Nanoparticles as an Efficient Adsorbent. *Water, Air, & Soil Pollution, 234*(4). https://doi.org/10.1007/s11270-023-06207-w.

CHAPTER 6

Spent Coffee Ground as a Source of Bioactive Compounds with Food and Agriculture Applications: A Review

KARLA A. LUNA, MÓNICA LIZETH CHÁVEZ-GONZÁLEZ, CRISTÓBAL N. AGUILAR, N. RAMÍREZ-GUZMAN, H. A. RUIZ, and L. M. HERNÁNDEZ

Bioprocesses and Bioproducts Group, Food Research Department, School of Chemistry, Universidad Autónoma de Coahuila, Saltillo, Coahuila, Mexico

ABSTRACT

Coffee is one of the most popular beverages worldwide due to its properties, aroma, and flavor. It is important to mention that in the processing of the beverage, only 5% of the fruit is used and the other 95% is agro-industrial waste. The coffee industry is responsible for large quantities of agro-industrial residues, among these residues is the coffee grounds that are obtained from the filters of the coffee machines. The coffee grounds are deposited in the environment without prior treatment which causes contamination to the environment. Coffee grounds can be valorized since it is a sustainable, renewable, and attractive raw material for research. In addition, the extracts have bioactive compounds that have problem-solving applications in the food industry in supplementation and agriculture as fertilizer and antimicrobial agents. They can also be harnessed in various ways to promote sustainable development and the circular economy.

Sustainable Agriculture and Global Environmental Health: Toward the Efficient Management of Resources. Mónica L. Chávez González, Cristóbal Noé Aguilar, Juan Carlos Contreras-Esquivel, and A. K. Haghi (Eds.)

6.1 INTRODUCTION

Coffee belongs to the Rubiaceae family, genus Coffea; the most important species, due to their flavor and aroma, are arabica (*Coffea arabica*), with approximately 75% of the cultivated area, and robusta (*C. canephora*), with approximately 24%. Currently, Mexico is among the top 10 producers worldwide, with a green bean production of around 239 million kilograms per year. Coffee consumption is more than 3,000 million cups per day worldwide, and more than 50 countries produce coffee commercially (Marín-Garza, Aguilar-Rivera, & Castañeda-Castro, 2019).

In recent years, the coffee industry has been responsible for the drastic increase in agro-industrial waste from its processing to obtain a beverage. It is important to note that coffee has become the second most popular product due to its aroma, taste, and properties. In the coffee industrialization process, only 5% of the weight of the fresh fruit is used in the preparation of the beverage, while the other 95% is represented by residues (Fernández-Cortés, Sotto-Rodríguez, & Vargas-Marín, 2020). The by-products of the coffee production chain include husks, mucilage, pulp, and coffee grounds, which are an important source of contamination and have been used as raw materials for some time. Coffee grounds are the dark brown solid residue that remains in the filter of coffee machines or presses, which typically has no added value and is usually deposited in the drainage system or on the ground without any treatment (Limón-Hernández et al., 2021). It is relevant to highlight that coffee grounds have valuable properties due to their high content of antioxidants and phenolic compounds such as chlorogenic acid, caffeine, and flavonoids. Their composition is related to high antioxidant, antibacterial, antiviral, and anti-inflammatory activity (Kovalcik, Obruca, & Marova, 2018). Therefore, this chapter aims to present a comprehensive review of recent literature on waste generated from the coffee industry, such as coffee grounds, its components, and proposals for the valorization of spent coffee grounds in related applications in the food industry and agriculture, as well as its impact on the environment and the economy.

6.2 COFFEE PRODUCTION

Coffee is grown in about 80 countries, of which 90% are developing countries, so its production has a relevant social importance (Dadi, Daba, Beyene, Luis, & Van der Bruggen, 2019). In addition, coffee beverage is a product found all over the world. Its production is obtained in tropical and subtropical zones in most underdeveloped countries (Bazán, del Ángel Zumaya, & Rivera, 2020). Currently, coffee consumption exceeds 3 billion cups per day globally, and more than 50 countries are coffee producers (Vegro & de Almeida, 2020). Arabica and robusta coffee production are important in the coffee industry due to their high quality, which accounts for approximately 56.6% of total production (95.7 million bags), while robusta coffee accounts for 43.4% (73.5 million bags). One of the characteristics of these species is their ease of production, which is why they have a lower commercialization price. On the other hand, world coffee production in 2019 is estimated at 174.5 million bags of approximately 60 kilograms, which is 10% higher than the previous year's estimated volume. As for the 2019/2020 production that started in October 2019 and reached a global production of 169.3 million bags, with a fall of 3% compared to 2018/2019 (Romero, 2020). More than 45 countries have purchased Mexican agribusiness, of which 13 have made constant purchases during the year, the rest only at least one month in the period (SIAP, 2019). In general, coffee is produced in developing or underdeveloped countries; it is transformed, commercialized, and distributed in industries considered developed. It is worth mentioning that the term "development" is understood under aspects of technology without considering other elements. As in other agrifood products, value-added is concentrated in the last links, benefiting mostly all the agents other than the producer. It is estimated that the most important intermediate consumer countries, also called importers, appropriate 84% of the final price of the product and only the rest goes to the producer (Velazquez Salazar, & Tenorio Noguéz, 2017). Coffee production is important in different countries since it has an increasing development, favoring the economy.

6.3 AGRO-INDUSTRIAL WASTE REVENUE GENERATION

A large amount of organic waste and related effluents is produced every year through food processing industries such as juice, chips, meat, confectionery, and fruit industries. These organic wastes can be used for different benefits; as the population is continuously increasing, the food requirements and their uses are also increasing (Alzate et al., 2011). Thus, in most countries, different food and beverage industries have increased remarkably to meet the need for food. Agriculture-based industries produce a large amount of waste every year. If these wastes are released into the environment without proper disposal procedures, they can cause environmental pollution and harmful health effects (Basanta, Delgado, Martínez, Vázquez, & Vázquez, 2007).

Most agro-industrial waste is not treated, so in most reports, it is disposed of either by burning, dumping, or unplanned deposit in landfills. Agricultural waste is rich in bioactive compounds. This waste can be used as an alternative source for the production of different products such as biogas, biofuel, fungi, and bioproducts. In addition, it can serve as raw material in various research and industries where it can help reduce the cost of production and also reduce the pollution load of the environment. In turn, it is also destined for composting or animal feed (options that generate low value-added products and increase the risks of spreading diseases, respectively). Many of the high-value-added functional molecules are lost or underutilized in these processes (Mahecha, 2021). Whereby a favorable application for agro-industrial wastes is in the manufacture of biofuels, enzymes, vitamins, antioxidants, antibiotics, and other chemicals through solid-state fermentation (SSF) among other bioprocesses (Kumar, Yadav, Kumar, Vyas, & Dhaliwal, 2017), ensuring the recovery of economically valuable waste materials. Thus, the wide range of novel applications of their bioactive compounds in the food industry and modern pharmaceutical and cosmetic industries (Rico, Gullón, Alonso, & Yáñez, 2020) makes them high-potential raw materials; the use of agro-industrial wastes can benefit both the economy and the environment by providing employment opportunities, reducing production costs, generating renewable energy, and reducing environmental pollution.

6.4 SPENT COFFEE GROUNDS

Spent coffee grounds are the remaining residues of ground coffee after extraction by infusion with hot water in the preparation of drinkable coffee. Usually, these residues remain in the filters of the coffee machine or are discarded, with a high moisture content (McNutt, 2019). From an environmental perspective, the disposal of coffee by-products (spent coffee grounds) represents a major challenge. These residues are often disposed of directly into the soil without prior treatment and contain large amounts of organic substances that require much oxygen to decompose. For this reason, SCGs are highly polluting and contribute to the waste management problem (Santos, Fonseca, Aires, Coutinho, & Trindade, 2017).

They consist of a dark brown solid residue with high moisture content, with first-origin residues being richer in chemical compounds. Compared to spent coffee from instant coffee industries. This is understandable since extraction is maximized for soluble coffee production to obtain the best yields (Mavria et al., 2022).

In addition, it is widely recognized that coffee grounds compounds contain a high content of polyphenols, especially chlorogenic acid, and flavonoids, which have biological functionalities (Lopez-Ramirez et al., 2018). Studies on coffee by-products such as SCG reveal an excellent profile of bioactive compounds with antioxidant properties. These compounds are related to human health, with antioxidant, antibacterial, antiviral, anti-inflammatory, and anticancer activities (Zabaniotou & Kamaterou, 2019).

6.4.1 SPENT COFFEE GROUNDS COMPOSITION

6.4.1.2 LIPIDS

Spent coffee grounds have a rich source of lipids, which can account for around 10–15% of their dry weight. These lipids are mainly unsaturated fatty acids, such as linoleic acid and oleic acid. The reported fatty acids are essential for the human organism as they have beneficial health properties by reducing cardiovascular diseases (Vargas Lazo, 2018).

In the chemical composition of coffee, the lipid fraction constitutes an important part as the lipids and fatty acids that compose it also favor the aromas and flavors. The lipids contained in coffee grounds generally range from 8 to 20% (Massaya et al., 2019). In addition, the lipids that are part of the coffee bean are not degraded in the roasting process. Some fatty acids release oxidation compounds such as aldehydes (Cardoso, Agnoletti, de Freitas, de Abreu Pinheiro, & Pereira, 2021). Lipids are used in the production of biopolymers, biofuels, and sunscreens (Rocha et al., 2014; Bhatia et al., 2017; Chiari-Andréo et al., 2014); based on the above, the use of lipids from coffee grounds promotes sustainable development by reducing waste, as it can be used to produce innovative and economically profitable products for the food, pharmaceutical, and energy industries.

6.4.1.2 PROTEINS AND AMINO ACIDS

Proteins are polymers resulting from the dehydration of amino acids and the union of these, using a specific type of covalent bond called peptide bond. Coffee grounds contain significant values of proteins present in their composition, which can reach up to 13–17% of their total weight (Battista et al., 2020; Coelho, Batista, Ávila, Franca, & Oliveira, 2021). Agro-industrial waste by-products, particularly seeds, are rich in proteins. These contain bioactive compounds that have a variety of biological properties, including antihypertensive, antioxidant, antimicrobial, and immunomodulatory (Leyva-López, Lizárraga-Velázquez, Hernández, & Sánchez-Gutiérrez, 2020).

It is important to mention that protein can be used as a nutrient supply and plays an important role in biotechnological and fermentative processes as a substrate. Studies have shown that protein and peptides from coffee wells have an antihypertensive and antioxidant effect, beneficial in reducing the risk of diabetes and hypertension (Ribeiro, de Souza Rocha, & Prudencio, 2021). In addition, coffee grounds protein contains beneficial nutrients such as dietary fiber and antioxidants and can therefore be used by the food industry in the production of beverages with nutritional value.

6.4.1.3 DIETARY FIBER AND FUNCTIONAL POLYSACCHARIDES

Spent coffee grounds are mainly a source of dietary fiber in the form of cellulose, hemicellulose, and resistant starch. Agro-industrial coffee by-products such as coffee pulp and spent coffee grounds can be considered an important source of dietary fiber, with applications in the food industry. The addition of fiber to food products would be a beneficial strategy to improve their nutritional, functional properties, and/or physical characteristics, helping in the prevention of diseases such as constipation, cholesterolemia, cancer, and diabetes (Ravani & Sharma, 2022). Spent coffee grounds with approximately 51% crude dietary fiber, previously classified as antioxidant dietary fiber, may be an innovative source for developing fortified foods rich in dietary fiber.

Most studies have focused on pectin extraction, considered one of the most prominent products obtained from waste, as coffee grounds pectin can contain small amounts of pectin and has various applications in the food industry, such as gelling agent, thickener, and stabilizer (Rakitikul & Nimmanpipug, 2016). Pectin also has beneficial health properties, such as lowering blood cholesterol levels, improving bowel function, and preventing constipation.

6.4.1.4 POLYPHENOLIC COMPOUNDS

Phenolic compounds or polyphenols comprise a broad group of chemical substances in the plant kingdom with different chemical structures encompassing more than 8,000 different compounds (Martinez-Gomez, Caballero, & Blanco, 2020). In addition, phenolic compounds, secondary plant metabolites abundant in fruits and their derivatives, exhibit antioxidant, antimutagenic, and anti-inflammatory properties, among others. It is important to highlight that antioxidant activity is usually associated with the phenolic content (Brglez Mojzer, Knez Hrnčič, Škerget, Knez, & Bren, 2016). Spent coffee grounds can contain a high number of phenolic compounds such as chlorogenic acid, caffeic acid, and feruloyl quinic acid, among others, due to the coffee bioactive substances reported in the literature (Saud & Salamatullah, 2021). Likewise, chlorogenic acid is a polyphenol found in coffee grounds that plays an important role in the gut microbiota as it reaches the small and large intestine undigested, where the gut microbiota can metabolize it to benefit health. Additionally, the coffee by-product contains melanoidins that act as dietary fibers and are fermented by the colonic microbiota, which also provides health benefits. The melanoidins present in SCG have been shown to have high antioxidant activity due to the presence of these polyphenols (Pérez-Burillo et al., 2020; Fogliano & Morales, 2011).

These compounds are of great interest since they provide a wide range of bioactivities as auxiliaries in inflammatory processes and viral and bacterial infections, and in addition, the phenolic content is a useful preliminary indicator of their antioxidant activity (Zabaniotou & Kamaterou, 2019).

It is important to mention that previous studies have shown that these secondary metabolites can exert various bioactivities, such as antioxidant, antiviral/antibacterial, anticancer, and anti-inflammatory activities (Zabaniotou & Kamaterou, 2019). Obtaining the extracts can be an alternative for the elaboration of different products due to the final cost since it is a waste that has a low cost.

6.5 APPLICATIONS IN INDUSTRY

6.5.1 FOOD APPLICATIONS

Studies have reported that coffee grounds contain bioactive compounds such as dietary fiber and phenolic compounds (Kovalcik, Obruca, & Marova, 2018). In addition, their biological properties may help reduce age-related diseases, making them of great interest in the formulation of functional foods, fortified foods, and dietary supplements (Martinez-Gomez, Caballero, & Blanco, 2020). It is important to mention that coffee grounds play an important role in the formulation of bakery products, as they are part of a dough ingredient up to 4% w/w without affecting the nutritional or sensory quality of the final product. Previous studies (Meerasri & Sothornvit, 2022) used oil extracted from coffee grounds as a partial substitute for butter in biscuits, with an acceptable substitution limit of 20%. Partial substitution of butter with SCG oil increased the phenolic content and antioxidant properties and improved the texture of the biscuits. In addition, partially substituting butter with alternative ingredients can reduce the final product's saturated fat and calorie content, which may also be beneficial for those wishing to reduce their saturated fat and calorie intake.

The reuse of coffee grounds as an alternative to producing innovative products can have benefits for the food industry by improving processes and reducing costs. At the same time, it can offer additional consumer benefits by improving the quality and health of the final products as shown in Table 6.1.

TABLE 6.1 Applications of Coffee Grounds in the Food Industry

Application	References
Bioactive peptides with antihypertensive and antioxidant potentials.	Ribeiro, de Souza Rocha, & Prudencio (2021)
Production of fermented and distilled alcoholic beverages improving organoleptic aspects and taste.	Machado, Mussatto, Teixeira, Vilanova, & Oliveira (2018)
Coffee-flavored liqueur.	Masino, Montevecchi, Calvini, Foca, & Antonelli (2022)
Antifungal and anti-mycotoxigenic material; Potential food ingredient with moderate cytotoxic and antibacterial activities.	Badr, El-Attar, Ali, Elkhadragy, Yehia, & Farouk (2022)
Spent coffee grounds as a potential ingredient to improve the nutritional and sensory quality of bread.	Martinez-Gomez, Caballero, & Blanco (2020)
Spent coffee grounds may be a valuable source of antioxidant and anti-inflammatory bioactive compounds.	Salazar-López et al. (2020)
It is used for cattle feed (dairy and beef) because of its high fiber and protein content.	San Martin et al. (2021)
It is a lignocellulosic composite material that is used for food packaging.	Garcia & Kim (2021)

6.5.2 AGRICULTURE APPLICATIONS

The valorization of coffee grounds provides opportunities for different areas of the industry to develop innovative products that are safe for the consumer, from cultivation to production. In agriculture, coffee grounds have advantages in promoting sustainability because this residue contains organic compounds such as polyphenols, as mentioned above, and minerals that can be used as a rich source of nutrients for the crop soil (Ciesielczuk et al., 2017).

Used coffee grounds are a source of nitrogen-rich biomass, containing approximately 30% protein by weight. From this biomass, glucose isomerization by-products can be obtained. In addition, the nitrogen content can be attributed in the form of alkaloids, amines, and free amino acids. Studies have shown that it has potential in agriculture as a fertilizer due to its nitrogen content (Anastopoulos, Karamesouti, Mitropoulos, & Kyzas, 2017). It is important to note that the nitrogen and other compounds present in this agro-industrial waste can be used as an alternative fertilizer, improving soil quality, and promoting crop growth.

It is important to mention that this residue can also bind to pesticide residues and other toxic heavy metals in soils, becoming less mobile in the soil and therefore less available to plants and other living organisms. This is because spent coffee grounds act as a "glue" that holds the contaminants together and prevents them from moving freely through the soil. In some cases, contaminants may even be adsorbed on the surface of the spent coffee grounds, making them even more difficult to remove from the soil (Kim et al., 2014). Recent studies using a mixture of eggshell and coffee grounds concluded that it had favorable results as it contained Ca, Mg, and K, had the right pH, and was also mixed with light clay ceramics for agricultural field drainage purposes and as a fertilizer for nursery production, as it retained water in the soil and improved leaf photosynthesis (Tombarkiewicz et al., 2022; Andreola et al., 2019; Ronga et al., 2020).

A problem in agriculture is the presence of plant pathogens that cause diseases in plant growth and affect the yield and quality of the crop in terms of size, color, texture, and very significantly, loss of commercial value. Moreover, controlling these plant diseases is costly, which is of interest for research, thanks to the compounds obtained from coffee grounds, an alternative for developing biocontrol agents that can solve problems in agriculture. Biocontrol agents are of great interest due to their ability to adapt to low-cost agro-industrial wastes, the enzymes they produce for antimicrobial activity are obtained through their biological control mechanisms and mass production (Das, Aguilar, Haridas, & Sabu, 2021). In the industry, different fungi are used one of them is *Trichoderma* which is the most used and effective bio fungicide of crop disease management through biological media.

Previous studies showed that Trichoderma and coffee residues as a substrate within a solid-state fermentation had an inhibitory effect against plant pathogens such as *Phytophthora* spp., *Rhizoctonia* spp., and *Fusarium* spp. (Das, Haridas, & Sabu, 2020). Other applications of coffee grounds in agriculture are shown in Table 6.2. Finally, coffee waste is an alternative solution to this problem that affects the different industries as well as the consumer. It opens new research for obtaining compounds from this waste to favor agriculture and have good quality crops and thus implement the circular economy.

TABLE 6.2 Applications of Coffee Grounds in Agriculture

Application	References
Fertilizer: coffee grounds are rich in essential nutrients for plant growth, which are used as a fertilizer to enrich the soil and improve the crop.	Cervera-Mata et al. (2019)
Spent coffee grounds are potential agents in the control of *Aedes aegypti* by reducing survival and blocking larval development.	Miranda, Fernandes, Bernardes, & Martins (2021)
Spent coffee grounds have demonstrated the ability to adsorb and remove pesticides.	Pavlovic et al. (2014)
Spent coffee grounds are used for composting to speed up the decomposition process and produce a nutrient-rich organic fertilizer for plants.	Ronga, Pane, Zaccardelli, & Pecchioni (2016)
Green synthesis of zinc oxide nanoparticles using coffee grounds with biocompatibility and antibacterial properties.	Abdelmigid, Hussien, Alyamani, Morsi, AlSufyani, & Kadi (2022)

6.6 CONCLUSION

Recently, the use of used coffee grounds for various value-added applications in the food industry and agriculture has been extensively researched. This has led to the discovery of new sustainable, cost-effective, and innovative alternatives to utilize coffee industry waste in favor of the circular economy and the reduction of environmental impact. In the food industry, it represents an opportunity for the development of value-added food and beverage products enriched with enhanced biological activities that exert potential health benefits with healthy and cost-effective products, which is becoming increasingly important nowadays. Spent coffee grounds also present promising solutions to different problems in agriculture, thanks to their properties as a fertilizer, antimicrobial agent, and substrate. It also has a favorable impact on farmers, improving the quality of plants and their production. The valorization of coffee grounds is a creative innovation to transform waste into valuable environmentally sustainable solutions.

KEYWORDS

- **agro-industrial waste**
- **bioactive compounds**
- **dietary fiber**
- **food applications**
- **polyphenolic compounds**
- **polysaccharides**
- **spent coffee grounds**

REFERENCES

Abdelmigid, H. M., Hussien, N. A., Alyamani, A. A., Morsi, M. M., AlSufyani, N. M., & Kadi, H. A. (2022). Green synthesis of zinc oxide nanoparticles using pomegranate fruit peel and solid coffee grounds vs. chemical method of synthesis, with their biocompatibility and antibacterial properties investigation. *Molecules, 27*(4), 1236.

Alzate Tamayo, L. M., Jimenez Cartagena, C., & Londoño Londoño, J. (2011). Use of agro-industrial waste to improve the sensory and nutritional quality of poultry products. *Clean+ Production, 6*(1), 108–127.

Anastopoulos, I., Karamesouti, M., Mitropoulos, A. C., & Kyzas, G. Z. (2017). A review for coffee adsorbents. *Journal of Molecular Liquids, 229*, 555–565.

Andreola, F., Borghi, A., Pedrazzi, S., Allesina, G., Tartarini, P., Lancellotti, I., & Barbieri, L. (2019). Spent coffee grounds in the production of lightweight clay ceramic aggregates in view of urban and agricultural sustainable development. *Materials, 12*, 3581.

Badr, A. N., El-Attar, M. M., Ali, H. S., Elkhadragy, M. F., Yehia, H. M., & Farouk, A. (2022). Spent coffee grounds valorization as bioactive phenolic source acquired antifungal, anti-mycotoxigenic, and anti-cytotoxic activities. *Toxins* 2022, *14*(2), 109; https://doi.org/10.3390/toxins14020109.

Basanta, R., Delgado, M. G., Martínez, J. C., Vázquez, H. M., & Vázquez, G. B. (2007). Sostenibilidad del reciclaje de residuos de la agroindustria azucarera: Una revisión [Sustainable recycling of waste from sugarcane agroindustry: A review]. *CYTA-Journal of Food, 5*(4), 293–305.

Battista, F., Barampouti, E. M., Mai, S., Bolzonella, D., Malamis, D., Moustakas, K., & Loizidou, M. (2020). Added-value molecules recovery and biofuels production from spent coffee grounds. *Renewable and Sustainable Energy Reviews, 131*, 110007.

Bazán, I. S., del Ángel Zumaya, J. A., & Rivera, O. J. (2020). Influence of roasting temperature of various blends of Coffea Arabica coffee and Coffea Robusta coffee. *Electronic Magazine on Technology, Education and Society, 7*(14), 237–255.

Bhatia, S. K., Kim, J. H., Kim, M. S., Kim, J., Hong, J. W., Hong, Y. G., Kim, H. J., Jeon, J. M., Kim, S. H., Ahn, J., Lee, H., & Yang, Y. H. (2017). Production of

(3-hydroxybutyrate-co-3-hydroxyhexanoate) copolymer from coffee waste oil using engineered Ralstonia eutropha. *Bioprocess Biosyst. Eng., 41*, 229–235.

Brglez Mojzer, E., Knez Hrnčič, M., Škerget, M., Knez, Ž., & Bren, U. (2016). Polyphenols: Extraction methods, antioxidative action, bioavailability and anticarcinogenic effects. *Molecules, 21*(7), 901.

Cardoso, W. S., Agnoletti, B. Z., de Freitas, R., de Abreu Pinheiro, F., & Pereira, L. L. (2021). Biochemical aspects of coffee fermentation. In *Quality Determinants in Coffee Production* (pp. 149–208). https://pubmed.ncbi.nlm.nih.gov/30711092/.

Cervera-Mata, A., Navarro-Alarcón, M., Delgado, G., Pastoriza, S., Montilla-Gómez, J., Llopis, J., & Rufián-Henares, J. Á. (2019). Spent coffee grounds improve the nutritional value in elements of lettuce (Lactuca sativa L.) and are an ecological alternative to inorganic fertilizers. *Food Chemistry, 282*, 1–8.

Chiari-Andréo, B. G., Trovatti, E., Pecoraro, É., Corrêa, M. A., Cicarelli, R. M. B., Ribeiro, S., & Isaac, V. L. B. (2014). Synergistic effect of green coffee oil and synthetic sunscreen for health care application. *Ind. Crop. Prod., 52*, 389–393.

Ciesielczuk, T., Rosik-Dulewska, C., Poluszyńska, J., Miłek, D., Szewczyk, A., & Sławińska, I. (2017). Acute Toxicity of Experimental Fertilizers Made of Spent Coffee Grounds. *Waste Biomass Valorization, 9*, 2157–2164.

Coelho, G. O., Batista, M. J., Ávila, A. F., Franca, A. S., & Oliveira, L. S. (2021). Development and characterization of biopolymeric films of galactomannans recovered from spent coffee grounds. *Journal of Food Engineering, 289*, 110083.

Dadi, D., Daba, G., Beyene, A., Luis, P., & Van der Bruggen, B. (2019). Composting and co-composting of coffee husk and pulp with source-separated municipal solid waste: A breakthrough in valorization of coffee waste. *International Journal of Recycling of Organic Waste in Agriculture, 8*, 263–277.

Das, M. M., Aguilar, C. N., Haridas, M., & Sabu, A. (2021). Production of bio-fungicide, Trichoderma harzianum CH_1 under solid-state fermentation using coffee husk. *Bioresource Technology Reports, 15*, 100708.

Das, M. M., Haridas, M., & Sabu, A. (2020). Process development for the enhanced production of bio-nematicide Purpureocillium lilacinum KU8 under solid-state fermentation. *Bioresource Technology, 308*, 123328.

Fernández-Cortés, Y., Sotto-Rodríguez, K. D., & Vargas-Marín, L. A. (2020). Environmental impacts of coffee production, and the sustainable use of the waste generated. *Producción+ limpia, 15*(1), 93–110.

Fogliano, V., & Morales, F. J. (2011). Estimation of dietary intake of melanoidins from coffee and bread. *Food Funct., 2*, 117–123.

Garcia, C. V., & Kim, Y. T. (2021). Spent Coffee Grounds and Coffee Silverskin as Potential Materials for Packaging: A Review. *Journal of Polymers and the Environment, 29*, 2372–2384.

Kim, M. S., Min, H. G., Koo, N., Park, J., Lee, S. H., Bak, G. I., & Kim, J. G. (2014). The effectiveness of spent coffee grounds and its biochar on the amelioration of heavy metals-contaminated water and soil using chemical and biological assessments. *Journal of Environmental Management, 146*, 124–130.

Kovalcik, A., Obruca, S., & Marova, I. (2018). Valorization of spent coffee grounds: A review. *Food and Bioproducts Processing, 110*, 104–119.

Kovalcik, A., Obruca, S., & Marova, I. (2018). Valorization of spent coffee grounds: A review. *Food and Bioproducts Processing, 110*, 104–119.

Kumar, K., Yadav, A. N., Kumar, V., Vyas, P., & Dhaliwal, H. S. (2017). Food waste: A potential bioresource for extraction of nutraceuticals and bioactive compounds. *Bioresources and Bioprocessing, 4*, 1–14.

Leyva-López, N., Lizárraga-Velázquez, C. E., Hernández, C., & Sánchez-Gutiérrez, E. Y. (2020). Exploitation of agro-industrial waste as potential source of bioactive compounds for aquaculture. *Foods, 9*(7), 843.

Limón-Hernández, R. A., López-Hernández, V., Aguilar-Aguilar, F. A., & Hernández-Martínez, I. (2021). Evaluation of the bioenergy potential of coffee grounds. *RINDERESU, 5*(1).

Lopez-Ramirez, N., Volke-Sepulveda, T., Gaime-Perraud, I., Saucedo-Castañeda, G., & Favela-Torres, E. (2018). Effect of stirring on growth and cellulolytic enzymes production by Trichoderma harzianum in a novel bench-scale solid-state fermentation bioreactor. *Bioresource Technology, 265*, 291–298.

Machado, E., Mussatto, S., Teixeira, J., Vilanova, M., & Oliveira, J. (2018). Increasing the sustainability of the coffee agro-industry: Spent coffee grounds as a source of new beverages. *Beverages, 4*, 105.

Mahecha, P. V. (2021). Agro-industrial chains: By-products and residues of agricultural activity. Alternatives for the use of waste in agro-industry.

Marín-Garza, T., Aguilar-Rivera, N., & Castañeda-Castro, O. (2019). Mexican coffee agribusiness, trends, challenges, and goals towards sustainability. *Current Politics and Economics of the United States, Canada, and Mexico 2019, 21*(2), 71–121.

Martinez-Gomez, A., Caballero, I., & Blanco, C. A. (2020). Phenols and Melanoidins as Natural Antioxidants in Beer. Structure, Reactivity and Antioxidant Activity. *Biomolecules, 10*, 400.

Masino, F., Montevecchi, G., Calvini, R., Foca, G., & Antonelli, A. (2022). Sensory evaluation and mixture design assessment of coffee-flavored liquor obtained from spent coffee grounds. *Food Qual. Prefer., 96*, 104427.

Massaya, J., Prates Pereira, A., Mills-Lamptey, B., Benjamin, J., & Chuck, C. J. (2019). Conceptualization of a Spent Coffee Grounds Biorefinery: A Review of Existing Valorisation Approaches. *Food and Bioproducts Processing, 118*, 149–166.

Mavria, A., Tsouko, E., Protonotariou, S., Papagiannopoulos, A., Georgiadou, M., Selianitis, D., Pispas, S., Mandala, I., & Koutinas, A. A. (2022). Sustainable Production of Novel Oleogels Valorizing Microbial Oil Rich in Carotenoids Derived from Spent Coffee Grounds. *Journal of Agricultural and Food Chemistry, 70*(35), 10807–10817.

McNutt, J. (2019). Spent coffee grounds: A review on current utilization. *Journal of Industrial and Engineering Chemistry, 71*, 78–88.

Meerasri, J., & Sothornvit, R. (2022). Novel development of coffee oil extracted from spent coffee grounds as a butter substitute in bakery products. *Journal of Food Process. Preserv., e16687*.

Miranda, F. R., Fernandes, K. M., Bernardes, R. C., & Martins, G. F. (2021). Biological, histological and immunohistochemical studies on the toxicity of spent coffee grounds and caffeine on the larvae of Aedes aegypti (Diptera: Culicidae). *Environmental Pollution, 271*, 116307.

Pavlovic, M. D., Buntic, A. V., Šiler-Marinkovic, S. S., Antonovic, D. G., Milutinovic, M. D., Radovanovic, N. R., & Dimitrijevic-Brankovic, S. I. (2014). Spent coffee grounds as adsorbents for pesticide paraquat removal from its aqueous solutions. In *International Conference on Civil, Biological and Environmental Engineering* (pp. 27–28). https://www.researchgate.

net/publication/378279412_Spent_Coffee_Grounds_Plant_Growth_Promoting_Bacteria_and_Medicinal_Plant_Waste_The_Biofertilizing_Effect_of_High-Value_Compost/fulltext/65d0ac63e51f606f997989f7/Spent-Coffee-Grounds-Plant-Growth-Promoting-Bacteria-and-Medicinal-Plant-Waste-The-Biofertilizing-Effect-of-High-Value-Compost.pdf.

Pérez-Burillo, S., Rajakaruna, S., Pastoriza, S., Paliy, O., & Rufián-Henares, J. Á. (2020). Bioactivity of food melanoidins is mediated by gut microbiota. *Food Chemistry, 316*, 126309.

Rakitikul, W., & Nimmanpipug, P. (2016). Degree of esterification and gelling properties of pectin structure in coffee pulp. In *Key Engineering Materials, 675*, 11–14.

Ravani, A., & Sharma, H. P. (2022). Meat Based Functional Foods. In *Functional Foods* (pp. 235–287).

Ribeiro, E., de Souza Rocha, T., & Prudencio, S. H. (2021). Potential of green and roasted coffee beans and spent coffee grounds to provide bioactive peptides. *Food Chemistry, 348*, 129061.

Rico, X., Gullón, B., Alonso, J. L., & Yáñez, R. (2020). Recovery of high value-added compounds from pineapple, melon, watermelon, and pumpkin processing by-products: An overview. *Food Research International, 132*, 109086.

Rocha, M. V. P., de Matos, L. J. B. L., de Lima, L. P., da Silva Figueiredo, P. M., Lucena, I. L., Fernandes, F. A. N., & Gonçalves, L. R. B. (2014). Ultrasound-assisted production of biodiesel and ethanol from spent coffee grounds. *Bioresource Technology, 167*, 343–348.

Romero, C. A. (2020). Commodities Observatory: Coffee. Lima, Peru: MINAGRI.

Ronga, D., Pane, C., Zaccardelli, M., & Pecchioni, N. (2016). Use of spent coffee ground compost in peat-based growing media to produce basil and tomato potting plants. *Communications in Soil Science and Plant Analysis, 47*(3), 356–368.

Ronga, D., Parisi, M., Barbieri, L., Lancellotti, I., Andreola, F., & Bignami, C. (2020). Valorization of spent coffee grounds, biochar, and other residues to produce lightweight clay ceramic aggregates suitable for nursery grapevine production. *Horticulturae, 6*, 58.

Salazar-López, N. J., López-Rodríguez, C. V., Hernández-Montoya, D. A., & Campos-Vega, R. (2020). Health benefits of spent coffee grounds. In *Food Wastes and By-Products: Nutraceutical and Health Potential* (pp. 327–351).

San Martin, D., Orive, M., Iñarra, B., García, A., Goiri, I., Atxaerandio, R., & Zufia, J. (2021). Spent coffee ground as second-generation feedstuff for dairy cattle. *Biomass Conversion and Biorefinery, 11*, 589–599.

Santos, C., Fonseca, J., Aires, A., Coutinho, J., & Trindade, H. (2017). Effect of different rates of spent coffee grounds (SCG) on composting process, gaseous emissions, and quality of end-product. *Waste Management, 59*, 37–47.

Saud, S., & Salamatullah, A. M. (2021). Relationship between the chemical composition and the biological functions of coffee. *Molecules, 26*(24), 7634.

SIAP. (2019). Agrifood and Fisheries Information System. Agri-food panorama. Agro-food and Fisheries Information System – Ministry of Agriculture and Rural Development, Mexico City. Retrieved from: https://nube.siap.gob.mx/gobmx-publicaciones-siap/pag/2019/atlas-agroalimentario-2019 (accessed on 28 May 2024).

Tombarkiewicz, B., Antonkiewicz, J., Lis, M. W., Pawlak, K., Trela, M., & Witkowicz, R., & Gorczyca, O. (2022). Chemical properties of the coffee grounds and poultry eggshells mixture in terms of soil improver. *Scientific Reports, 12*, 2592.

Vargas Lazo, A. M. (2018). Study of the production of pellets from coffee grounds. *Chemical Engineering.*

Vegro, C. L. R., & de Almeida, L. F. (2020). Global coffee market: Socio-economic and cultural dynamics. In *Coffee Consumption and Industry Strategies in Brazil* (pp. 3–19). Woodhead Publishing.

Velazquez Salazar, M., & Tenorio Noguéz, A. (2017). Global value chains: A methodological proposal for the analysis of coffee chains in Mexico. In *Rural Perspectives Magazine New Era* (pp. 13–41). National University, Costa Rica.

Zabaniotou, A., & Kamaterou, P. (2019). Food waste valorization advocating Circular Bioeconomy-A critical review of potentialities and perspectives of spent coffee grounds biorefinery. *Journal of Cleaner Production, 211*, 1553–1566.

Zabaniotou, A., & Kamaterou, P. (2019). Food waste valorization advocating Circular Bioeconomy-A critical review of potentialities and perspectives of spent coffee grounds biorefinery. *Journal of Cleaner Production, 211*, 1553–1566.

CHAPTER 7

Sustainable Use of Semi-Desert Plants

ZAIDA M. ARANDA,[1] MÓNICA LIZETH CHÁVEZ-GONZÁLEZ,[1] CRISTIAN TORRES-LEÓN,[2] JOSÉ LUIS MARTÍNEZ HERNÁNDEZ,[1] and MAYELA GOVEZA SALAS[1]

[1]*Bioprocesses and Bioproducts Group, Food Research Department, School of Chemistry, Universidad Autónoma de Coahuila, Saltillo, Coahuila, Mexico*

[2]*Research Center and Ethnobiological Garden (CIJE-UAdeC), Universidad Autonoma de Coahuila, Unidad Torreón, Viesca, Coahuila, Mexico*

ABSTRACT

The semi-desert region zones cover a considerable territorial extension in the world. Regions such as the Southwestern United States, the Northern states of Mexico, the Gran Chaco in South America, Europe mostly in southeastern Spain, parts of Greece, Italy, Portugal, and Cyprus, Africa, Australia, and South Asia primarily in India and Pakistan have a wealthy semi-arid environment. Those areas are covered with endemic species that have outstanding medicinal potential, which has recently been studied due to their sustainable applications in the food and pharmaceutical areas, bringing important economic and social benefits. Therefore, the bio-compound extraction of semi-desert plants is an important concern nowadays, and emerging or green technologies have been intended to cover sustainable and high-quality performance.

7.1 INTRODUCTION

The semi-desert flora of the area is characterized by having great potential in the agro-industrial and pharmaceutical fields, as well as various applications

Sustainable Agriculture and Global Environmental Health: Toward the Efficient Management of Resources. Mónica L. Chávez González, Cristóbal Noé Aguilar, Juan Carlos Contreras-Esquivel, and A. K. Haghi (Eds.)

in the food area, due to the fact that the plants have phytochemicals with high biological potential and high added value. Among the secondary metabolites are tannins, terpenoids, alkaloids, and flavonoids which have notable anti-microbial, antioxidant, and antitumor activities and can be valuable in medical treatments (Al-Saleem, Awaad, Alothman, & Alqasoumi, 2018).

The extraction of these bioactive compounds is fundamental in the investigation of plants with medicinal potential, as their efficiency impacts the final result (Azmir et al., 2013). In terms of extraction of bioactive compounds, there is a wide range of techniques, among the most common, also known as conventional, are Soxhlet extraction, maceration, and hydro distillation regularly used for the extraction of phenolic compounds. However, the trend of recent years has promoted the development of new sustainable, efficient, and fast extractive technologies, based on a green extraction approach such as ultrasonic extraction, pulsed electric field, enzymatic digestion, extrusion, microwave heating, ohmic heating, supercritical fluids, and accelerated solvents.

7.2 SEMI-DESERT IMPORTANCE IN SUSTAINABILITY

The semi-desert is a dry, hot, and arid environment. These climates tend to have extremely hot summers and warm to cool winters, with minimal rainy periods. These zones cover about one-fifth of Earth's surface and are mostly located around the fringes of subtropical deserts. The soil in these regions can be sandy, gravelly, or stony, depending on the type of desert. As mentioned before, the precipitation rates are very low, and normally, they receive at most 50 centimeters (20 inches) of rainfall a year. For this reason, the organisms that live in semi-deserts are adapted to this extremely dry climate. However, semi-arid deserts are considerably cooler than hot deserts. The dry summers in semi-arid deserts are preceded by winters with some rain (Beck, Zimmermann, McVicar, Vergopolan, Berg, & Wood, 2018).

Unfortunately, future climate changes foreshadow an abrupt change in the ecosystem's vegetation. In this regard, semi-desert environments with weak vegetation have been considered the principal target of global climate change. At the same time, the notable rise in temperature affects plant species immensely, which could be exacerbated if precipitation decreases by some degrees. Thus, tracking down the changes in the vegetation of semi-desert zones has been the most important matter of recent research. The vegetation plays a major role in the food chain because plants not only provide balance to terrestrial ecosystems via carbon sequestration and photosynthesis, reduce the undesirable effects of worldwide climate change, and offer scenic attractions with major tourism potential, but also serve as medical sources for humans and have outstanding drug potential (Bagheri, Fordoei, Mousavi, & Tahmasebi, 2021).

This region is home to diverse plants and animals that have adaptive traits to survive in harsh conditions like dryness, extremely high temperatures, and practically no precipitation. Small plants, like grasses, shrubs, and small trees, prevail in the semi-desert zones. Most of the plants in these areas have developed some specific characteristics to survive extreme conditions. For instance, their thorny branches or waxy cuticles help to reduce evaporation and water loss through their leaves. More specifically, the cacti have enlarged stems to store water, as well as spines to protect these water reserves from animals. The animals that inhabit the semi-desert also have advanced organisms to live easily in this ecosystem. Some species present in these regions are bison, antelope, gazelle, zebra, wolves, lions, jackals, or coyotes, most of the time. For example, the black-tailed jackrabbit (Lepus californicus) is highly adapted to its environment with its extra-long ears helping to transfer excess heat from its body into the air (Eldridge, Koen, Killgore, Huang, & Whitford, 2012).

Around the world, the more prominent semi-arid deserts are found in North America, usually in the Southwestern United States, in the Northern states of Mexico and the ABC Islands, South America in zones such as the Gran Chaco, Greenland, Europe mostly in southeast Spain and parts of Greece, Italy, Portugal, and Cyprus, Africa, Australia, and South Asia primarily in India and Pakistan (Bagheri, Fordoei, Mousavi, & Tahmasebi, 2021).

7.2.1 MEXICAN SEMI-DESERT

The Mexican semi-desert has a wide variety of herbaceous flora and woody species. The vegetation pattern is spatially diverse across the landscape. It is influenced by highly variable patterns of precipitation, as well as subtle changes in edaphic factors, which also impact the regimes of natural and anthropogenic disturbances (Villavicencio-Gutiérrez et al., 2021).

In Mexico, the arid and semi-arid regions are home to around 6,000 species of vascular plants, of which approximately 60% are endemic. The Chihuahuan Desert is considered the main biological reservoir in the country, due to its extension, as well as its remarkable diversity and endemism, mainly favored by environmental gradients that form mosaics (Hernández-Magaña, Guadalupe Hernández-Oria, & Chávez, 2012).

The semi-arid region of the country is located primarily in the north, comprising 70.79 million hectares, and is characterized by its high vulnerability to desertification where its inhabitants (around 5 million people) suffer extreme marginalization and increased migration, for which they see themselves in need to use non-timber forest resources (RFNM) of various species in semi-desert areas to obtain raw material that impacts on their food safety, health, well-being, and economic income. Coahuila is the state that reports the highest RFNM production numbers, with 280,680 tons annually (Villavicencio-Gutiérrez et al., 2021).

7.2.2 SUSTAINABLE USE OF SEMI-DESERT PLANTS

Semi-desert species contain secondary metabolites predominantly: tannins, terpenoids, alkaloids, and flavonoids which have remarkable antimicrobial, antioxidant, and antitumor properties turning them into highly valuable compounds in treatments. Present research has found about 7,000 species of medicinal plants, and their inner value involves diverse chemical substances that produce a specific physiological action on the human body (Al-Saleem, Awaad, Alothman, & Alqasoumi, 2018).

Endemic species contain exceptional amounts of bioactive compounds and phytochemicals that provide innumerable benefits to our health. These compounds have a high biological potential and therefore an important added value, even when they are present in semi-arid zones where the climate tends to be dry, and hot, and rainfall is scarce (Hernández-Magaña, Guadalupe Hernández-Oria, & Chávez, 2012). For this reason, the members of the communities that are located in the semi-desert have used them for the treatment of various diseases as an alternative, due to the lack of formal medical attention. The communities are aware of the benefits that plants provide them and that they have known over time, from generation to generation, it is part of their ancestral knowledge, however, they are unaware that these semi-desert species contain bioactive compounds. In this way, science and technology, through botanical, horticultural, and chemical research, seek to generate biotechnological transformations of these semi-desert species into products with added value to significantly boost the economic growth of these regions.

The use of the numerous plants of the semi-desert is utilized by the locals to alleviate some types of diseases or ailments, thus compensating for the lack of specialized medical attention. In recent years, efforts have been made to significantly reduce these existing problems by increasing scientific knowledge about semi-desert plants, thus favoring their development and use to improve the quality of life of rural communities. The scientific validation through laboratory tests of the medicinal properties reported by the communities is a priority for their subsequent applications in the food industry or pharmacology. Using science, technology, and culture as tools, numerous works have been reported on the characterization, extraction, and optimization of bioactive compounds from semi-desert plants, usually to evaluate their biological activities, or for the generation of a food additive. Under this scenario, the economic growth of many communities has benefited, without neglecting the sustainability of resources and the constant disclosure of information on semi-desert species with potential applications in the industry (Torres-León et al., 2023) (Figure 7.1).

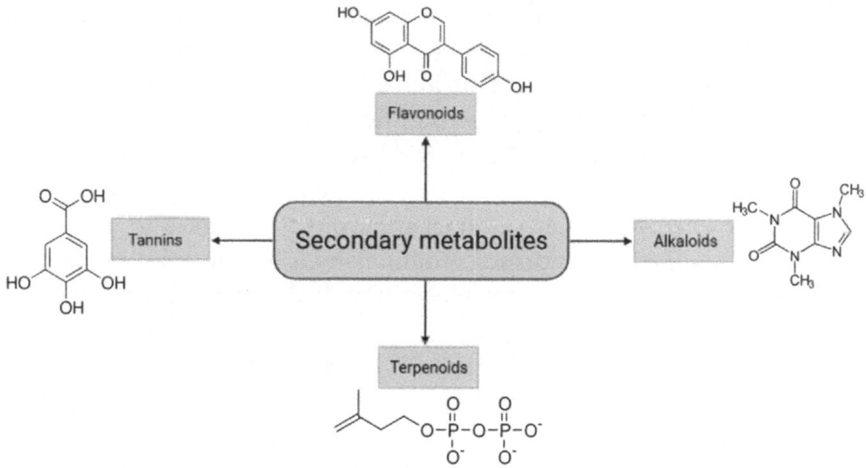

FIGURE 7.1 Secondary metabolite's chemical structure.

7.3 TYPES OF PLANTS REPORTED

Various semi-desert species have been previously reported, but the research on medicinal potential plants has increased in the last few years. Recently, 77 species of medicinal plants belonging to 36 botanical families were cited (Torres-León et al., 2023), mostly because of their viable application in medicine and auspicious use that promotes the sustainable production of medicinal plant-based products. In this way, it could be possible to avoid the irrational generation of synthetic drugs. The positive application of this medicinal species promotes the sustainable use of biodiversity and consequently brings notable improvements in the social and economic areas of nations. Also, floristic diversity and traditional knowledge have facilitated the advancement of several bioactive natural products to clinical trials (Calzada & Bautista, 2020).

Furthermore, special interest has been directed to herbal-based products, and recently into species of semi-arid zones. Their importance relies on the potential to be included as sustainable medication to decrease the pharmacological damage to the environment. The plant-derived drugs provide reduced side effects, and affordability due to their natural availability, thus making them suitable over other conventional medications (Al-Saleem, Awaad, Alothman, & Alqasoumi, 2018). Several plant extracts and naturally isolated compounds were reported to have potential anti-ulcerogenic activity, anti-inflammatory, as well as analgesic activities, and hepatoprotective capacity (El-Meligy, Awaad, Soliman, Kenawy, & Alqasoumi, 2017). For instance, some species of the Calendula family have been applied in the traditional medicine field as an anti-inflammatory, immunomodulator, antimicrobial, antimalarial, antiallergic, antidiabetic, antipyretic, disinfectant, antispasmodic, diuretic in the treatment of kidney and gall stones, emmenagog, diaphoretic, sedative, healing properties, hepatoprotective, and for treating burns (Al-Saleem, Awaad, Alothman, & Alqasoumi, 2018).

Hence, the species identified and classified as traditional medicinal plants could undergo modern ethnopharmacological studies to determine their specific biological properties. Additionally, it is important to mention some of the species that have been recently studied in semi-desert regions to further contribute to this field.

The Parthenium hysterophorus species belongs to the Asteraceae family. Its reproduction primarily occurs through seeds. It can grow up to 2.6 meters long and has lobed leaves that measure between 8 and 20 centimeters long and between 4 and 8 centimeters wide. Its leaves and stems are characterized by having short and fluffy trichomes. The inflorescences tend to be fuzzy, containing relatively small, capitular white flowers (Bashar et al., 2021). In general, each head has between five to eight ray florets. They produce compressed black seeds of approximately 2 mm in size and can produce up to 10,000 to 14,000 seeds, which are dispersed through various mechanisms.

Parthenium hysterophorus is present throughout the year. It supports extreme climates, particularly hot areas where droughts and high temperatures predominate. It is also capable of growing in sandy soils and can withstand saline soils. The plant has deep roots that allow it to survive in soils with little moisture, but it also thrives in moist soils (Iqbal et al., 2022). It germinates in temperatures ranging from 10°C to 25°C and requires approximately 500 mm of rainfall in summer (Singh, n.d.).

The plant was first located in northern Mexico (Iqbal et al., 2022). However, it is estimated that this species currently extends to around 46 countries, mainly in regions of the southern United States, Central America, northern Argentina, southern Bolivia, southwestern Brazil, and Western India, among others. Its rapid propagation causes significant economic losses because its high toxicity affects numerous crops.

Recent studies show a different perspective regarding its potential as an anticancer and antitumor agent, mainly due to its antioxidant properties (Bashar et al., 2021). Likewise, Parthenium hysterophorus contains high levels of nitrogen, magnesium, calcium, and potassium, among other nutrients. Extracts of the plant have been used as pesticides to control diseases in different species, improve seed germination, and promote the growth of seedlings, biomass, and crop yield.

In recent years, studies such as the one by Atkins & Shabbir (2014) have determined that Parthenium hysterophorus is extremely invasive and resistant to a wide range of conditions due to a series of allelochemicals that allow it to suppress or deprive the growth of species that develop close to it. Allelopathy describes the beneficial or harmful effects of a plant (Bashar et al., 2021). Along this line, even though normally the allelopathic characteristic of this species was considered inherently negative and harmful, a new approach arises that aims to develop its potential benefits, especially within the area of sustainable herbicides.

Its compounds: Parthenin, hysterin, hymenin, and ambrosin have allelopathic activity, favoring the production of phytochemicals, which are synthesized for the development of potential and innovative herbicides of natural origin. According to Motmainna et al. (2021), the presence of 82 known compounds in the Parthenium hysterophorus extract was confirmed, most of them previously reported in a large number of studies, allowing entry into the field of bioherbicides for the efficient control of weeds as a sustainable alternative, which presents significant increases in crop yields. On the other hand, the metabolomic identification, the isolation of potential allelopatins, and the formulation techniques of various surfactants or nanoformulations are relevant in the penetration and absorption of active compounds.

Within the positive impact that this species can achieve, Parthenium hysterophorus presents favorable characteristics to venture into the area of medicine. In the past, it has been an ally in treating diseases such as neurological disorders, inflammation, fever, malaria, gynecological ailments, herpes, diarrhea, emmenagog, and urinary tract infections, to name just a few. According to what was established by Singh & Kaushik (n.d.), it serves as a pharmacologically active substance capable of treating muscular rheumatism and hepatic amoebiasis. In the same way, it has anti-cancer properties that present exceptional performance in terms of tumor reduction. The flower, when subjected to a leaching process in methanol, presented antitumor activities. Parthenine, on the other hand, showed a cytotoxic character against T cells and the cancer cell line (Singh, n.d.).

Along the same lines, in the in vivo study of glucose levels in the blood of rats with type II diabetes, the aqueous extract of the species showed a strong hypoglycemic action (Bashar et al., 2021). It is important to mention the impact that this species represents for the inhabitants of regions where medical service is not immediately available, highlighting its use to treat itching, skin conditions, rheumatic pain, eczema, heart complications, and reproductive problems.

Ephedra antisyphilitica is a plant similar to a shrub, with evergreen leaves that generally grows in rocky areas or cracked terrain between 1,000 and 1,800 meters above sea level. Its fruits are reddish with an approximate length of 6–7 mm and 4–6 mm long. It flowers in the months of May and June and bears fruit in July and August (Bashenova & Aidarbayeva, 2022).

When addressing the species Ephedra antisyphilitica, we find that it has a long history in the field of medicine since its knowledge within this goes back to the origins of humanity; from the genus Ephedraceae, there are approximately 69 species, four subspecies, and two varieties distributed around the world, being found in greater quantity in arid and semi-arid areas, mainly in southwestern North America, South America, Europe, Asia, and North Africa.

González-Juárez et al. (2020) refer that the orogenetics and climatic changes typical of the North American regions favor the diversity of ephedra species, reporting the presence of 13 species, the closest being E. antisyphilitica Berland. ex C.A.Mey., E. aspera Engelm ex S.Watson, E. californica S.Watson, E. compacta Rose, E. nevadensis S.Watson, E. pedunculata Engelm ex S.Watson, E. torreyana S.Watson and E. trifurca Torr. ex S. Watson, located in northern Mexico.

As previously established, its applications in treating diseases date back to the year 5000 B.C. within traditional Chinese medicine. However, a special focus is made on the ancient civilizations of American Indians who used it for treatment against syphilis and gonorrhea. In the same way, they are used for relatively common diseases such as cough, flu, allergies, bronchial asthma, colds, chills, edema, fever, headache, and nasal congestion. Its application is so broad that currently, some species of ephedra have been used for weight loss (González-Juárez et al., 2020). In the Mexican semi-desert, the natives consume this species as an infusion to relieve stomach aches, kidney problems, and gonorrhea, or they grind the seeds to produce a drink similar to coffee. Similarly, they also make ointments for wound healing with the help of crushed stems and leaves.

Ephedra is characterized by having small leaves and photosynthetic shoots; they usually grow in the form of shrubs. However, they also grow as climbing plants that reach up to four meters in length. Pollination occurs through the wind, or with the collaboration of insects, there are also specimens that present both types of pollination and are known as ambophiles (González-Juárez et al., 2020).

This species is dioecious, with unisexual cones. The ovulated strobila contains three ovules surrounded by one or two bracts, also known as envelopes, and the integument forms a micropylar projection to produce a pollination drop (Zumajo-Cardona & Ambrose, 2022).

Chemically, this species has 26 alkaloids, and 75 phenolic compounds, which include aromatic compounds, flavonoids, lignans, proanthocyanidins, and seven amino acid derivatives. It is possible to produce essential oils from Ephedra, which can be obtained by extraction with supercritical CO2, producing 70 volatile compounds, or by hydrodistillation, resulting in 98 volatile compounds (González-Juárez et al., 2020). Ephedra antisyphilitica has promising bioactive compounds with potential applications in the pharmaceutical industry, cosmetics, food products, or within the agricultural sector. In the same way, there is also a wide variety of flavonoids and phenolic compounds present in this species; we find benzoic acid, p-hydroxybenzoic acid, vanillic acid, protocatechuic acid, trans-cinnamic acid, and p-coumaric acid.

Heterotheca inuloides is an endemic species that grows wild, sometimes as a weed, located primarily in the center and north of Mexico, mostly in temperate forests, tropical forests, grasslands, and xerophytic scrub (Rodríguez-Chávez et al., 2017). The species settle in mountainous terrain, at altitudes between 2000 and 3000 meters above sea level, mainly in three regions: the Sierra Madre Occidental in the west, the Sierra Madre Oriental in the east, and the Volcanic Belt Trans-Mexican in the south of the country. It is a diploid, herbaceous, and perennial plant. It can measure up to 1 meter in height, has alternate, lanceolate leaves, and is pubescent with serrated margins. Its smell is almost imperceptible, and it has a bitter taste. The inflorescence consists of a head surrounded by ray florets (Rodríguez-Chávez et al., 2017).

Also known as "Mexican arnica," it is inherent to traditional Mexican medicine and has been widely used as an infusion in the treatment of bruises and muscle pain (Carmona-Aparicio et al., 2019). Its application in medicine is mainly due to its abundant components, with approximately 140 compounds calculated, including sesquiterpenoids, triterpenoids, and sterols found in greater quantity (Guadarrama-Reyes, n.d.).

Previous studies have shown the antioxidant properties of dried flowers. It was found that they can inhibit lipid peroxidation and confirmed their role as a cytoprotective agent. Currently, various studies have focused on the isolation of flavonoids, terpenes, triterpenoids, and phytosterols. The study carried out by Carmona-Aparicio et al. (2019) demonstrated a new medical approach as an anticonvulsant treatment proposal, tested in vivo models subjecting male Wistar rats to analysis. Heterotheca inuloides is recognized by the World Health Organization (WHO) for its antimicrobial activity, antifungal, cytotoxic, and antioxidant properties. It has also been used to treat dental disease and gastrointestinal disorders (Guadarrama-Reyes, n.d.).

Likewise, Rodríguez-Chávez et al. (2017) report that its use extends to treating dermatological diseases such as erysipelas, topical skin inflammation, pimples, rashes, etc., bites of dogs, burns, and infections caused by insects or spider bites. In addition, it serves as an ally in treating cancer and diabetes.

In the economic part, this species has had a high demand throughout the years, being an important source of income in many families of rural communities in the country. According to Rodríguez-Chávez et al. (2017), the Heterotheca inuloides contains abundant chemical compounds with different structural patterns. It is possible to obtain essential oils and organic extracts from flowers or roots. Through techniques such as gas chromatography, column chromatography, thin-layer chromatography, mass spectrometry with gas chromatography, and HPLC, it is possible to isolate and purify these compounds; in the same sense, ultrasound has recently been used as an alternative extraction method, obtaining favorable results in terms of processing time, rate of extraction, and solvent consumption. Among the most abundant compounds of this species are the phenolics identified from spectrophotometry and evaluated with the Folin-Ciocalteu method. Its flowers contain the compounds phenolics: kaempferol 3β-glucopyranoside,

3β-galactopyranoside, 3β-rutinoside, 3β-arabinobioside, quercetin 3β-glucopyranoside, 3β-galactopyranoside, 3β-glucuronide, 3β-glucuronide-6"-methyl ester, 3-α-L-arabinoside, 3β-rutinoside, 3-O-glucuronide-7-O-galactopyranoside, 3-O-glucuronide-7-O-glucopyranoside, 3-O-rhamnosylgalactopyranoside, and 3-O-caffeoylglucopyranoside.

In the same sense, Heterotheca inuloides presents a promising performance regarding its antimicrobial capacity. The study carried out by Rodríguez Chávez et al. (2017), where the methanolic extract showed a greater inhibitory capacity than metronidazole for the bacterium Helicobacter pylori. In the field of development of sustainable and environmentally friendly insecticides, this species has demonstrated efficient results in the elimination of pests such as the spider mite, honey bees, larvae, or the corn weevil (Sitophilus zeamais). The species Parthenium incanum, also commonly known as Mariola, is found within the genus Parthenium; approximately 16 species are estimated to be different from shrubs, herbaceous perennials, and annual plants. As for its location, they are found primarily in the southwestern United States and northern Mexico. It has grayish-green leaves and reaches up to 2–3 feet tall. It has also been widely used in traditional medicine, forming part of treatments for indigestion, constipation, and fatty liver (Carmona-Aparicio et al., 2019) (Table 7.1).

TABLE 7.1 Extraction Techniques and Bioactive Compounds Reported in Semi-Desert Plants

Plant	Extraction Technique	Bioactive Compounds	References
Parthenium hysterophorus	Soxhlet extraction	Alkaloids, terpenes, flavonoids, lignans, plant steroids, curcumins, saponins, phenolics, and glucosides.	Bashar et al. (2021)
Ephedra antisyphilitica	Supercritical CO_2 fluid extraction.	Alkaloids (ephedrine-type), Phenolic compounds (aromatic compounds, flavonoids, lignans, and proanthocyanidins), and amino acid derivatives.	González-Juárez et al. (2020)
Heterotheca inuloides	Ultrasound-assisted extraction	Sesquiterpene compounds, polyphenol compounds, triterpenes, and phytosterols.	Rodriguez-Chávez et al. (2017)
Parthenium incanum	Soxhlet and maceration extraction	Alkaloids, terpenes, flavonoids, plant steroids, curcumins, tannins, and sesquiterpene lactones.	Hernández-Marín et al. (2018)

7.3.1 REPORTED BIOACTIVE COMPOUNDS OF SEMI-DESERT PLANTS

Flavonoids are compounds found in countless plants, including semi-desert species. They are part of an important group called phenolic compounds. Flavonoids have a positive impact on human health, especially due to their antioxidant and free radical scavenging properties. Recent studies have shown that flavonoids play a significant role in reducing cardiovascular diseases, neurodegeneration, and cancer. These compounds help protect the body from damage caused by oxidizing agents. Since the body cannot produce these chemicals on its own, they must be obtained through diet or supplements.

Semi-desert plants generally contain polyphenols, which exhibit a series of reduction and oxidation reactions due to the resonance effects of their phenyl rings. These compounds have strong antioxidative activity because of their chemical composition. Therefore, research has focused on analyzing extracts after specific extraction treatments. The most commonly used methods include ABTS, Trolox equivalence antioxidant capacity (TEAC), ferric reducing antioxidant power (FRAP), 2,2-diphenyl-1-picrylhydrazyl (DPPH), and oxygen radical absorbance capacity (ORAC). These assays are based on the ability of extracts to inhibit free radical activity (Dzah, Duan, Zhang, Wen, Zhang, Chen, & Ma, 2020).

It is important to establish that every plant matrix is different from another. Therefore, the polyphenol composition varies immensely according to their environment. The polyphenols give the plant important protection mechanisms mostly against predators, but also work as an auto-oxidative compound, inducing oxidative stress, drought stress, infections, and extreme weather conditions. They attract pollinators, allow the conservation of enzymes, and permit the preservation of cell membranes. In the same way, the extraction techniques, solvents used, temperature, solvent-to-material ratio, time, frequency, intensity, and power are parameters that could negatively affect the composition of extracts in any extraction process (Dzah, Duan, Zhang, Wen, Zhang, Chen, & Ma, 2020). In addition, the medicinal use of semi-desert plants has been an important part of numerous traditional communities, primarily for disease treatment.

7.4 SUSTAINABLE TECHNIQUES FOR THE EXTRACTION OF BIOACTIVE COMPOUNDS

Nowadays, the extraction of these outstanding compounds has become an important issue because of consumers' trend to practice a healthier lifestyle and have a fulfilled diet, including antioxidants. For this reason, the implementation of diverse extracting techniques has been performed recently, so reaching desirable bioactive compounds and optimal yields has become the principal focus (Carpentieri, Mazza, Nutrizio, Jambrak, Ferrari, & Pataro, 2021).

The extraction process must always be the first step in the investigation of plants with medicinal potential. It is directly related to the efficiency of its final result, taking into account that the properties of the matrix, temperature, pressure, time, and type of solvent have a significant impact on the process (Azmir et al., 2013).

Regarding the extraction of bioactive compounds, there is a wide range of techniques. Among the most common are percolation, maceration, and Soxhlet extraction, usually used for the extraction of phenolic compounds (Carpentieri, Mazza, Nutrizio, Jambrak, Ferrari, & Pataro, 2021). The latter is the reference method to compare the efficiency of the new extraction techniques (Azmir et al., 2013). The trend in recent years to reduce the solvents involved in these processes, excessive energy expenditure, high temperatures, prolonged extraction, and the formation of undesirable compounds has promoted the development of new sustainable, efficient, and fast extractive technologies based on a green extraction approach. Under this action, the introduction of physical cell disruption pretreatments, such as pulsed electric fields (PEF) and ultrasound (US), which favor the weakening of the cell membrane and wall, has shown good performance, reduces solvent consumption, energy, time, and simultaneously improves the extraction kinetics of the compound (Carpentieri, Mazza, Nutrizio, Jambrak, Ferrari, & Pataro, 2021).

During the last 50 years, multiple research has focused on the reduction of synthetic and organic solvents with a sustainable approach, without neglecting yield, extract quality, and selectivity, as well as the optimization of the operation time (Azmir et al., 2013).

7.4.1 EMERGING TECHNOLOGIES

The extraction process is always the first step in the research of plants with medicinal potential. Its efficiency impacts the final result. Regarding the extraction of bioactive compounds, there is a wide range of techniques. Among the most common, also known as conventional, are Soxhlet extraction, maceration, soaking, water percolation, and hydro distillation usually used for the extraction of phenolic compounds. However, these techniques involve long extraction time and degradation of the bioactive compounds. Hence, the trend of recent years has promoted the development of new sustainable, efficient, and fast extractive technologies based on a green extraction approach such as microwave-assisted extraction, ultrasonic-assisted extraction, supercritical fluid extraction, enzyme and microorganism-assisted extraction, high-pressure processing, pulsed electric fields processing, cold plasma, and ohmic heating (Figure 7.2).

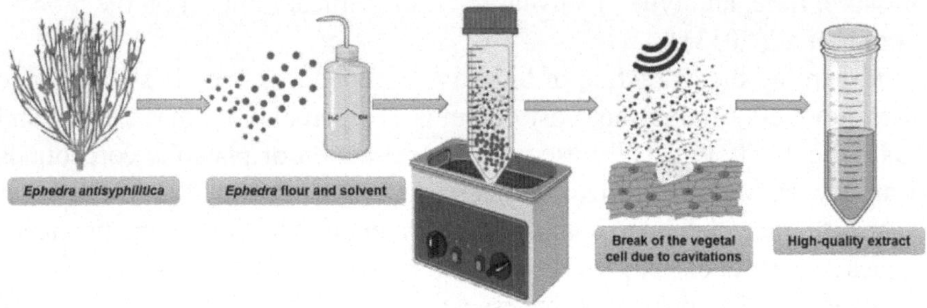

FIGURE 7.2 Comparison between conventional and emerging technologies.

7.4.1.1 MICROWAVE-ASSISTED EXTRACTION

Microwave-assisted extraction is an effective extraction method where microwaves promote cellular matrix breakup to obtain valuable bioactive compounds. The principle consists of heating the molecules of a certain sample, both internally and externally, without a thermal gradient, for better energy absorption (Osorio-Tobón, 2020).

MAE is an emerging technology based on radiation principles, usually working in a frequency range of 300 MHz to 300 GHz classified as microwaves. The energy of the heat generated by the microwave facilitates the extraction process mainly because of the dipole polarization effect and the ion conduction of molecules. Therefore, in this technology, ionic conduction and dipole movement warm the sample in the presence of microwaves (Bagade & Patil, 2021).

This technique breaks the sample wall to interact with the polar components, generating heat. The heat present in the microwave acts on the targeted molecules through ionic conduction and dipole rotation. For the technique to be efficient, the dissipation factor must be considered because it absorbs microwave energy and dissipates heat to the surroundings (Chan, Yusoff, Ngoh, & Kung, 2011).

The correct extraction of bioactive compounds requires a particular amount of energy, which can be supplied randomly. This occurs when the microwaves penetrate the sample irregularly. This is also called a multi-mode of energy. On the other hand, the focused mode of the applied energy is related to a single mode, where microwaves hit a specific or restricted zone. The first type of energy is a closed and high-pressurized system, while the second one is an open system that operates at atmospheric pressure. Nevertheless, both systems ought to include a magnetron, a waveguide, an applicator, and a circulator (Osorio-Tobón, 2020).

There are two types of systems in Microwave-Assisted Extraction technology: the multi-mode and the focused-mode. In the first one, the dispersion of microwaves occurs randomly by a stirrer, while in the focused mode, the microwave radiation occurs in a specific, focused, and restricted zone. The two kinds of systems are also classified as 'closed system' and 'open system' to categorize them according to their pressure operation (Chan, Yusoff, Ngoh, & Kung, 2011).

The microwave that operates as a closed system carries out extractions in a sealed vessel, under uniform heating. In this process, two factors are involved to reach an efficient extraction: high pressure and an adequate temperature. In the first case, the pressure must be controlled carefully because it must not exceed the working pressure of the vessel, while the temperature must be above the boiling point of the solvent involved in the process (Chan, Yusoff, Ngoh, & Kung, 2011).

On the contrary, in an open system, the operation conditions are milder, with a higher sample throughput and more solvent. It also operates at atmospheric pressure in a vessel that is only exposed to mono-mode microwave radiation. This kind of system is usually used for extracting thermolabile bioactive compounds (Chan, Yusoff, Ngoh, & Kung, 2011).

For correct operation of the microwave, there are some factors that must be taken into accounts such as solvent nature, solvent-to-feed ratio, extraction time, microwave power, temperature, sample characteristics, and the effect of stirring (Chan, Yusoff, Ngoh, & Kung, 2011).

The main goal of an extraction process is to separate a matrix into two parts: the dissolvable material from the insoluble material or residue. To carry out this separation successfully, an adequate solvent should be involved. It must be taken into account that many secondary metabolites are present in a single plant; therefore, finding the correct conditions for the extraction process is the major aim in this case (Bagade & Patil, 2021).

At first, the microwave-assisted extraction (MAE) technique was used as a digestion method for different samples: biological, environmental, and geological matrices. However, in recent years, this technique has been widely used for the extraction of bioactive compounds from various plants, leading to the research and development of a new field. Besides, this green technology reaches optimal extraction yields compared to conventional technologies, additionally has been sustainable because of their low solvent consumption, minimal waste generation, and a significant decrease in the use of hazardous solvents that harm the environment (Bagade & Patil, 2021).

Normally, polar solvents such as ethanol and methanol mixed with water are used in this technology since their affinity to react with the microwaves produced in the process allows the heating of the particles, also called ionic conduction, where their movement is affected due to an alteration of the electric field. Also, their polarity allows satisfactory extraction yields and effective flavonoid extraction (Chávez-González et al., 2020). On the other hand, less polar solvents like hexane and chloroform remain unaffected by heat and thereby don't react with microwaves. The correct selection of the solvent is an important element that should be taken into account because of the direct effect on the extraction yield and elution of plant components from the matrix. Besides, the solvent involved in the process is a critical element that should be taken into account mostly due to its dielectric properties (Bagade & Patil, 2021).

One of the main factors for a successful extraction by MAE is the moisture present in the matrix or plant cell. In this way, the goal of microwaves is to heat uniformly and deeply with high pressure to promote evaporation and remove the moisture inside the plant. During this process, the cell wall of the plant breaks due to the high pressure, supporting the correct extraction of phytoconstituents. Other factors that should not be overlooked to achieve a high and optimal extraction yield are high temperatures, dielectric susceptibility of the solvent, extraction time, and microwave irradiation energy, according to the specific characteristics of the plant matrix, mostly due to its thermal sensitivity (Bagade & Patil, 2021).

Additionally, the operational cost of the entire process must be taken into consideration, focusing on reducing extraction times and the use of solvents as the main aspects in terms of economics (Chávez-González et al., 2020).

However, flavonoids can lead to a wide variety of negative effects, making the optimization stage the most important. The parameters or factors mentioned before can be optimized with an experimental design, mainly to reduce the number of trials needed to assess the influence of these factors on the MAE process and to determine the optimal extraction conditions. Usually, a response surface methodology (RSM) is used to optimize this type of process (Llompart, Celeiro, & Dagnac, 2019).

7.4.1.2 ULTRASOUND-ASSISTED EXTRACTION

Ultrasound-assisted extraction is a method typically used to obtain an optimal quantity of bioactive compounds, mostly from species with medicinal and pharmaceutical applications.

Most species with medicinal potential contain secondary metabolites, also known as phenolic compounds, that are present in plants in the form of phenolic acids and flavonoids. In recent years, these compounds have gained a positive and outstanding reputation mainly because of their health benefits such as antioxidative, anti-inflammatory, anticancer, and antimicrobial properties. Due to these benefits to human health, research on them has become more relevant than ever, and studies on various extraction technologies for the biological activities involved have been increasing recently (Dzah, Duan, Zhang, Wen, Zhang, Chen, & Ma, 2020).

The principle of Ultrasound-Assisted Extraction consists of acoustic cavitations, which produce small bubbles, also considering cavitation waves that break the plant wall to extract the bioactive compounds by allowing the solvent into the matrix and promoting the release of the compounds; in the same way, this stage also enables a significant particle size reduction involving the solvent-substrate interaction. It is essential to take into account specific variables that have a major impact on flavonoid extraction. Under this statement, the optimization of these variables is a key stage in the performance of this extraction technique, where an experimental matrix is the most common path to the maximum recovery of bioactive compounds (Chávez-González et al., 2020).

There are two categories of UAE technology according to the type of interaction between the extraction medium and the ultrasound source. First, direct ultrasonication allows better contact and interaction between sound waves and the extraction medium. Besides, this type of ultrasound reaches intense cavitations improving the extraction of bioactive compounds. On the other hand, indirect ultrasonication occurs when the sound waves pass through two media, usually between the water bath and a sample container, before arriving at the extraction medium. In this stage, a significant reduction of the cavitations and extraction rate takes place, leaving direct ultrasonication better positioned (Dzah, Duan, Zhang, Wen, Zhang, Chen, & Ma, 2020).

The frequency rate is another matter that is constantly analyzed in this process; there are different working rates according to its purpose. For example, ranges between 20 kHz and 100 kHz are mostly applied in cleaning, plastic welding, and sonochemistry. Ranges of 100 kHz–2 MHz are also used in sonochemistry, while ranges between 2 MHz and 10 MHz are usually applied in the medical field. Finally, low ranges between 20 and 40 kHz are used for the extraction of bioactive compounds mainly from diverse plant species with medicinal potential for the recovery of polyphenols. These low-frequency rates allow the formation of large cavitation bubbles present in the solvents, which implode violently, producing high shear while creating microjets that degrade the cells significantly, increasing solvent penetration and reaching a higher extraction rate (Dzah, Duan, Zhang, Wen, Zhang, Chen, & Ma, 2020).

Besides, the temperature of the extraction process is directly involved in the polyphenol yield. In a recent study conducted by Maran et al. (2017), the extracts of Nephelium lappaceum demonstrated that at higher temperatures, the yield is significantly increased due to growth in material porosity, higher solvation, and suitable mass transfer. Similarly, an increased temperature reduces surface tension and viscosity in extracts, which also promotes extraction yield, aids in the penetration of the matrix, and thus has a positive effect on the extraction of polyphenols. However, some of these compounds are susceptible to oxidative degradation and hydrolysis at high temperatures, usually over 60°C, which leads to a decrease in phenolic compound extraction. For this reason, the correct selection of extraction temperatures is one of the main subjects that must be taken into account in UAE technology to guarantee the stability of phenolic compounds present in the matrix (Dzah, Duan, Zhang, Wen, Zhang, Chen, & Ma, 2020).

Furthermore, the selection of a suitable solvent is important in this process. There are factors that interfere with the quality of the extraction: polarity, melting point, boiling point, density, specific gravity, affinity for the target component, as well as the effect on the purity and activity of the extracted compound, which set the limits to ensure the stability of polyphenols and the efficiency of the process. Frequently mixtures between organic and aqueous phases are made for the extraction of bioactive plant compounds. Ethanol, methanol, acetone, and isopropanol are mixed with different proportions of water, with the most common being water. However, it's also possible to extract with 100% of either organic solvent or water. The specific characteristics of solvents about treated materials, their affinities, proportions, and extraction conditions are relevant parameters that are helpful in economic terms (Dzah, Duan, Zhang, Wen, Zhang, Chen, & Ma, 2020).

Optimizing the parameters involved in Ultrasound-Assisted Extraction leads to reaching acceptable extraction yields as well as reliable preservation of the biological properties present in the extracts. Some factors implicated in the process, such as time, frequency, power, temperature, and solvent-to-material ratio, must also be taken into account for the efficiency and success of the polyphenol extraction (Dzah, Duan, Zhang, Wen, Zhang, Chen, & Ma, 2020).

To increase the yield of bioactive compounds, usually a surface methodology is applied. At the same time, the extraction technique and the particular characteristics of the plant directly impact the recovery of bioactive compounds and flavonoids. In addition, for an acceptable quality of the extract, the reduction of extraction time, frequency, solvent selection, solvent concentration, solid-solvent ratio, and temperature contribute to achieving better biological properties of the extract involved (Chávez-González et al., 2020) (Figure 7.3).

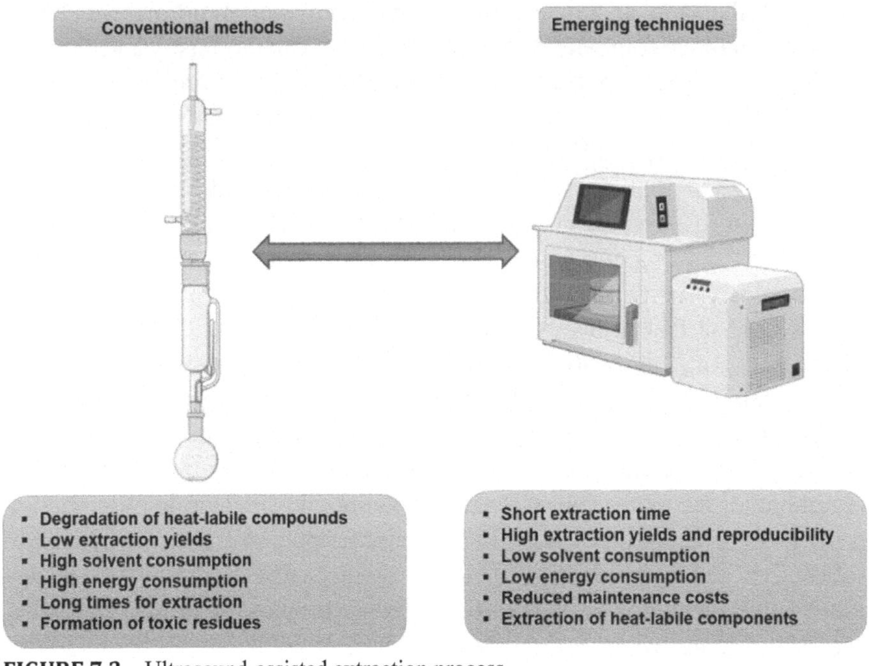

FIGURE 7.3 Ultrasound-assisted extraction process.

7.4.1.3 SUPERCRITICAL FLUID EXTRACTION

In more recent years, the application of supercritical fluid extraction has been increasing due to the latest trends in sustainable and ecological technologies.

This is because of its selectivity, ease of use, and separation capacity to recover diverse organic compounds, many of which are difficult or nonviable to extract through conventional methods. In addition, this process avoids organic solvents that have a negative impact on the environment. Some of the most relevant areas that apply SFE are food, drug, toxicology, chemistry, environment, textile, petrochemicals, and polymers (Ahmad, Masoodi, Rather, Wani, & Gull, 2019). The versatility of supercritical fluid extraction is an outstanding characteristic that enables the extraction of natural therapeutic compounds (Da Silva, Rocha-Santos, & Duarte, 2016).

The main purpose of this process is to achieve high selectivity, short extraction times, increased pollution prevention, and the use of non-toxic organic solvents. Supercritical fluid extraction is based on the principle of fluids that includes concepts of density, diffusivity, dielectric constant, and viscosity, as well as temperature and pressure. All these concepts allow the fluid to fluctuate between gas and liquid because it is similar to liquid density and much the same as gas viscosity. Moreover, a superfluid presents better transport qualities compared to a normal liquid due to its density and capacity to adjust easily to abrupt changes in temperature and pressure, usually above the critical point of a compound, mixture, or element (Ahmad, Masoodi, Rather, Wani, & Gull, 2019).

Before getting into the principle of Supercritical Fluid Extraction (SFE), it is relevant to elaborate on the term "supercritical," which refers to a fluid in a non-condensing and single-phase state when brought above its critical temperature (Tc) and critical pressure (Pc). The principal characteristic of this fluid is that it shares both gas and liquid properties: high density, intermediate diffusivity, low viscosity, and specific surface tension. In terms of how this process is carried out, SFE works with a mixture of carbon dioxide and pressure to destroy microorganisms without compromising the nutritional and organoleptic characteristics. In this way, the technique is favorable for the extraction of bioactive compounds, reduction of extraction time, less consumption of organic solvents, adaptation for thermo-sensitive substances, and the production of cleaner extracts. The environmental benignity of this technology situates it as one of the most promising (Ahmad, Masoodi, Rather, Wani, & Gull, 2019).

To extract and inactivate the vegetative cell of a plant matrix, the supercritical fluid, due to its physicochemical properties such as low viscosity and intermediate values of diffusivity combined with the absence of surface tension, allows penetration into the matrix, breaking the cell wall, and promoting the extraction of its bioactive compounds (Ahmad, Masoodi, Rather, Wani, & Gull, 2019).

It is important to mention that the SFE process takes place in two stages: initially, the bioactive compounds inside the plant matrix are solubilized and separated into the supercritical solvent. In this stage, the solvent flows through the packed bed solubilizing the compounds of the vegetable matrix. Finally, the second stage consists of the exit of the solvent with the solubilized compounds, which afterward become solvent-free due to a decrease in pressure or an increase in temperature (Da Silva, Rocha-Santos, & Duarte, 2016).

SFE possesses multiple advantages over diverse conventional methods, one of them being the ability to modify the density of the fluid by altering its pressure and/or temperature. With this, the solvent strength of the fluid can be modified. In addition, another advantage is the application of solvents GRAS (generally recognized as safe), as well as better efficiency regarding higher extraction yields and lower operation times (Da Silva, Rocha-Santos, & Duarte, 2016).

In terms of the use of carbon dioxide, when the extraction is completed and the system is decompressed, the elimination of CO2 is achieved, yielding a solvent-free extract. However, because of its low polarity, CO2 is less effective in extracting polar compounds, so the use of modifiers or cosolvents comes into play to overcome this problem (Da Silva, Rocha-Santos, & Duarte, 2016). Among the most popular cosolvents are hexane, methanol, ethanol, isopropanol, and dichloromethane, with ethanol being optimal for this process due to its lower toxicity and miscibility in CO2 (Ahmad, Masoodi, Rather, Wani, & Gull, 2019).

Some factors that undoubtedly affect the final composition of the extracts are pressure, time, CO2 flow rate, and the use of a co-solvent (Gallego, Bueno, & Herrero, 2019). For this reason, the optimization of this process becomes necessary, and it is essential to have knowledge of the thermodynamic and kinetic characteristics of the compounds involved, with a special emphasis on solubility, selectivity, and mass transfer coefficients (Da Silva, Rocha-Santos, & Duarte, 2016).

7.4.1.4 ENZYMATIC-ASSISTED EXTRACTION

The basis of Enzyme-assisted extraction (EAE) is the degradation of a plant cell wall, with the aim of releasing phenolic compounds. This process achieves high efficiency in terms of extraction due to the specificity of these biomolecules for their substrate. Additionally, the hydrolysis of high molecular weight compounds to lower molecular weight allows the extracts to increase their bioactivity (Cascaes et al., 2020). The function of enzymes in EAE is to use their hydrolytic properties to break down the plant matrix to access the cytosolic spaces where bioactive compounds are located. Most cell walls are composed of complex polysaccharides such as cellulose, hemicellulose, pectin, lignin, and proteins (Miękus, Iqbal, Marszałek, Puchalski, & Świergiel, 2019).

To start with the main advantages of this process, the EAE operates at low temperatures, shortens the processing time, eliminates waste by using the whole plant material, requires fewer steps, uses a substrate specificity that allows the extraction of a large number of bioactive compounds, and finally reduces production costs significantly in contrast with conventional techniques (Gligor et al., 2019).

There are some factors that influence the efficiency of this process. For instance, the enzymatic mixtures heavily impact hydrolytic properties through composition, types of enzymatic activity of their constituents, and their respective proportions. Similarly, extraction time and concentration affect the operation, particularly because bioactive compounds may undergo disintegration or oxidation processes with long extraction times. Temperature is another factor that must be taken into account due to its effect on enzymatic activity and rheological properties of solubilized compounds, which can halt the hydrolysis process and break compounds. Finally, the influence of pH during extraction also has an important effect on plant cell walls and enzymatic efficacy. For example, an acidic extraction medium leads to the destabilization of hydrogen bonds, resulting in increased cellular wall plasticity, changes in the catalytic activity of proteins, and their capacity to bind to substrates (Alexandre et al., 2019).

EAE has a superior disruption of the vegetal walls from many plants, in comparison with other emerging technologies due to its lower energy waste. Besides, the addition of specific enzymes into the process enhances the extraction of compounds by breaking the cell wall usually of polysaccharides that hydrolyze and are retained in the polysaccharide-lignin network by hydrogen or hydrophobic bonding and are hardly accessible with a solvent extraction process (Miękus, Iqbal, Marszałek, Puchalski, & Świergiel, 2019).

EAE is present in food processing, nutraceutical and cosmeceutical manufacturing, as well as biodiesel production. The most common enzymes used in this process are proteases, pectinase, cellulase, tannase, or carbohydrases (Miękus, Iqbal, Marszałek, Puchalski, & Świergiel, 2019).

7.4.1.5 HIGH-PRESSURE PROCESSING

The basic principle of the high-pressure (HP) process is the application of a certain pressure, which normally operates up to 600 MPa at room temperature or a value close to this. Three processing factors are involved to achieve desirable results: pressure, temperature, and processing time. On the other hand, this process is used in the inactivation of enzymes and microorganisms to ensure acceptable quality, prevention of food deterioration, preservation of sensory attributes, increase the shelf life of food products, modify the bioaccessibility and bioavailability of nutrients as well as the bioactive compounds avoiding the development of toxic substances (Chemat et al., 2020).

In general, the equipment of the HP process consists of a treatment vessel, closures or plugs, a yoke/wire-wound steel frame, a pressure pump, and a control system. The process is divided into the following stages: 1. Loading the packaged product into the vessel, 2. Filling up with water, 3. Selection of the operating pressure and temperature, 4. Maintaining the pressure for a particular time, 5. Discharging water, and 6. Unloading the treated product (Chemat et al., 2020).

The concepts that explain the function of the HP process are Le Chatelier's principle, microscopic ordering, and the Arrhenius relationship (Chemat et al., 2020). Le Chatelier's principle states that when any kind of disturbance is applied to a system, it reacts by trying to come back into a balanced state. For instance, when high pressure is applied to activate various phenomena, it induces a reaction promoting a decrease in volume that improves the extraction efficiency. Essentially, this high pressure causes the rupture of ionic bonds to reduce the volume due to electrostriction forces acting on water. As a response, the cell structure is modified internally, which decreases the mass transfer in the cell, and consequently, the bioactive compounds are extracted, leaving behind other components such as vitamins, alkaloids, saponins, flavonoids, pigments, or fragrance ingredients unaffected (Khan, Aslam, & Makroo, 2019).

Studies have shown a number of advantages of the HP technique, such as reduced processing time, lower power consumption, increased extraction yields, and the reduction of impurities. Besides, HP technology operates at a low heat rate, reducing the thermal degradation and structure of bioactive compounds. It also operates at reduced temperatures of about 60 °C and high pressures to extract the compounds almost immediately with a low quantity of organic solvents (Khan, Aslam, & Makroo, 2019). Another advantage of this technique is that almost any product of any shape can be processed (Chemat et al., 2020).

This technique has been widely used for the extraction of high-added value compounds: polysaccharides, anthocyanins, ginsenosides, rutin, flavonoids, and phenolic compounds (Khan, Aslam, & Makroo, 2018). It can be classified as an effective sustainable technology because it uses green solvents that reach excellent extraction yields (Chemat et al., 2020).

In terms of optimization, this process applies algorithmic designs: Factorial central composite design and orthogonal array designs are the most popular in research. The parameters that affect the extraction recovery most of the time are solvent nature and volume, pressure, extraction time, number of extraction cycles, and temperature (Khan, Aslam, & Makroo, 2019).

7.4.1.6 PULSED ELECTRIC FIELDS PROCESSING

PEF treatment is regarded as environmentally friendly compared with conventional techniques. Therefore, PEF technology has been positioned as a sustainable option due to its multiple benefits, such as lower energy and water depletion, additionally contributing to sustainable food processing without compromising the safety or quality of products, and promoting an economic advantage for the food industry (Arshad et al., 2021).

Pulsed electric field (PEF) technology is a sustainable technique that allows the recovery of many bioactive compounds, usually applied in the processing of various liquid, semi-liquid, solid, and muscle foods. Certainly, this technique has improved the efficiency and yield of the extraction process, mainly because PEF allows the permeabilization of cell walls with the aid of the following parameters: electric field strength, treatment time, specific energy, pulse shape, pulse width, frequency, temperature, pH, conductivity, size, and shape of the targeted cell (Arshad e al., 2020).

The process consists of gradually increasing the transmembrane potential to initiate the formation of pores in the membrane of the plant cell. As soon as pores are formed in an order of radius of 0.5 nm, they start to grow due to exposure to the electrical field, disrupting the flow of material into and out of the cell (Arshad et al., 2020). Usually, plant cells require a lower critical electric field strength (0.5–2 kV cm-1) for electroporation due to their larger cell size compared to microbial cells (Arshad et al., 2021).

Notably, the damage and loss of the membrane liberate the cell contents into their surroundings and cause its death. This stage is also known as irreversible electroporation, while reversible electroporation, even after the removal of the external electric field, maintains the cell alive (Arshad et al., 2020). Besides, these two kinds of electroporation make this technique able to be widely implemented in the food-processing industries, mainly for the bio compounds extraction of plant matrices (Arshad et al., 2021).

PEF treatment is perceived as an environmentally friendly and economical technology that improves mass transfer in a wide range of applications within the food field. To start with some advantages of this technique, it is important to mention the following: effective preservation of liquids with low ionic strength and low conductivity, high yields, enhanced mass transfer, a softening effect that reduces the energy needed for cutting, minimizes temperatures, reduces processing times, energy-efficient, environmentally friendly, and reduces processing costs (Arshad e al., 2020).

Most PEF systems include a treatment chamber, pulsed power supply, and control and monitoring system (Gómez et al., 2019). Specifically, a pulse power generator is used to apply high-voltage pulsations across the treatment chamber. This equipment consists of a combination of passive discrete elements (capacitive, inductive, and resistive), transformers, and power switches that must transfer the energy stored in the capacitors or magnetic fields of inductors. The stage of energy transfer is a delicate matter in economic terms and can influence the electrical design of this technology (Arshad et al., 2020). Also, one of the benefits of this process is the application of short electrical pulses at high voltages creates full control of thermal effects to remain low (Gómez et al., 2019).

On the other hand, the parameters that influence the processing conditions for targeted applications and the efficiency of this technology at a commercial level are electric field strength, flow rate, pulse waveform, pulse repetition rate, exposure time, specific energy density, and temperature variation in the sample involved (Arshad e al., 2020). In addition, the design of the treatment chambers is probably the most significant element in terms of the quality of the results. The shape of the chamber has a big influence on the electric field distribution, the total resistance, and the discharge circuit (Gómez et al., 2019).

PEF, as a non-thermal technique, is characterized by its low energy consumption, its continuous operability with short processing times, and is a waste-free process, allowing the development of brand new, cost-effective, and sustainable processes in the food and drink industry, as well as in the biotechnology and pharmaceutical industry (Gómez et al., 2019).

7.4.1.7 COLD PLASMA

Cold plasma technology, also known as nonthermal plasma, contains electrons at a high temperature, unlike ions and neutrons. However, the gas temperature of cold plasma stays low in a state of thermodynamic non-equilibrium, and it is possible to operate both at atmospheric or low pressure. With that being stated, this emerging technology is ideal for temperature-sensitive biological materials (Rashid, Bao, Ahmed, & Huang, 2020).

Plasma is a substance that is a partially ionized gas and contains free electrons, ions, radicals, excited and non-excited molecules, and photons. In the same way, cold plasma works at a lower temperature (under 40°C) and at atmospheric pressure conditions. While cold plasma (CP) consists of ions, electrons, ultraviolet photons, atoms, molecules, and free radicals that break covalent bonds and produce different chemical reactions (Keshavarzi, Najafi, Ahmadi Gavlighi, Seyfi, & Ghomi, 2020). It is used in diverse areas such as polymer modification and functionalization, the biomedical field, and food processing, among others. Also, it functions as a microbial inactivator due to the strong oxidative stress and cell membrane leakage caused by plasma reactive species to targeted microorganisms. Finally, in recent years, cold plasma has been widely used in the agricultural field for the extraction of bioactive compounds and to raise the quality of food products, regarding their availability and antioxidant capacity of bio compounds. (Bao, Reddivari, & Huang, 2020).

This technology has been applied for the alteration of biological materials, for example, surface rupture, wettability, and roughness, which are the characteristics closely related to the disposition of the migration of their substances from the interior to the surface during extraction (Rashid, Bao, Ahmed, & Huang, 2020).

Some factors that can cause phenolic compound degradation are plasma source type, time, and power of treatment, varied active species, type of food, and plant matrix. The principal reason is that CP produces many reactive oxygen species that cause the oxidation of phenolic compounds and a decrease in antioxidant activity. However, using an adequate CP source that reduces the number of reactive oxygen species and increases the extraction rate of phenolic compounds from plants is essential for their safe application in the food and pharmaceutical industry (Keshavarzi, Najafi, Ahmadi Gavlighi, Seyfi, & Ghomi, 2020).

In terms of sustainability, cold plasma has been considered a green technology in the field of food. For instance, when it is being operated at atmospheric conditions, it is economically viable because it does not require a costly reaction chamber to regulate pressure and temperature conditions (Rashid, Bao, Ahmed, & Huang, 2020).

7.4.1.8 OHMIC HEATING

Ohmic heating is a homogeneous, efficient, and rapid heating process. Its basis is the use of current alternative electrical heating techniques widely applied in the food industry (Çilingir, Goksu, & Sabanci, 2021). It is based on Ohm's law, an important concept in ohmic heating, which states that the voltage across a conductor is proportional to the electrical current that passes through the conductor, and the ratio between these two directly proportional quantities is described as the resistance (Gavahian, Sastry, Farhoosh, & Farahnaky, 2020).

Basically, the principle under which this technology operates is to apply a determined amount of heat through an alternating current that passes the liquid/solid material between the two electrodes. The application of this technique is outstanding for improving solute extraction with lower energetic costs and superior product quality (El Darra, Grimi, Vorobiev, Louka, & Maroun, 2013). Some advantages of this process include providing homogeneous, efficient, and rapid heating with minimal intervention in the process control. It turns off by a power cut and turns on again if the supply of electricity is restored. Moreover, it has been used in food engineering areas such as heating liquid foods, extraction, pre-drying, evaporation, cooking, and distillation (Çilingir, Goksu, & Sabanci, 2021).

Recently, some studies reported that an ohmic extraction process can be affected by the following factors: frequency, electric field strength, extraction duration, the electrical conductivity of the extraction medium, the physicochemical characteristics of the biomass, and the extraction temperature. In this way, the adequate control of these parameters ensures a desirable extraction yield (Gavahian, Sastry, Farhoosh, & Farahnaky, 2020).

In conclusion, Ohmic heating technology is a non-thermal process with the capacity to provide fast, homogeneous, and precise heating. The direct application of electrical energy to the food ensures highly efficient energy transfer, making it one of the better options in the food processing field as an emerging and sustainable technique (Çilingir, Goksu, & Sabanci, 2021).

7.5 CONCLUSION

Semi-desert plants, in particular, hold great potential in the food and pharmaceutical fields, mainly due to their secondary metabolites that offer significant health benefits to humans. Tannins, terpenoids, alkaloids, and flavonoids are the main bioactive compounds that several studies aim to extract. Recent research has been gradually shifting from traditional techniques such as Soxhlet extraction, maceration, soaking, water percolation, and hydro distillation, mainly due to their low extraction yields, long extraction times, formation of toxic residues, heat degradation of compounds, high energy and solvent consumption, to more sustainable technologies: microwave-assisted extraction, ultrasonic-assisted extraction, supercritical fluid extraction, enzyme and microorganism-assisted extraction, high-pressure processing, pulsed electric fields processing, cold plasma, and ohmic heating. These technologies offer multiple benefits in terms of the efficiency and quality of the extracted compounds. However, determining the optimal parameters and conditions for these emerging technologies has become the primary focus of current studies.

Bentonite was modified and used to remove Congo red from dye effluent water. It is proven to be an effective adsorbent. The equilibrium time required is determined as 4 hours. Univariate parametric studies were conducted, and an ineffective variable (pH) was eliminated. Dye uptake increased with temperature, indicating the endothermic nature of the process. The effects of various parameters and their interactions were studied using response surface methodology. Effluent concentration and dosage were identified as the most influential parameters, and the second-order effects of the same are also influencing the process. The statistical quadratic model was generated, and its robustness was tested as per the standards given in the literature.

KEYWORDS

- **bioactive compounds**
- **cold plasma**
- **emerging technologies**
- **extraction process**
- **microwave-assisted extraction**
- **second-order effects**
- **semi-desert species**
- **sustainability**

REFERENCES

Adkins, S., & Shabbir, A. (2014). Biology, ecology and management of the invasive parthenium weed (Parthenium hysterophorus L.). Pest Management Science, 70(7), 1023–1029. https://doi.org/10.1002/ps.3708.

Ahmad, T., Masoodi, F. A., Rather, S., Wani, S. M., & Gull, A. (2019). Supercritical Fluid Extraction: A Review. Journal of Biological and Chemical Chronicles, 5(1), 114–122. https://doi.org/10.33980/jbcc.2019.v05i01.019.

Alexandre, E. M. C., Silva, S., Santos, S. A. O., Silvestre, A. J. D., Duarte, M. F., Saraiva, J. A., & Pintado, M. (2019). Antimicrobial activity of pomegranate peel extracts performed by high pressure and enzymatic assisted extraction. Food Research International, 115, 167–176. https://doi.org/10.1016/j.foodres.2018.08.044.

Al-Saleem, M. S., Awaad, A. S., Alothman, M. R., & Alqasoumi, S. I. (2018). Phytochemical standardization and biological activities of certain desert plants growing in Saudi Arabia. Saudi Pharmaceutical Journal, 26(2), 198–204. https://doi.org/10.1016/j.jsps.2017.12.011.

Al-Saleem, M. S., Awaad, A. S., Alothman, M. R., & Alqasoumi, S. I. (2018). Phytochemical standardization and biological activities of certain desert plants growing in Saudi Arabia. Saudi Pharmaceutical Journal, 26(2), 198–204. https://doi.org/10.1016/j.jsps.2017.12.011.

Arshad, R. N., Abdul-Malek, Z., Munir, A., Buntat, Z., Ahmad, M. H., Jusoh, Y. M. M., Bekhit, A. E. D., Roobab, U., Manzoor, M. F., & Aadil, R. M. (2020). Electrical systems for pulsed electric field applications in the food industry: An engineering perspective. Trends in Food Science and Technology, 104, 1–13. https://doi.org/10.1016/j.tifs.2020.07.008.

Arshad, R. N., Abdul-Malek, Z., Roobab, U., Munir, M. A., Naderipour, A., Qureshi, M. I., El-Din Bekhit, A., Liu, Z. W., & Aadil, R. M. (2021). Pulsed electric field: A potential alternative towards a sustainable food processing. Trends in Food Science and Technology, 111, 43–54. https://doi.org/10.1016/j.tifs.2021.02.041.

Azmir, J., Zaidul, I. S. M., Rahman, M. M., Sharif, K. M., Mohamed, A., Sahena, F., Jahurul, M. H. A., Ghafoor, K., Norulaini, N. A. N., & Omar, A. K. M. (2013). Techniques for extraction of bioactive compounds from plant materials: A review. Journal of Food Engineering, 117(4), 426–436. https://doi.org/10.1016/j.jfoodeng.2013.01.014.

Azmir, J., Zaidul, I. S. M., Rahman, M. M., Sharif, K. M., Mohamed, A., Sahena, F., Jahurul, M. H. A., Ghafoor, K., Norulaini, N. A. N., & Omar, A. K. M. (2013). Techniques for extraction of bioactive compounds from plant materials: A review. Journal of Food Engineering, 117(4), 426–436. https://doi.org/10.1016/j.jfoodeng.2013.01.014.

Bagade, S. B., & Patil, M. (2021). Recent Advances in Microwave Assisted Extraction of Bioactive Compounds from Complex Herbal Samples: A Review. Critical Reviews in Analytical Chemistry, 51(2), 138–149. https://doi.org/10.1080/10408347.2019.1686966.

Bagheri, R., Fordoei, A. R., Mousavi, H., & Tahmasebi, P. (2021). Climate-driven abrupt changes in plant communities of desert and semi-desert region. Theoretical and Applied Climatology, 146(1–2), 331–348. https://doi.org/10.1007/s00704-021-03720-y.

Bao, Y., Reddivari, L., & Huang, J. Y. (2020). Enhancement of phenolic compounds extraction from grape pomace by high voltage atmospheric cold plasma. LWT, 133. https://doi.org/10.1016/j.lwt.2020.109970.

Bashar, H. M. K., Juraimi, A. S., Ahmad-Hamdani, M. S., Uddin, M. K., Asib, N., Anwar, M. P., & Rahaman, F. (2021). A mystic weed, parthenium hysterophorus: Threats, potentials and management. Agronomy, 11(8). MDPI AG. https://doi.org/10.3390/agronomy11081514.

Bashenova, M., & Aidarbayeva, D. (2022). Biological sciences bioecological feature and medicinal value of Eohedra L.

Beck, H. E., Zimmermann, N. E., McVicar, T. R., Vergopolan, N., Berg, A., & Wood, E. F. (2018). Present and future köppen-geiger climate classification maps at 1-km resolution. Scientific Data, 5. https://doi.org/10.1038/sdata.2018.214.

Calzada, F., & Bautista, E. (2020). Plants used for the treatment of diarrhoea from Mexican flora with amoebicidal and giadicidal activity, and their phytochemical constituents. In Journal of Ethnopharmacology (Vol. 253). Elsevier Ireland Ltd. https://doi.org/10.1016/j.jep.2020.112676.

Carmona-Aparicio, L., Cárdenas-Rodríguez, N., Delgado-Lamas, G., Pedraza Chaverri, J., Montesinos-Correa, H., Rivera-Espinosa, L., Torres-Espíndola, L. M., Hernández, M. E., López-Aceves, T., Pérez-Lozano, D. L., Hernández-Velasco, N., Narváez-Delgado, O., Gutiérrez-Alejandre, A. P., Fuentes-Mejía, M., Bello-Robles, E., Martínez-Ponce, K., Sánchez-Valle, V., Sampieri, A., Granados-Rojas, L., & Coballase-Urrutia, E. (2019). Dose-Dependent Behavioral and Antioxidant Effects of Quercetin and Methanolic and Acetonic Extracts from Heterotheca inuloides on Several Rat Tissues following Kainic Acid-Induced Status Epilepticus. Oxidative Medicine and Cellular Longevity, 2019. https://doi.org/10.1155/2019/5287507.

Carpentieri, S., Mazza, L., Nutrizio, M., Jambrak, A. R., Ferrari, G., & Pataro, G. (2021). Pulsed electric fields- and ultrasound-assisted green extraction of valuable compounds from Origanum vulgare L. and Thymus serpyllum L. International Journal of Food Science and Technology, 56(10), 4834–4842. https://doi.org/10.1111/ijfs.15159.

Cascaes Teles, A. S., Hidalgo Chávez, D. W., Zarur Coelho, M. A., Rosenthal, A., Fortes Gottschalk, L. M., & Tonon, R. V. (2020). Combination of enzyme-assisted extraction and high hydrostatic pressure for phenolic compounds recovery from grape pomace. Journal of Food Engineering, 288. https://doi.org/10.1016/j.jfoodeng.2020.110128.

Chan, C. H., Yusoff, R., Ngoh, G. C., & Kung, F. W. L. (2011). Microwave-assisted extractions of active ingredients from plants. Journal of Chromatography A, 1218(37), 6213–6225. https://doi.org/10.1016/j.chroma.2011.07.040.

Chávez-González, M. L., Sepúlveda, L., Verma, D. K., Luna-García, H. A., Rodríguez-Durán, L. V., Ilina, A., & Aguilar, C. N. (2020). Conventional and emerging extraction processes of flavonoids. Processes, 8(4). https://doi.org/10.3390/PR8040434.

Chemat, F., Abert Vian, M., Fabiano-Tixier, A. S., Nutrizio, M., Režek Jambrak, A., Munekata, P. E. S., Lorenzo, J. M., Barba, F. J., Binello, A., & Cravotto, G. (2020). A review of sustainable and intensified techniques for extraction of food and natural products. Green Chemistry, 22(8), 2325–2353. https://doi.org/10.1039/c9gc03878g.

Çilingir, S., Goksu, A., & Sabanci, S. (2021). Production of Pectin from Lemon Peel Powder Using Ohmic Heating-Assisted Extraction Process. Food and Bioprocess Technology, 14(7), 1349–1360. https://doi.org/10.1007/s11947-021-02636-9.

Da Silva, R. P. F. F., Rocha-Santos, T. A. P., & Duarte, A. C. (2016). Supercritical fluid extraction of bioactive compounds. TrAC – Trends in Analytical Chemistry, 76, 40–51. https://doi.org/10.1016/j.trac.2015.11.013.

Dzah, C. S., Duan, Y., Zhang, H., Wen, C., Zhang, J., Chen, G., & Ma, H. (2020). The effects of ultrasound assisted extraction on yield, antioxidant, anticancer and antimicrobial activity of polyphenol extracts: A review. Food Bioscience, 35. https://doi.org/10.1016/j.fbio.2020.100547.

El Darra, N., Grimi, N., Vorobiev, E., Louka, N., & Maroun, R. (2013). Extraction of Polyphenols from Red Grape Pomace Assisted by Pulsed Ohmic Heating. Food and Bioprocess Technology, 6(5), 1281–1289. https://doi.org/10.1007/s11947-012-0869-7.

Eldridge, D. J., Koen, T. B., Killgore, A., Huang, N., & Whitford, W. G. (2012). Animal foraging as a mechanism for sediment movement and soil nutrient development: Evidence from the semi-arid Australian woodlands and the Chihuahuan Desert. Geomorphology, 157–158, 131–141. https://doi.org/10.1016/j.geomorph.2011.04.041.

El-Meligy, R. M., Awaad, A. S., Soliman, G. A., Kenawy, S. A., & Alqasoumi, S. I. (2017). Prophylactic and curative anti-ulcerogenic activity and the possible mechanisms of action of some desert plants. Saudi Pharmaceutical Journal, 25(3), 387–396. https://doi.org/10.1016/j.jsps.2016.10.008.

Gallego, R., Bueno, M., & Herrero, M. (2019). Sub- and supercritical fluid extraction of bioactive compounds from plants, food-by-products, seaweeds and microalgae – An update. TrAC – Trends in Analytical Chemistry, 116, 198–213. https://doi.org/10.1016/j.trac.2019.04.030.

Gavahian, M., Sastry, S., Farhoosh, R., & Farahnaky, A. (2020). Ohmic heating as a promising technique for extraction of herbal essential oils: Understanding mechanisms, recent findings, and associated challenges. Advances in Food and Nutrition Research, 91, 227–273. https://doi.org/10.1016/bs.afnr.2019.09.001.

Gligor, O., Mocan, A., Moldovan, C., Locatelli, M., Crişan, G., & Ferreira, I. C. F. R. (2019). Enzyme-assisted extractions of polyphenols – A comprehensive review. Trends in Food Science and Technology, 88, 302–315. https://doi.org/10.1016/j.tifs.2019.03.029.

Gómez, B., Munekata, P. E. S., Gavahian, M., Barba, F. J., Martí-Quijal, F. J., Bolumar, T., Campagnol, P. C. B., Tomasevic, I., & Lorenzo, J. M. (2019). Application of pulsed electric fields in meat and fish processing industries: An overview. Food Research International, 123, 95–105. https://doi.org/10.1016/j.foodres.2019.04.047.

González-Juárez, D. E., Escobedo-Moratilla, A., Flores, J., Hidalgo-Figueroa, S., Martínez-Tagüeña, N., Morales-Jiménez, J., Muñiz-Ramírez, A., Pastor-Palacios, G., Pérez-Miranda, S., Ramírez-Hernández, A., Trujillo, J., & Bautista, E. (2020). A review of the Ephedra genus: Distribution, ecology, ethnobotany, phytochemistry and pharmacological properties. Molecules, 25(14). MDPI AG. https://doi.org/10.3390/molecules25143283.

Guadarrama-Reyes, S. C., Morales-Luckie, R. A., Sánchez-Mendieta, V., González-Pedroza, M. G., Lara-Carrillo, E., Velazquez-Enriquez, U., Toral-Rizo, V., & Scougall-Vilchis, R. (n.d.). Green Synthesis of Silver Nanoparticles Using Heterotheca inuloides and Its Antimicrobial Activity in Catgut Suture Threads.

Hernández-Magaña, R., Guadalupe Hernández-Oria, J., & Chávez, R. (2012). Datos para la conservación florística en función de la amplitud geográfica de las especies en el semidesierto queretano, méxico. In Acta Botanica Mexicana, 99.

Iqbal, J., Khan, A. A., Aziz, T., Ali, W., Ahmad, S., Rahman, S. U., Iqbal, Z., Dablool, A. S., Alruways, M. W., Almalki, A. A., Alamri, A. S., & Alhomrani, M. (2022). Phytochemical Investigation, Antioxidant Properties and In Vivo Evaluation of the Toxic Effects of Parthenium hysterophorus. Molecules, 27(13). https://doi.org/10.3390/molecules27134189.

Keshavarzi, M., Najafi, G., Ahmadi Gavlighi, H., Seyfi, P., & Ghomi, H. (2020). Enhancement of polyphenolic content extraction rate with maximal antioxidant activity from green tea leaves by cold plasma. Journal of Food Science, 85(10), 3415–3422. https://doi.org/10.1111/1750-3841.15448.

Khan, S. A., Aslam, R., & Makroo, H. A. (2019). High pressure extraction and its application in the extraction of bio-active compounds: A review. Journal of Food Process Engineering, 42(1). https://doi.org/10.1111/jfpe.12896.

Llompart, M., Celeiro, M., & Dagnac, T. (2019). Microwave-assisted extraction of pharmaceuticals, personal care products and industrial contaminants in the environment. TrAC – Trends in Analytical Chemistry, 116, 136–150. https://doi.org/10.1016/j.trac.2019.04.029.

Męekus, N., Iqbal, A., Marszałek, K., Puchalski, C., & Świergiel, A. (2019). Green chemistry extractions of carotenoids from daucus carota L.-Supercritical carbon dioxide and enzyme-assisted methods. Molecules, 24(23). https://doi.org/10.3390/molecules24234339.

Motmainna, M., Juraimi, A. S., Uddin, M. K., Asib, N. B., Islam, A. K. M. M., Ahmad Hamdani, M. S., & Hasan, M. (2021). Phytochemical constituents and allelopathic potential of parthenium hysterophorus l. In comparison to commercial herbicides to control weeds. Plants, 10(7). https://doi.org/10.3390/plants10071445.

Osorio-Tobón, J. F. (2020). Recent advances and comparisons of conventional and alternative extraction techniques of phenolic compounds. Journal of Food Science and Technology, 57(12), 4299–4315. https://doi.org/10.1007/s13197-020-04433-2.

Pereira, R. N., Rodrigues, R. M., Genisheva, Z., Oliveira, H., de Freitas, V., Teixeira, J. A., & Vicente, A. A. (2016). Effects of ohmic heating on extraction of food-grade phytochemicals from colored potato. LWT, 74, 493–503. https://doi.org/10.1016/j.lwt.2016.07.074.

Rashid, F., Bao, Y., Ahmed, Z., & Huang, J. Y. (2020). Effect of high voltage atmospheric cold plasma on extraction of fenugreek galactomannan and its physicochemical properties. Food Research International, 138. https://doi.org/10.1016/j.foodres.2020.109776.

Rodríguez-Chávez, J. L., Egas, V., Linares, E., Bye, R., Hernández, T., Espinosa García, F. J., & Delgado, G. (2017). Mexican Arnica (Heterotheca inuloides Cass. Asteraceae: Astereae): Ethnomedical uses, chemical constituents and biological properties. Journal of Ethnopharmacology, 195, 39–63. https://doi.org/10.1016/j.jep.2016.11.021.

Singh, I. (n.d.). Parthenium hysterophorus a Threat or Beneficial Weed and Management: A Review. https://www.researchgate.net/publication/347354811 (accessed on 28 May 2024).

Torres-León, C., Rebolledo Ramírez, F., Aguirre-Joya, J. A., Ramírez-Moreno, A., Chávez-González, M. L., Aguillón-Gutierrez, D. R., Camacho-Guerra, L., Ramírez-Guzmán, N., Hernández Vélez, S., & Aguilar, C. N. (2023). Medicinal plants used by rural communities in the arid zone of Viesca and Parras Coahuila in northeast Mexico. Saudi Pharmaceutical Journal, 31(1), 21–28. https://doi.org/10.1016/j.jsps.2022.11.003.

Villavicencio-Gutiérrez, E. E., Cano-Pineda, A., Castillo-Quiroz, D., Hernández Ramos, A., & Martínez-Burciaga, O. U. (2021). Sustainable forest management of non-wood resources in the semi-desert of northern Mexico. In Revista Mexicana de Ciencias Forestales (pp. 31–63). National Institute of Forestry, Agricultural and Livestock Research. https://doi.org/10.29298/rmcf.v12iEspecial-1.1083.

Zumajo-Cardona, C., & Ambrose, B. A. (2022). Fleshy or dry: transcriptome analyses reveal the genetic mechanisms underlying bract development in Ephedra. EvoDevo, 13(1). https://doi.org/10.1186/s13227-022-00195-4.

CHAPTER 8

Streptomyces sp. as a Producer of Bioactive Compounds for the Generation of Organic Herbicides

VÍCTOR NAVARRO-MACIAS,[1] OLGA B. ALVAREZ-PÉREZ,[2] MAYELA GOVEA-SALAS,[1] ELAN LAREDO-ALCALÁ,[3] MÓNICA LIZETH CHÁVEZ-GONZÁLEZ,[1] RODOLFO RAMOS-GONZALES,[1] and ROBERTO ARREDONDO-VALDÉS[1]

[1]*Facultad de Ciencias Químicas, Universidad Autónoma de Coahuila, Saltillo, Coahuila, Mexico*

[2]*Microbiological Research and Development Department, Greencorp Biorganiks de Mexico S.A. de C.V., Saltillo, Coahuila, Mexico*

[3]*Centro de Investigación para la Conservación de la Biodiversidad y Ecología de Coahuila (CICBEC), Universidad Autónoma de Coahuila, Saltillo, Coahuila, Mexico*

ABSTRACT

The environmental and economic problems generated by chemical herbicides in recent years and the resistance they cause in weeds have led to the search for alternatives to control these plants. In this sense, formulating new products based on natural and biological sources has become essential. Herbicides from metabolites produced by microorganisms can be an attractive solution due to their lower toxicity and residuality at the time of application. Among the genera of microorganisms with the most significant potential to produce compounds with herbicidal activity is *Streptomyces*, a genus of actinobacteria widely used to produce metabolites of industrial

Sustainable Agriculture and Global Environmental Health: Toward the Efficient Management of Resources.
Mónica L. Chávez González, Cristóbal Noé Aguilar, Juan Carlos Contreras-Esquivel, and A. K. Haghi (Eds.)

interest. Among the metabolites with herbicidal potential are amino acids, amides, and phytohormones, which have different mechanisms of action depending on the species and the compound generated. These metabolites are produced through biotechnological processes, such as fermentation in the liquid and solid states. However, saving state actors can affect or promote output, so searching for the best conditions remains a topic of great industrial interest, significantly when proposed alternatives reduce environmental damage and costs in the production process. This review focuses on the metabolites produced by *Streptomyces* and their herbicidal effect on weeds.

8.1 INTRODUCTION

Herbicides are agrochemicals that prevent the growth and development of unwanted plants (weeds). Weeds can be annual or perennial, and their control is difficult because they can grow from seeds, which are produced in large quantities, or underground structures such as roots or rhizomes. These plants generate significant economic losses (Camargo et al., 2019; Mesnage, Székács, & Zaller, 2021; Xu et al., 2019). The WHO mentions that, in developed countries, weeds reduce agricultural production by 5 to 10%, while in developing countries, the losses can be higher than 30%. Applying chemical herbicides has solved this problem because methodologies such as tillage and crop rotation have ceased to be used. However, the uncontrolled use of these products has triggered adverse environmental effects, such as damage to biodiversity, decreased agricultural soil fertility, affectation to flora and fauna, high residuality remaining longer in aquifers, and damage to health (Hassaan & El-Nemr, 2020). Currently, 50% of the pesticides used are herbicides, those of a chemical nature that predominate in the market (Zhou et al., 2020). Due to this, herbicides of biological origin are an alternative to weed control and can be important for sustainable agriculture (Bo et al., 2019). Natural herbicides originate naturally from living organisms or their metabolites, which can be selective, degradable, and safer for the environment (Radhakrishnan et al., 2018). Metabolites are produced during growth and development. However, in microorganisms, secondary metabolites are made in the stationary and cell death phase and are usually under metabolic stress; limitation or excess of nutrients, light, and pH are some environmental factors that influence their production (Lajtai-Szabó et al., 2022). Among the microbial metabolites reported as herbicides or herbicidal adjuvants are amino acids, peptides, nucleosides, analogs, cyclic esters, acyclic esters, and organic acids, among others; these compounds act through different mechanisms of action (Shi et al., 2020). One of the microorganisms with the most significant potential to produce compounds with herbicidal activity is *Streptomyces*; approximately 45% of the microbial metabolites identified to date come from this genus. In addition, its filamentous morphology helps it penetrate the plant better (Amelia-Yap et al., 2022). *Streptomyces* is characterized by its life cycle and different forms of growth in solid and liquid cultures. In this sense, solid and liquid fermentation as a biotechnological process to produce metabolites with herbicidal activity plays a significant role since combining factors makes obtaining a good process possible.

8.2 HERBICIDE TOXICITY

With the increase in population, there has been a need to satisfy the demand for food. For this reason, the use of chemical herbicides has become indispensable not just to improve food production but also in some crops such as wheat, oats, barley, maize, and sorghum where herbicides can act as desiccants. Despite the advantages of chemical herbicides, there are drawbacks, such as damage to biodiversity (Hassaan & El-Nemr, 2020). Among the most toxic herbicides are those belonging to the bipyridyl group, such as diquat and paraquat. Paraquat (1,1-dimethyl-4,4-bipyridyl dichloride) is a highly effective non-selective herbicide against broadleaf weeds. Poisoning with this herbicide mainly occurs through ingestion. Once ingested, it is absorbed quickly, mainly by the small intestine; 1–5% is absorbed in the first 6 hours of poisoning (Dinis-Oliveira et al., 2008). It has been shown that in shallow doses, it can also cause health damage. Paraquat goes through a process called the redox cycle; in the cell, this compound can be metabolized by enzymes such as NADPH-cytochrome P450 reductase, and NADH-ubiquinone oxidoreductase. Through oxidation reactions, NADPH is consumed, promoting the formation of free oxygen radicals, which will affect the membrane through the oxidation of lipids (Kevin Baer n.d.; Liu, Yang, & Liu, 2022). Paraquat has been associated with Parkinson's disease caused by mitochondrial dysfunction and induces apoptosis in organs such as lungs, liver, and kidneys due to oxidative stress (Franco et al., 2022; Liu et al., 2022) and at high concentrations and prolonged contact, it has a mortality rate of 60–70% (Fallahi & Dalfardi, 2018). Reported pulmonary fibrosis in a patient ingesting paraquat and that 15 mL to 20% paraquat can cause lethal poisoning.

Glyphosate is another herbicide that is attributed to toxic effects. Over 8,00,000 tons of glyphosate are used worldwide every year (van Bruggen et al., 2018). The expiration of Monsanto's patent in 1991 (Roundup®)

allowed the production of other herbicides based on this compound. In addition, the introduction of Roundup-ready crops (genetically modified to resist glyphosate) led to glyphosate becoming the most used herbicide in the world in 2010 (Linz & Homan, 2011; Mesnage & Antoniou, 2017). What led to this herbicide has generated different problems of contamination.

Several reports indicate the toxicity of this compound in different organisms. This is present in soil, water, air, and food, so it is present practically throughout the food chain. The primary exposure is by contact during its application. However, the intake of water and food contaminated with this herbicide has generated problems. In foods, glyphosate is found mainly in grains and cereals such as oats, corn, wheat, barley, rye, and soybeans, as well as in products derived from these such as cakes, cookies, tortillas, and beer, among others (Xu et al., 2019), and from these, it is absorbed by humans and animals. The International Agency for Research on Cancer (IARC) has classified Glyphosate as a carcinogenic category 2A (Probably carcinogenic to humans). This herbicide may inhibit mitochondrial succinate dehydrogenase activity, which causes the death of human beings' umbilical, embryonic, and mobile cells (Zhao et al., 2015). In addition, constant exposure causes heart, respiratory, and hepatorenal damage and can alter the structure of DNA (Mesnage & Antoniou, 2017; Sandoval-Gío et al., 2022).

On the other hand, there are triazines; some of this family's molecules are associated with diseases such as cancer and hormonal dysfunctions in the human body. This herbicidal compound group is selective and effective against broadleaf and grass weeds (Liu et al., 2022). Cyanazine is an herbicide of the triazine group; according to the World Health Organization, it is one of the herbicides with the highest degree of toxicity. Endocrine disorders, regular disruptions of hormonal function, and reproductive tumors are some of the effects it causes on human health (Zhang et al., 2013). On the other hand, atrazine, other herbicides of the same group, have been reported as a compound with toxic effects on human health; these can be acute from eye and skin irritation, chronic such as those related to reproductive damage, it also has high diabetogenic potential. (Gupta et al., 2023) They mention that exposure to atrazine coupled with a moderately high diet promotes type 2 diabetes because the glucose mechanism is disrupted, generating insulin resistance; in addition, there is the early appearance of diabetic complications such as cardiopathy. The toxicity of herbicides can be low and cause only irritation or so severe that it can lead to death; however, the toxicity and harmful health effects are not the only problems; another problem that has been generated because of their excessive use is the resistance of weeds to these products (Table 8.1).

TABLE 8.1 Toxicity of Some Herbicides on The Market

Herbicide	Herbicide Type	Action Mechanism	Toxicity Grade *WHO	Possible Health Risks	References
Diquat	Bipyridyls	Photosystem I inhibitor	High	Systemic toxicity	Ness-Cochinwala et al. (2022)
Paraquat	Bipyridyls	Photosystem I inhibitor	High	Induces oxidative stress, affects DNA, and cytotoxicity in lymphocytes	Ahmadian et al. (2020); Alizadeh et al. (2022)
Bromoxynil	Foliar action miscellaneous	Photosystem II inhibitor	High	Classified as a group C possible human carcinogen.	*EPA
Cyanazine	Triazines	Photosystem II inhibitor	High	Possible cancer risks	AbdEl-Aty & El-Dib (2009)
2,4-D	Aryloxy-alkanoic acids	Auxin-type herbicides	Moderate	Eye irritation	*EPA
Metolachlor	Thiol carbamates	Lipid biosynthesis inhibitors	Moderate	Possible cancer risks	Gupta (2018)
Atrazyne	Triazines	Photosystem II inhibitor	Moderate	Induce ROS production	Karadayian et al. (2022)
Alachlor	Chloroacetamides	Cell division inhibitors	Moderate	Possible cancer risks	Gupta (2018)
Aciflurofen	Diphenyl ethers	Chlorophyll synthesis inhibitors	Low	Irreversible eye damage.	Rodriguez (2014)
Glyphosate	Glyphosate	Amino acid synthesis inhibitors	Low	Problems in pregnancy, affecting human placental JEG3 cells.	Richard et al. (2005)

8.3 HERBICIDE RESISTANCE

Implementing sustainable alternatives for controlling weed populations worldwide continues to be a significant issue. Weeds, in general, infest various crops, affecting production yields, handling and harvesting operations, and increasing field decontamination costs. For this reason, good management and control practices have been developed and implemented since before 1970 (Avila et al., 2007; Powles et al., 1997). Since then, herbicide resistance has been defined as "the heritable ability of a weed population to survive the application of a herbicide that is normally lethal to most individuals of that species" (Powles et al., 1997). Different factors contribute to generating resistance in a population; among them, we find the initial frequency with which resistant weeds appear, the number of treated individuals, the hereditary genes between generations that cause resistance, and the nature and extent of herbicide use (Friesen, Ferguson, & Hall, 2000). This resistance is caused by genetic mutations that can affect herbicide activity, site of action, metabolic systems, translocation, and absorption. However, resistant individuals must appear first for a population to become resistant. Usually, resistance occurs over a single chemical site, or it can happen over several areas at the same time, being called multiple resistance, for example, through a combination of target insensitivity and enhanced metabolism (Bailly et al., 2012).

An attempt has been made to define how to explain these phenomena, and for this reason, two theories were proposed; the first is the well-known "Gene-pool," which mentions that the process occurs due to a genetic mutation at a specific time in plant life, as a response to external stimuli such as the application of weed control products (Gaines et al., 2020). The second theory is that of selection, where it is mentioned that due to natural variation, there will be at least one individual in a resistant population. For this reason, it is not possible to avoid the phenomenon of resistance; it can occur randomly at any time (Friesen et al., 2000). With this second theory, the effect known as herbicide selection pressure is presented, which is referred to as the effectiveness with which the herbicide selects resistant plants in a population and is directly proportional to the efficacy of the herbicide. Different models can explain resistance, such as the one proposed by Gressel & Segel (1990), which mentions that three variables generate resistance: selection pressure, seed bank process(es), and relative plant fitness. Following these bases, it is established that those species that die quickly due to the effect of herbicides will probably be the first to be resistant (Gaines et al., 2020).

Belz et al. (2022) also mention the existence of two categories of resistance mechanisms, target-site resistance (TSR) and non-target-site resistance (NTSR). In the TSR mechanism, the aim is to dilute the effect of the herbicide or increase the target molecule to dilute herbicide activity, which often results in a high level of resistance, as in the case of resistance to acetolactate synthetase (Li et al., 2019) is caused by a mutation in the encoding gene, while by the NTSR mechanism (Matzrafi et al., 2014). The amount of herbicide that reaches the target molecule is minimal, thus reducing its effect or causing translocation, increasing metabolic degradation, or moving the herbicide away from the target, thus generating low levels of resistance, such as that generated towards propanil, where high enzymatic levels of aryl acrylamides hydrolyze the compound making it non-phytotoxic. Some examples are mentioned by Délye et al. (2013) in their Table 8.2.

Several management strategies have been proposed in recent years, such as herbicide rotation, using mixtures, strictly following herbicide use instructions, including non-selective herbicides, good weed control practices, and sound monitoring of population changes. Many studies have found that the emergence of herbicide resistance causes substantial economic losses (Beltran, Pannell, & Doole, 2012).

There are studies of resistance generated by herbicides; however, the type of weeds varies depending on the location, area, and crop. Despite this, these data also allow for establishing resistance prediction in each zone (Hulme, 2023). Additionally, weeds differ in their resistance due to different modes of action of applied herbicides and may not respond to external factors in the same way. Hulme (2023) mentions five variables that may explain the variation in resistant plants worldwide. Trends indicate that the intensification of agriculture is associated with increased numbers of resistant weeds worldwide. This is mainly due to the destructive practices carried out, leading to underestimation of the problem and failure to manage this problem correctly in the future.

8.4 MICROBIAL HERBICIDES

Natural herbicides entered the market in 1980, with the United States and some European countries being the first to use them (Radhakrishnan et al., 2018). These products originate naturally from living organisms or their natural metabolites during growth and development. They can be selective, degradable, and safer for the environment to help maintain sustainable agricultural production (Bo et al., 2019).

Secondary metabolites are not associated with the growth and development of microorganisms; these metabolites are produced in the stationary and cell death phases and usually under metabolic stress. Limitation or excess of nutrients, temperature, and pH are some of the environmental factors that influence their production (Lajtai-Szabó et al., 2022). Among the microbial metabolites that have been reported as herbicides are amino acids, peptides,

TABLE 8.2 Effects Associated with Resistance to Herbicides by Délye, Jasieniuk, & Le Corre (2013)

Resistance Mechanism	Herbicide Group	Species	The Effect Associated with Herbicide Resistance			
			Seed Germination	Vegetative Biomass	Seed production	Competitive Ability
TSR, Ile-1781-Leu	A	*Alopecurus myosuroides*	Delay	None detected	None detected	Na
TSR, Ile-2041-Asn	A	*Alopecurus myosuroides*	None detected	None detected	None detected	Na
TSR, Gly-2078-Ala	A	*Alopecurus myosuroides*	Acceleration	Reduction	Reduction	Na
TSR, Ile-1781-Leu	A	*Lolium* sp.	Delay	None detected	None detected	None detected
TSR, Ile-1781-Leu	A	*Setaria* sp.	Na	Increase	Increase	Increase
TSR, Pro-197-Ala	B	*Lolium* sp.	Na	None detected	Na	Na
TSR, Pro-197-Arg	B	*Lolium* sp.	Na	None detected	Na	Na
TSR, Pro-197-Gln	B	*Lolium* sp.	Na	None detected	Na	Na
TSR, Pro-197-Ser	B	*Lolium* sp.	Na	None detected	Na	Na
TSR, Trp-574-Leu	B	*Lolium* sp.	Na	None detected	Na	Na
TSR, Pro-197-His	B	*Lactuca serriola*	Na	Reduction	Reduction	Na
TSR, Trp-574-Leu	B	*Amaranthus powellii*	Na	Reduction	Reduction	Na
NTSR	A, B	*Lolium* sp.	None detected	Reduction	Reduction	Reduction
NTSR	O	*Sinapis arvensis*	Na	Reduction	Reduction	Reduction
NTSR	G	*Lolium* sp.	Na	Reduction	Reduction	Na

nucleosides, analogs, cyclic esters, acyclic esters, and organic acids, which act as toxins for weeds (Shi et al., 2020). However, some beneficial metabolites for plants, such as phytohormones, in high concentrations have a detrimental effect on their growth and development. Indole-3-acetic acid (IAA) is a compound that acts as a regulator, favoring the growth of plants; however, at high concentrations, it stimulates aminocyclopropane-1-carboxylate (ACC), which is a precursor of ethylene. This causes inhibition of growth and elongation of the roots (Phukan et al., 2021). In addition to some toxic volatile compounds, they inhibit plant growth and seed germination. These metabolites are produced by bacterial and fungal cells (Radhakrishnan et al., 2018). Some native soil bacteria, such as *Bacillus, Pseudomonas, and Streptomyces*, can produce them (Table 8.3).

The compounds produced by these microorganisms can act through different modes of action, such as inhibition of photosynthesis, inhibition of photosynthetic pigments, cell wall inhibition, and hormonal effects. In the mechanism of inhibition of photosynthesis, the transport of electrons is blocked. This leads to the formation of activated oxygen, causing lipid peroxidation and tissue destruction (Kim et al., 2020a). The inhibition of glutamine is another mechanism that has been described as a mechanism of action of herbicides; in this mechanism, the activity of the enzyme glutamine synthetase is inhibited, preventing the entry of glutamate and ammonium into the complex formed by ATP and the enzyme. Therefore, glutamine is not included, which avoids ammonia assimilation (Bo et al., 2019).

Amino acids are the basis for protein synthesis, which has various functions in the plant, such as catalytic capacity and transport of compounds in and out of the plant (El-Sobki et al., 2021). Therefore, the inhibition of these is also a mechanism of action of herbicides.

Pigments such as chlorophyll and carotenoids are involved in photosynthesis; inhibition of the biosynthesis of these pigments occurs by inhibiting their precursors, δ-aminolevulinic acid and phytoene respectively (Bordin et al., 2021; Jiao et al., 2022). Pigment inhibition can lead to plant death. Lipid inhibitors Herbicides block the enzyme ACCase involved in synthesizing fatty acids, which is essential to maintaining cell structure (Székács, 2021). These mechanisms cause symptoms such as necrosis, chlorosis, deformation, and atrophy. It has also been shown that some adjuvants can increase herbicidal activity (Bo et al., 2019).

8.5 STREPTOMYCES

Streptomyces is a gram-positive bacterium of the actinomycetes family. Currently, more than 1,200 species of this genus have been identified. They are ubiquitous in nature but are mainly found in soil and are characterized by their complex morphological cycle, divided into two phases: vegetative and reproductive (Amelia-Yap et al., 2022; Del Carratore et al., 2022; Cuozzo et al., 2023). Its cell wall component is the peptide glucan, similar to any other

TABLE 8.3 Bacterial Metabolites with Herbicidal Activity

Metabolite	Microorganism	Action Mode	References
Hydrogen cyanide	*Pseudomonas aeruginosa*	Hormonal	Blier et al. (2012)
Hydrogen cyanide	*Bacillus megaterium*	Hormonal	Faraji et al. (2022)
IAA	*Bacillus pumilus* *Bacillus subtilis* *Pseudomonas fluorescens*	Hormonal	Ait Bessai et al. (2022)
IAA	*Enterobacter* spp.	Hormonal	Kumar Ghosh et al. (2015)
Romidepsin	*Burkholderia rinojensis*	Inhibition of plant histone deacetylases	Owens et al. (2020)
Unidentified substance	*Bacillus cereus* *Bacillus pumilus*	Inhibition of seed germination	Carvalho et al. (2007)
Volatiles metabolites	*Pseudomonas* sp., *Enterobacter* sp., *Bacillus* sp.	Inhibition of seed germination	Verdugo-Navarrete et al. (2021)
Unidentified substance	*Bacillus subtilis*	Inhibition of seed germination	Martínez-Mendoza (2012)
Herbicidin A Herbicidin B	*Streptomyces scopuliridis*	Disruption cell membrane	Won et al. (2016)
N-phenylpropanamide	*Streptomyces* sp.	Inhibition of seed germination	Priyadharsini, Dhanasekaran, & Kanimozhi (2013)

bacterium. However, the life cycle of this genus of bacteria is like that of fungi since it also produces mycelium and spores (Salwan & Sharma, 2020). The process begins with the germination of the spores; from these emerge the germinative tubes, and later branched hyphae will develop, forming vegetative mycelium (Xu et al., 2022). In a solid culture, the mycelium grows deeply after programmed cell death occurs, during which some cells of the substrate mycelium undergo a degradation process (Lajtai-Szabó et al., 2022). Subsequently, the formation of aerial branched hyphae begins in this stage, where the production of antibiotics and secondary metabolites occurs. Hyphae differ into multinucleated hyphae, where each segment contains genetic material. Finally, a second programmed cell death event is followed by compartmentalization and spore formation (Amelia-Yap et al., 2022; Xu et al., 2022). On the other hand, in a liquid medium, after germination and mycelium development, three characteristic structures can be formed: dispersed mycelium, open mycelium networks, and mycelium granules. In the latter, programmed cell death occurs in the granule's central part (Lajtai-Szabó et al., 2022).

Streptomyces can use different carbon sources such as sorbitol, dextrose, fructose, inositol, lactose, maltose, raffinose, rhamnose, sucrose, and xylose (Gopalakrishnan, Srinivas, & Prasanna, 2020). However, it can produce enzymes to degrade organic material such as starch, cotton textiles, cellulose, lignocellulose, and chitin, and use them as a substrate. The presence of these bacteria in the environment is influenced by environmental conditions (pH, temperature, humidity, salinity, soil, and texture, among others) (El-Naggar, 2021). *Streptomyces* strains require inorganic nitrogen sources, organic carbon, and mineral salts. Optimum growing temperatures are 30°C and 37°C, while the pH range is 6.5 to 8.0 (El-Naggar, 2021). Devi et al. (2022) mention that the optimum growth temperature of *Streptomyces* spp. is 30°C and that they can also grow at a concentration of 5% NaCl. Other reports indicate that the temperature range is 30–40C, the pH can be 5–11, and the percentage of NaCl can be up to 10% (Gopalakrishnan et al., 2020). This genus of bacteria can grow in various environmental conditions; in addition, the ability to produce a wide variety of metabolites makes *Streptomyces* a beneficial microorganism in the pharmaceutical, food, medicine, and agriculture industries.

8.5.1 BIOTECHNOLOGICAL APPLICATIONS

The wide variety of metabolites that Streptomyces can produce gives these bacteria significant applications in various areas. Different species of Streptomyces have been widely used to produce antibiotics; among the best known is streptomycin, a compound discovered in 1944, and isolated from *S. griseus*. This antibiotic stops the growth of bacteria by inhibiting proteins. Spiramycin production from *S. ambofaciens* has been used mainly against *Staphylococcus aureus* (Grosset & Singer, 2013).

In addition to antibiotics, some strains of *Streptomyces* also produce pigments. The pigments that Streptomyces can produce can be used in the textile industry or the food industry due to each pigment's properties. Zhu et al. (2020) produced a new blue pigment (4,8,13-trihydroxy-6,11-dione-trihydrogranaticins A) with antioxidant properties by checking this activity through testing DPPH and ABTS. They mention that this compound remained stable at high temperatures of 100°C and acidic and alkaline pH levels (3–10). Due to its properties, it can be an attractive natural pigment to use as an additive in food. Karuppiah et al. (2013) optimized the production of a red pigment to which anticancer properties were attributed.

In agriculture, the Streptomyces genus can synthesize metabolites that act as growth promoters or biocontrol agents. Some enzymes produced by this microorganism are cellulases, chitinase, proteases, and lipases, which also make phytohormones and siderophores that benefit plant development (Gopalakrishnan et al., 2020).

These metabolites help fight insect pests, weeds, and microorganisms. In this sense, Deng et al. (2021) mention that it is possible to produce phenazine; this compound has antifungal properties. Gorantla et al. (2014) observed that phenazine managed to inhibit *Rhizoctonia solani* and *Fusarium oxysporum,* two of the fungi that most affect crops. Similarly, they tended to inhibit the growth of *Fusarium oxysporum, Alternaria,* and *Sclerotinia* (Luo et al., 2015) (Table 8.4).

8.5.2 APPLICATION AS HERBICIDES

Several authors have reported the preemergence and post-emergence herbicidal capacity of metabolites produced from *Streptomyces strains*. Among the best-known compounds with herbicidal activity is bialaphos, which is produced by *S. hygroscopicus*. These molecules gave rise to glufosinate, one of the herbicides currently on the market; glufosinate is a broad-spectrum weed control herbicide. Its mechanism of action is provided by inhibition of the enzyme glutamine synthetase, which intervenes in the assimilation of ammonia. This, in large quantities, causes yellowing and death of the plants (Puertas de Freitas Silva et al., 2016).

Other metabolites with a similar mechanism of action are nigericin, hydanthocidin, geldanamycin, and blasticidin. These affect plant growth and development and are produced by *S. hygroscopicus, Streptomyces griseohrmogenes,* and *S. saganonensis*. (Won et al., 2016) compared the capacity of the herbicidin A and B produced by *S. scopuliridis* on *Echinochloa oryzoides, Digitaria ciliaris, Abutilon theophrasti*, and *Amaranthus retroflexus* observing necrosis at three days of application at a dose of 4,000 ppm. They mention that the herbicidal effect was similar to that obtained with glufosinate and lower than that shown by paraquat and glyphosate.

Kim et al. (2020b) report that *Streptomyces* sp. causes whitish and chlorosis on leaves treated with this bacterium. They also mention that the metabolites produced by this strain managed to inhibit 80–90% of the growth of *Digitaria ciliaris*. In their evaluation of electron loss, the

TABLE 8.4 Metabolites with Antimicrobial Potential Produced by *Streptomyces* sp.

Microorganism	Metabolite	Application	References
Streptomyces cacaoi	Polyether-type polyketides	Antimicrobial	Khan et al. (2019)
Streptomyces angustmyceticus	Enzymes that break down cell walls.	Antifungal: *Colletotrichum* sp. *Curvularia* sp.	Wonglom et al. (2019)
Streptomyces sp.	Unidentified metabolites	Antifungal	Marimuthu et al. (2020)
Streptomyces cuspidosporus	1,2-Benzenedicarboxylic acid, bis(2-Methylpropyl)	Antimicrobial: *Escherichia coli*, *Klebsiella pneumonia*, *Staphylococcus aureus*, *Salmonella typhi*	Sholkamy et al. (2020)
Streptomyces sp. *FXP04*	Piericidin A	Antifungal: *P. infestans*	Fu et al. (2022)
Streptomyces sp.	Unidentified antibiotic	Antimicrobial	Al-Ansari et al. (2019)
Streptomyces strain GLD22	2-tert-butyl-4,6-bis(3,5-di-tert-butyl-4-hydroxy benzyl) phenol Dibutyl phthalate Cyclo(leucyloprolyl)	*Klebsiella pneumoniae*, *Pseudomonas aeruginosa*, and *Escherichia coli*	Djebbah et al. (2021)
Streptomyces enissocaesilis	Compound group	*Pseudomonas* sp., *Vibrio anguillarum*, *Aeromonas hydrophilia*, *Vibrio parahaemolyticus*, and *Vibrio harveyi*	Kumaran et al. (2020)
Streptomyces sp. KLBMP5084	Volatile organic compounds	Antifungal, *Alternaria solani*	Gong et al. (2020)

metabolites showed similar results to those of paraquat at 10, 50, and 100 μg mL^{-1}. Umurzokov et al. (2022) produced an anthraquinone (rubiginone D2) from *Streptomyces gardneri*; this compound contained to inhibit 100% of the growth of *Digitaria ciliaris* by inhibiting photosynthetic pigments (chlorophyll) at 500 μg mL-1.

On the other hand, Bo et al. (2019) found that *S. Anulatus* isolated from soil produced two compounds with the herbicidal activity of more than 90%

inhibition in preemergence and 100% in post-emergence causing necrosis and chlorosis on grasses and broadleaf weeds. The symptoms observed are characteristic of photosynthesis inhibitor herbicides. *Streptomyces* also produces enzymes and other metabolites that degrade plant biopolymers, such as cellulose, lignin, and chitin. Loria et al. (2008) mention that *S. scabies* has thaxtomin; a dipeptide that mainly affects during growth in the formation of the cell wall. Albucidine is another secondary metabolite that reduces cell growth and inhibits cellulose synthesis. Species such as *Streptomyces scabies, Streptomyces acidiscabies,* and *Streptomyces turgidiscabies* produce the same effect.

The effect of preemergence has also been evaluated (Priya Dharsini et al., 2017). They produced herbicidal metabolites from *Streptomyces sp.,* which showed an impact on germination, shoot length, root, sub-root, fresh, and dry weight at the concentration of 1.5 mg/ml of the herbicide compound. The type of application time mode, the Streptomyces species, and the production conditions of the metabolites will determine the herbicidal activity of the metabolites.

8.6 PRODUCTION OF SECONDARY METABOLITES BY SUBMERGED FERMENTATION

Fermentation is a process where metabolites of interest are obtained from a nutrient-rich substrate. The metabolic processes of growth and reproduction of the microorganism are carried out thanks to the primary metabolites, while the secondary metabolites are not essential for growth. However, these are produced as a mechanism of defense and survival in the stationary phase or death of the microorganism, usually in a state of stress generated by factors such as temperature, pH, light, oxygen, some components of the culture medium, or by the presence of another antagonistic microorganism (Lajtai-Szabó et al., 2022). Secondary metabolites are essential in the economy and human health since they are of great interest to medicine, agriculture, and the pharmaceutical and food industries.

Streptomyces spp. produce large amounts of metabolites, such as antibiotics and pesticides. Different species of *Streptomyces* that produce some agricultural pesticides (fungicides or herbicides) have been reported. The amount and type of metabolite are related to the production process and the conditions under which it is produced. Submerged fermentations are biotechnological processes that have been reported to produce compounds from *Streptomyces*.

8.6.1 SUBMERGED FERMENTATION

Submerged fermentation is the technique of microbial growth with a high content of free water, where only 5% is solutes; due to a large amount of free water and the nutrients being dissolved in the water, the absorption of them is faster in the same way as bacterial growth (Belorkar & Jogaiah, 2022). This technique has advantages over other metabolite production processes; it is easier to recover and purify when microorganisms excrete metabolites. It is also possible to monitor and modify different factors that directly influence the microorganism's growth and the production of compounds (Barbosa et al., 2021).

An essential factor for the microorganism's metabolism is oxygen; this is closely related to agitation; agitation favors oxygen dissolution in the culture medium during fermentation. (Umurzokov et al., 2022) Mention that the metabolites produced at 200 and 250 rpm had a better herbicidal effect (100%). Similarly, Sholkamy et al. (2020) obtained the highest concentration of metabolites at 200 rpm by *Streptomyces cuspidosporus* (Kim et al., 2020b). They report no significant difference in bacterial growth at 150 and 200 rpm; however, the herbicidal activity was better when 150 rpm was used, obtaining up to 100%. In both cases, greater oxygenation favored the production of secondary metabolites (Sholkamy et al., 2020).

Kim et al. (2020b) mention that pH 5.5 is ideal for the production of metabolites with herbicidal activity with *Streptomyces olivochromogenes*; however, Umurzokov et al. (2022) evaluated a pH range of 3 to 10 and observed that the growth of *Streptomyces gardneri* was only affected at pH 3. They mention that from 5.5 to 10, the development of the bacterium was similar; however, they report that the highest concentration of herbicide metabolite was obtained at pH 10. In the cases mentioned above, the optimal production values are different because they are different species adapted to different environments. Sholkamy et al. (2020) found that *Streptomyces cuspidosporus* obtained the highest production of biomass and active compound at pH 9, while at pH seven, the growth of the microorganism was lower. *Streptomyces olivochromogenes* produces metabolites with 98% herbicidal activity at 27°C (Kim et al., 2020b). On the other hand, the study realized by Umurzokov et al. (2022) found that the optimum temperature for the production of metabolites with herbicidal activity was 37°C, tripling the concentration obtained with the optimal growth temperature of *Streptomyces* at 27°C. Sholkamy et al. (2020) received the highest concentration of metabolites at a temperature of 30°C.

The components of the medium play perhaps the most essential role in the production of metabolites; it is necessary to find a culture medium that promotes high yields. In producing metabolites, inducers can be a strategy to achieve optimal production (Loria et al., 2008). They mention that cellobiose encourages the biosynthesis of taxtomine because the polymers of the plant cell wall stimulate production; however, glucose repression is also necessary (Singh et al., 2018). Supplement the culture medium with starch as the only carbon source to make the propagation more significant. They mention that with this component in the medium, a more significant herbicidal effect was achieved, reporting 60–80% inhibition of seed germination. Adding tryptophan promotes indoleacetic acid production, and supplementing the medium with carboxymethyl cellulose will help the production of cellulases; both compounds are beneficial but can also have herbicidal effects at high concentrations.

8.7 CONCLUSION

The application of products generated from microorganisms such as Streptomyces offers an alternative to reduce synthetic herbicides polluting the environment and toxic to humans. Natural herbicides can help prevent weeds from developing resistance and thus reduce the use of chemical herbicides. However, it is essential to develop biological herbicides with high rates of effectiveness against weeds that can compete with chemical herbicides. Therefore, it is crucial to develop a biotechnological process that allows this. Additionally, bioherbicides must be produced to generate little waste, be non-toxic, and not affect the environment.

KEYWORDS

- **carboxymethyl cellulose**
- **fermentation**
- **herbicides**
- **metabolites**
- **microorganisms**
- **seed germination**
- **streptomyces**
- **toxicity**

REFERENCES

AbdEl-Aty, A. M., & El-Dib, M. A. (2009). Uptake and the effects of cyanazine on Scenedesmus obliquus and Anabaena flos-aquae. *Desalination, 249*(3), 1294–1297. https://doi.org/10.1016/j.desal.2009.09.003.

Ahmadian, E., Eftekhari, A., Kavetskyy, T., Khosroushahi, A. Y., Turksoy, V. A., & Khalilov, R. (2020). Effects of quercetin loaded nanostructured lipid carriers on the paraquat-induced toxicity in human lymphocytes. *Pesticide Biochemistry and Physiology, 167*. https://pubmed.ncbi.nlm.nih.gov/32527420/.

Ait Bessai, S., Bensidhoum, L., & Nabti, E. H. (2022). Optimization of IAA production by telluric bacteria isolated from Northern Algeria. *Biocatalysis and Agricultural Biotechnology, 41*. https://doi.org/10.1016/j.bcab.2022.102319.

Al-Ansari, M., Alkubaisi, N., Vijayaragavan, P., & Murugan, K. (2019). Antimicrobial potential of Streptomyces sp. to the Gram positive and gram negative pathogens. *Journal of Infection and Public Health, 12*(6), 861–866. https://doi.org/10.1016/j.jiph.2019.05.016.

Alizadeh, S., Anani-sarab, G., Amiri, H., & Hashemi, M. (2022). Paraquat induced oxidative stress, DNA damage, and cytotoxicity in lymphocytes. *Heliyon, 8*(7). https://doi.org/10.1016/j.heliyon.2022.e09895.

Amelia-Yap, Z. H., Azman, A. S., AbuBakar, S., & Low, V. L. (2022). Streptomyces derivatives as an insecticide: Current perspectives, challenges and future research needs for mosquito control. *Acta Tropica, 229*(February), 106381. https://doi.org/10.1016/j.actatropica.2022.106381.

Avila, W., Bolaños, A., & Valverde, B. E. (2007). Characterization of the cross-resistance mechanism to herbicides inhibiting acetyl coenzyme-A carboxylase in itchgrass (*Rottboellia cochinchinensis*) biotypes from Bolivia. *Crop Protection, 26*(3), 342–348. https://doi.org/10.1016/j.cropro.2006.03.019.

Bailly, G. C., Dale, R. P., Archer, S. A., Wright, D. J., & Kaundun, S. S. (2012). Role of residual herbicides for the management of multiple herbicide resistance to ACCase and ALS inhibitors in a black-grass population. *Crop Protection, 34*, 96–103. https://doi.org/10.1016/j.cropro.2011.11.017.

Barbosa, J. R., Da Silva, S. B., & De Carvalho, R. N. (2021). Biosurfactant production by solid-state fermentation, submerged fermentation, and biphasic fermentation. In *Green Sustainable Process for Chemical and Environmental Engineering and Science: Microbially-Derived Biosurfactants for Improving Sustainability in Industry* (pp. 155–171). Elsevier.

Beltran, J. C., Pannell, D. J., & Doole, G. J. (2012). Economic implications of herbicide resistance and high labour costs for management of annual barnyardgrass (*Echinochloa crus-galli*) in Philippine rice farming systems. *Crop Protection, 31*(1), 31–39. https://doi.org/10.1016/j.cropro.2011.09.012.

Belz, R. G., Carbonari, C. A., & Duke, S. O. (2022). The potential influence of hormesis on evolution of resistance to herbicides. *Current Opinion in Environmental Science and Health, 27*.

Blier, A. S., Vieillard, J., Gerault, E., Dagorn, A., Varacavoudin, T., Le Derf, F., Orange, N., Feuilloley, M., & Lesouhaitier, O. (2012). Quantification of *Pseudomonas aeruginosa* hydrogen cyanide production by a polarographic approach. *Journal of Microbiological Methods, 90*(1), 20–24. https://doi.org/10.1016/j.mimet.2012.04.005.

Bo, A. B., Kim, J. D., Kim, Y. S., Sin, H. T., Kim, H. J., Khaitov, B., Ko, Y. K., Park, K. W., & Choi, J. S. (2019). Isolation, identification and characterization of Streptomyces metabolites as a potential bioherbicide. *PLoS ONE, 14*(9), 1–18. https://doi.org/10.1371/journal.pone.0222933.

Bordin, E. R., Camargo, A. F., Stefanski, F. S., Scapini, T., Bonatto, C., Zanivan, J., Modkovski, T. A., Reichert Júnior, F., Mossi, A. J., Fongaro, G., Ramsdorf, W. A., & Treichel, H. (2021). Current production of bioherbicides: Mechanisms of action and technical and scientific challenges to improve food and environmental security. *Biocatalysis and Biotransformation, 39*(5), 346–359. https://doi.org/10.1080/10242422.2020.1833864.

Camargo, A. F., Stefanski, F. S., Scapini, T., Weirich, S. N., Ulkovski, C., Carezia, C., Bordin, E. R., Rossetto, V., Reichert Júnior, F., Galon, L., Fongaro, G., Mossi, A. J., & Treichel, H. (2019). Resistant weeds were controlled by the combined use of herbicides and bioherbicides. *Environmental Quality Management, 29*(1), 37–42. https://doi.org/10.1002/tqem.21643.

Carvalho, D. D. C., Oliveira, D. F., Corrêa, R. S. B., Campos, V. P., Guimarães, R. M., & Coimbra, J. L. (2007). Rhizobacteria able to produce phytotoxic metabolites. *Brazilian Journal of Microbiology, 38*(4), 759–765. https://doi.org/10.1590/S1517-83822007000400032.

Cuozzo, S., De Moreno De Leblanc, A., Leblanc, J. G., Hoffmann, N., & Tortella, G. R. (2023). Streptomyces genus as a source of probiotics and its potential for its use in health. *Microbiological Research, 266*(August 2022), 127248. https://doi.org/10.1016/j.micres.2022.127248.

Del Carratore, F., Hanko, E. K., Breitling, R., & Takano, E. (2022). Biotechnological application of Streptomyces for the production of clinical drugs and other bioactive molecules. *Current Opinion in Biotechnology, 77*, 102762. https://doi.org/10.1016/j.copbio.2022.102762.

Délye, C., Jasieniuk, M., & Le Corre, V. (2013). Deciphering the evolution of herbicide resistance in weeds. *Trends in Genetics, 29*(11), 649–658.

Deng, R. X., Zhang, Z., Li, H. L., Wang, W., Hu, H. B., & Zhang, X. H. (2021). Identification of a novel bioactive phenazine derivative and regulation of PhoP on its production in *Streptomyces lomondensis* S015. *Journal of Agricultural and Food Chemistry, 69*(3), 974–981. https://doi.org/10.1021/acs.jafc.0c06498.

Devi, A. P., Jesumaharaja, G. L., Balasundaram, K., Sahana, N., Battacharya, P. M., Roy, A., Bandyopadhyay, S., & Khalko, S. (2022). *Streptomyces sp.*: A feasible biocontrol agent for sustainable management of crop diseases. In *Microbes and Microbial Biotechnology for Green Remediation* (pp. 377–388). Elsevier.

Dinis-Oliveira, R. J., Duarte, J. A., Sánchez-Navarro, A., Remião, F., Bastos, M. L., & Carvalho, F. (2008). Paraquat poisonings: Mechanisms of lung toxicity, clinical features, and treatment. *Critical Reviews in Toxicology, 38*(1), 13–71.

Djebbah, F. Z., Belyagoubi, L., Abdelouahid, D. E., Kherbouche, F., Al-Dhabi, N. A., Arasu, M. V., & Ravindran, B. (2021). Isolation and characterization of novel *Streptomyces* strain from Algeria and its in-vitro antimicrobial properties against microbial pathogens. *Journal of Infection and Public Health, 14*(11), 1671–1678. https://doi.org/10.1016/j.jiph.2021.09.019.

El-Naggar, N. E. A. (2021). *Streptomyces*-based cell factories for production of biomolecules and bioactive metabolites. In *Microbial Cell Factories Engineering for Production of Biomolecules* (pp. 183–234). Elsevier.

El-Sobki, A. E., Saad, A. M., El-Saadony, M. T., El-Tahan, A. M., Taha, A. E., Aljuaid, B. S., El-Shehawi, A. M., & Salem, R. E. M. E. (2021). Fluctuation in amino acids content in *Triticum aestivum* L. cultivars as an indicator on the impact of post-emergence herbicides in controlling weeds. *Saudi Journal of Biological Sciences, 28*(11), 6332–6338. https://doi.org/10.1016/j.sjbs.2021.06.097.

Fallahi, M. J., & Dalfardi, B. (2018). Early lung fibrosis after accidental ingestion of paraquat herbicide. *Visual Journal of Emergency Medicine, 13*, 66–67. https://doi.org/10.1016/j.visj.2018.09.008.

Faraji, F., Mahandra, H., Wang, J., & Ghahreman, A. (2022). Evaluation of medium and applicability of corn steep waste for biogenic cyanide production by *Bacillus megaterium* for sustainable gold recovery. *Journal of Cleaner Production, 372*. https://doi.org/10.1016/j.jclepro.2022.133691.

Franco, D. S. P., Georgin, J., Lima, E. C., & Silva, L. F. O. (2022). Advances made in removing paraquat herbicide by adsorption technology: A review. *Journal of Water Process Engineering, 49*.

Friesen, L. J. S., Ferguson, G. M., & Hall, J. C. (2000). Management strategies for attenuating herbicide resistance: Untoward consequences of their promotion.

Fu, X., Liu, S., Ru, J., Tang, B., Zhai, Y., Wang, Z., & Wang, L. (2022). Biological control of potato late blight by *Streptomyces sp.* FXP04 and potential role of secondary metabolites. *Biological Control, 169*. https://doi.org/10.1016/j.biocontrol.2022.104891.

Gaines, T. A., Duke, S. O., Morran, S., Rigon, C. A. G., Tranel, P. J., Küpper, A., & Dayan, F. E. (2020). Mechanisms of evolved herbicide resistance. *Journal of Biological Chemistry, 295*(30), 10307–10330.

Gong, Y., Chen, L. J., Pan, S. Y., Li, X. W., Xu, M. J., Zhang, C. M., Xing, K., & Qin, S. (2020). Antifungal potential evaluation and alleviation of salt stress in tomato seedlings by a halotolerant plant growth-promoting actinomycete *Streptomyces sp.* KLBMP5084. *Rhizosphere, 16*. https://doi.org/10.1016/j.rhisph.2020.100262.

Gopalakrishnan, S., Srinivas, V., & Prasanna, S. L. (2020). *Streptomyces*. In *Beneficial Microbes in Agro-Ecology: Bacteria and Fungi* (pp. 55–71). Elsevier.

Gorantla, J. N., Kumar, S. N., Nisha, G. V., Sumandu, A. S., Dileep, C., Sudaresan, A., Kumar, M. M. S., Lankalapalli, R. S., & Kumar, B. S. D. (2014). Purification and characterization of antifungal phenazines from a fluorescent *Pseudomonas* strain FPO4 against medically important fungi. *Journal de Mycologie Médicale, 24*(3), 185–192. https://doi.org/10.1016/j.mycmed.2014.02.003.

Grosset, J. H., & Singer, T. (2013). Streptomycin. In *Brenner's Encyclopedia of Genetics: Second Edition* (pp. 568–569). Elsevier Inc.

Gupta, H. P., Fatima, M. u., Pandey, R., & Ram, K. R. (2023). Adult exposure of atrazine alone or in combination with carbohydrate diet hastens the onset/progression of type 2 diabetes in *Drosophila*. *Life Sciences, 316*. https://doi.org/10.1016/j.lfs.2023.121370.

Gupta, P. K. (2018). Toxicity of herbicides. In *Veterinary Toxicology: Basic and Clinical Principles: Third Edition* (pp. 553–567). Elsevier.

Hassaan, M. A., & el Nemr, A. (2020). Pesticides pollution: Classifications, human health impact, extraction and treatment techniques. *Egyptian Journal of Aquatic Research, 46*(3), 207–220. https://doi.org/10.1016/j.ejar.2020.08.007.

Hulme, P. E. (2023). Weed resistance to different herbicide modes of action is driven by agricultural intensification. *Field Crops Research, 292*. https://doi.org/10.1016/j.fcr.2023.108819.

Jiao, Q., Mu, Y., Deng, J., Yao, X., Zhao, X., Liu, X., Li, X., Jiang, X., & Zhang, F. (2022). Direct toxicity of the herbicide florasulam against *Chlorella vulgaris*: An integrated physiological and metabolomic analysis. *Ecotoxicology and Environmental Safety, 246*(October), 114135. https://doi.org/10.1016/j.ecoenv.2022.114135.

Karuppiah, V., Aarthi, C., Sivakumar, K., & Kannan, L. (2013). Statistical optimization and anticancer activity of a red pigment isolated from *Streptomyces* sp. PM4. *Asian Pacific Journal of Tropical Biomedicine, 3*(8), 650–656. https://doi.org/10.1016/S2221-1691(13)60131-8.

Kevin Baer. Paraquat. Retrieved from: https://pubchem.ncbi.nlm.nih.gov/compound/Paraquat (accessed on 28 May 2024).

Khan, N., Yılmaz, S., Aksoy, S., Uzel, A., Tosun, Ç., Kirmizibayrak, P. B., & Bedir, E. (2019). Polyethers isolated from the marine actinobacterium *Streptomyces cacaoi* inhibit autophagy and induce apoptosis in cancer cells. *Chemico-Biological Interactions, 307*, 167–178. https://doi.org/10.1016/j.cbi.2019.04.035.

Kim, H. J., Bo, A. B., Kim, J. D., Kim, Y. S., Khaitov, B., Ko, Y. K., Cho, K. M., Jang, K. S., Park, K. W., & Choi, J. S. (2020a). Herbicidal characteristics and structural identification of the potential active compounds from *Streptomyces* sp. KRA17-580. *Journal of Agricultural and Food Chemistry, 68*(52), 15373–15380. https://doi.org/10.1021/acs.jafc.0c01974.

Kim, H. J., Bo, A. B., Kim, J. D., Kim, Y. S., Khaitov, B., Ko, Y. K., Cho, K. M., Jang, K. S., Park, K. W., & Choi, J. S. (2020b). Herbicidal characteristics and structural identification of the potential active compounds from *Streptomyces* sp. KRA17-580. *Journal of Agricultural and Food Chemistry, 68*(52), 15373–15380. https://doi.org/10.1021/acs.jafc.0c01974.

Kumar Ghosh, P., Sen, S. K., & Maiti, T. K. (2015). Production and metabolism of IAA by *Enterobacter* spp. (Gammaproteobacteria) isolated from root nodules of a legume *Abrus precatorius* L. *Biocatalysis and Agricultural Biotechnology, 4*(3), 296–303. https://doi.org/10.1016/j.bcab.2015.04.002.

Kumaran, S., Bharathi, S., Uttra, V., Thirunavukkarasu, R., Nainangu, P., Krishnan, V. G., Renuga, P. S., Wilson, A., & Balaraman, D. (2020). Bioactive metabolites produced from *Streptomyces enissocaesilis* SSASC10 against fish pathogens. *Biocatalysis and Agricultural Biotechnology, 29*. https://doi.org/10.1016/j.bcab.2020.101802.

Lajtai-Szabó, P., Hülber-Beyer, É., Nemestóthy, N., & Bélafi-Bakó, K. (2022). The role of physical support in secondary metabolite production by *Streptomyces* species. *Biochemical Engineering Journal, 185*(February), 0–3. https://doi.org/10.1016/j.bej.2022.108495.

Li, J., Gao, X., Li, M., & Fang, F. (2019). Resistance evolution and mechanisms to ALS-inhibiting herbicides in *Capsella bursa-pastoris* populations from China. *Pesticide Biochemistry and Physiology, 159*, 17–21. https://doi.org/10.1016/j.pestbp.2019.05.010.

Linz, G. M., & Homan, H. J. (2011). Use of glyphosate for managing invasive cattail (*Typha* spp.) to disperse blackbird (*Icteridae*) roosts. *Crop Protection, 30*(2), 98–104. https://doi.org/10.1016/j.cropro.2010.10.003.

Liu, L., Tu, H. C., Liu, F., Qiao, L. Q., Wang, X. X., Lin, J. M., Wang, M. L., Wu, Y. N., & Zhao, R. S. (2022). Hydroxyl group-enriched microporous organic network for high-performance solid-phase extraction of triazine herbicides: Experiment and DFT calculation on adsorption behavior. *Chemical Engineering Journal, 442*(P1), 136171. https://doi.org/10.1016/j.cej.2022.136171.

Liu, X., Yang, H., & Liu, Z. (2022). Signaling pathways involved in paraquat-induced pulmonary toxicity: Molecular mechanisms and potential therapeutic drugs. *International Immunopharmacology, 113*.

Loria, R., Bignell, D. R. D., Moll, S., Huguet-Tapia, J. C., Joshi, M. V., Johnson, E. G., Seipke, R. F., & Gibson, D. M. (2008). Thaxtomin biosynthesis: The path to plant pathogenicity in the genus *Streptomyces*. *Antonie van Leeuwenhoek, International Journal of General and Molecular Microbiology, 94*(1), 3–10. https://doi.org/10.1007/s10482-008-9240-4.

Luo, Q., Hu, H., Peng, H., Zhang, X., & Wang, W. (2015). Isolation and structural identification of two bioactive phenazines from *Streptomyces griseoluteus* P510. *Chinese Journal of Chemical Engineering, 23*(4), 699–703. https://doi.org/10.1016/j.cjche.2015.01.001.

Marimuthu, S., Karthic, C., Mostafa, A. A., Al-Enazi, N. M., Abdel-Raouf, N., & Sholkamy, E. N. (2020). Antifungal activity of *Streptomyces sp.* SLR03 against tea fungal plant pathogen *Pestalotiopsis theae*. *Journal of King Saud University – Science, 32*(8), 3258–3264. https://doi.org/10.1016/j.jksus.2020.08.027.

Martínez-Mendoza, E. K. (2012). Effects of *Bacillus subtilis* extracts on weed seed germination of *Sorghum halepense* and *Amaranthus hybridus*. *African Journal of Microbiology Research, 6*(9). https://doi.org/10.5897/ajmr11.554.

Matzrafi, M., Gadri, Y., Frenkel, E., Rubin, B., & Peleg, Z. (2014). Evolution of herbicide resistance mechanisms in grass weeds. *Plant Science, 229*, 43–52. https://doi.org/10.1016/j.plantsci.2014.08.013.

Mesnage, R., & Antoniou, M. N. (2017). Roundup ready! Glyphosate and the current controversy over the world's leading herbicide (Vols. 1–5). Elsevier Inc.

Ness-Cochinwala, M., Proaño, J. S., Bernstein, J. N., Martinez, P., Ladd, H., & Totapally, B. (2022). A case of a lethal diquat ingestion in a toddler. *Journal of Emergency Medicine, 62*(2), e16–e19. https://doi.org/10.1016/j.jemermed.2021.10.007.

Owens, D. K., Bajsa-Hirschel, J., Duke, S. O., Carbonari, C. A., Gomes, G. L. G. C., Asolkar, R., Boddy, L., & Dayan, F. E. (2020). The contribution of romidepsin to the herbicidal activity of *Burkholderia rinojensis* biopesticide. *Journal of Natural Products, 83*(4), 843–851. https://doi.org/10.1021/acs.jnatprod.9b00405.

Phukan, J., Deka, J., Kurmi, K., & Kalita, S. (2021). Deleterious rhizobacteria as a potential bioherbicide-A review. *International Journal of Agriculture & Environmental Science, 8*(2), 1–5. https://doi.org/10.14445/23942568/ijaes-v8i2p101.

Powles, S. B., Preston, C., Bryan, I. B., & Jutsum, A. R. (1997). Herbicide resistance: Impact and management.

Priya dharsini, P., Dhanasekaran, D., Gopinath, P. M., Ramanathan, K., Shanthi, V., Chandraleka, S., & Biswas, B. (2017). Spectroscopic identification and molecular modeling of diethyl 7-hydroxytrideca-2, 5, 8, 11-tetraenedioate: A herbicidal compound from *Streptomyces* sp. *Arabian Journal for Science and Engineering, 42*(6), 2217–2227. https://doi.org/10.1007/s13369-016-2401-2.

Priyadharsini, P., Dhanasekaran, D., & Kanimozhi, B. (2013). Isolation, structural identification and herbicidal activity of N-phenylpropanamide from *Streptomyces* sp. KA1-3. *Archives of Phytopathology and Plant Protection, 46*(3), 364–373. https://doi.org/10.1080/03235408.2012.758418.

Puertas de Freitas Silva, I., Carbonari, C. A., Velini, E. D., Silva Jr, J. F., Tropaldi, L., Gomes, G. L. G. C., & Autor responsable. (2016). Absorption velocity of glufosinate and its effects on weeds and cotton velocidad de absorción del glufosinato y sus efectos en malezas y algodón (Vol. 50).

Radhakrishnan, R., Alqarawi, A. A., & Abd Allah, E. F. (2018). Bioherbicides: Current knowledge on weed control mechanism. *Ecotoxicology and Environmental Safety, 158*(November 2017), 131–138. https://doi.org/10.1016/j.ecoenv.2018.04.018.

Richard, S., Moslemi, S., Sipahutar, H., Benachour, N., & Seralini, G. E. (2005). Differential effects of glyphosate and Roundup on human placental cells and aromatase. *Environmental Health Perspectives, 113*(6), 716–720. https://doi.org/10.1289/ehp.7728.

Rodriguez, Y. R. (2014). Acifluorfen, sodium salt. In *Encyclopedia of Toxicology: Third Edition* (pp. 59–62). Elsevier.

Salwan, R., & Sharma, V. (2020). Bioactive compounds of *Streptomyces*: Biosynthesis to applications. In *Studies in Natural Products Chemistry* (Vol. 64, pp. 467–491). Elsevier B.V.

Shi, L., Wu, Z., Zhang, Y., Zhang, Z., Fang, W., Wang, Y., Wan, Z., Wang, K., & Ke, S. (2020). Herbicidal secondary metabolites from actinomycetes: Structure diversity, modes of action, and their roles in the development of herbicides. *Journal of Agricultural and Food Chemistry, 68*(1), 17–32. https://doi.org/10.1021/acs.jafc.9b06126.

Sholkamy, E. N., Muthukrishnan, P., Abdel-Raouf, N., Nandhini, X., Ibraheem, I. B. M., & Mostafa, A. A. (2020). Antimicrobial and antinematicidal metabolites from *Streptomyces cuspidosporus* strain SA4 against selected pathogenic bacteria, fungi and nematode. *Saudi Journal of Biological Sciences, 27*(12), 3208–3220. https://doi.org/10.1016/j.sjbs.2020.08.043.

Singh, H., Naik, B., Kumar, V., & Bisht, G. S. (2018). Screening of endophytic actinomycetes for their herbicidal activity. *Annals of Agrarian Science, 16*(2), 101–107. https://doi.org/10.1016/j.aasci.2017.11.001.

Székács, A. (2021). Herbicide mode of action.

Umurzokov, M., Lee, Y. M., Kim, H. J., Cho, K. M., Kim, Y. S., Choi, J. S., & Park, K. W. (2022). Herbicidal characteristics and structural identification of a potential active compound produced by *Streptomyces* sp. KRA18-249. *Pesticide Biochemistry and Physiology, 187*(August), 105213. https://doi.org/10.1016/j.pestbp.2022.105213.

van Bruggen, A. H. C., He, M. M., Shin, K., Mai, V., Jeong, K. C., Finckh, M. R., & Morris, J. G. (2018). Environmental and health effects of the herbicide glyphosate. *Science of the Total Environment, 616–617*, 255–268. https://doi.org/10.1016/j.scitotenv.2017.10.309.

Verdugo-Navarrete, C., Maldonado-Mendoza, I. E., Castro-Martínez, C., Leyva-Madrigal, K. Y., & Martínez-Álvarez, J. C. (2021). Selection of rhizobacteria isolates with bioherbicide potential against Palmer amaranth (*Amarathus palmeri* S. Wats.). *Brazilian Journal of Microbiology, 52*(3), 1443–1450. https://doi.org/10.1007/s42770-021-00514-2.

Won, O. J., Kim, Y. T., Choi, J. S., Oh, T. K., Shinogi, Y., & Park, K. W. (2016). Herbicidal activity and mode of action of *Streptomyces scopuliridis* metabolites. *Journal of the Faculty of Agriculture, Kyushu University, 61*(1), 47–51. https://doi.org/10.5109/1564081.

Wonglom, P., Suwannarach, N., Lumyong, S., Ito, S., Matsui, K., & Sunpapao, A. (2019). *Streptomyces angustmyceticus* NR8-2 as a potential microorganism for the biological control of leaf spots of *Brassica rapa* subsp. *Pekinensis* caused by *Colletotrichum sp.* and *Curvularia lunata*. *Biological Control, 138*. https://doi.org/10.1016/j.biocontrol.2019.104046.

Xu, J., Smith, S., Smith, G., Wang, W., & Li, Y. (2019). Glyphosate contamination in grains and foods: An overview. *Food Control, 106*(March). https://doi.org/10.1016/j.foodcont.2019.106710.

Xu, Z., Ji, L., Tang, W., Guo, L., Gao, C., Chen, X., Liu, J., Hu, G., & Liu, L. (2022). Metabolic engineering of *Streptomyces* to enhance the synthesis of valuable natural products. *Engineering Microbiology, 2*(2), 100022. https://doi.org/10.1016/j.engmic.2022.100022.

Zhang, G., Zhang, Y., Zhang, Y., & Li, Y. (2013). Spectroscopic studies of cyanazine binding to calf thymus DNA with the use of ethidium bromide as a probe. *Sensors and Actuators, B: Chemical, 182*, 453–460. https://doi.org/10.1016/j.snb.2013.03.038.

Zhao, H., Tao, K., Zhu, J., Liu, S., Gao, H., & Zhou, X. (2015). Bioremediation potential of glyphosate-degrading *Pseudomonas spp.* strains isolated from contaminated soil. *Journal of General and Applied Microbiology, 61*(5), 165–170. https://doi.org/10.2323/jgam.61.165.

Zhou, C., Luo, X., Chen, N., Zhang, L., & Gao, J. (2020). C-P natural products as next-generation herbicides: Chemistry and biology of glufosinate. *ACS Applied Materials and Interfaces*. https://doi.org/10.1021/acs.jafc.0c00052.

Zhu, Y., Shang, X., Yang, L., Zheng, S., Liu, K., & Li, X. (2020). Purification, identification and properties of a new blue pigment produced from *Streptomyces* sp. A1013Y. *Food Chemistry, 308*. https://doi.org/10.1016/j.foodchem.2019.125600.

CHAPTER 9

Biological Control as an Environmental Sustainable Strategy for Application in Postharvest Fruit Packaging

LAIHSA V. RODRÍGUEZ-GONZÁLEZ,[1]
MÓNICA LIZETH CHÁVEZ-GONZÁLEZ,[1] NATHIELY RAMÍREZ-GUZMÁN,[2]
LILIANA FLORES-LÓPEZ,[3] HÉCTOR S. VILLADA-CARRILLO,[1] and
CRISTÓBAL N. AGUILAR[1]

[1]*Department of Food Research, Universidad Autónoma de Coahuila, Mexico*

[2]*Center for Interdisciplinary Studies and Research, Universidad Autónoma de Coahuila, Mexico*

[3]*Center for Scientific and Technological Research and Innovation, Universidad Autónoma de Coahuila, Mexico*

ABSTRACT

Plastics are the most widely used packaging materials in the food industry because they offer millions of advantages. However, in the last decades, the awareness to help the planet has awakened in the consumer and the demand for perishable food has increased, so partially or replacing non-biodegradable plastics is an everyday task. This bibliographic review describes the use of biotechnologies such as submerged fermentation and systems such as co-cultures, together with biopolymers, which are polymers obtained from organic materials of vegetable origin, that are better for packaging products. This is a potential alternative to extend the life of fruits since they are perishable foods very susceptible to damage by pathogens, which translates into a great economic loss for the producer.

Sustainable Agriculture and Global Environmental Health: Toward the Efficient Management of Resources. Mónica L. Chávez González, Cristóbal Noé Aguilar, Juan Carlos Contreras-Esquivel, and A. K. Haghi (Eds.)

9.1 INTRODUCTION

The demand for fruits and vegetables has been increasing in recent years because they provide high nutritional value and their consumption reduces the risk of cancer, cardiovascular diseases, asthma, and diabetes, among other ailments (Robinson et al., 2002). The consumption of vegetables is highly dependent on their physical attributes including appearance, color, and texture, but above all on ensuring microbial safety (Ruelas et al., 2013). Fresh fruits and vegetables are perishable foods with relatively short postharvest storage lives (Herrera-Cebreros et al., 2022). Due to this, the application of different technologies has been implemented to extend shelf life and facilitate transportation, such as the use of edible films and coatings (Moncayo Martínez, 2013; Fernández Valdés et al., 2015), low temperatures (refrigeration and freezing), controlled atmospheres, immersion in additives, vacuum packaging (Anaya-Esparza et al., 2020), ultraviolet radiation (Villarroel et al., 2014) among others. The purpose of these technologies is to minimize losses during food handling, transportation, and storage, without neglecting the use of packaging, since the main priority is the preservation and protection of products. In recent years, there has been growing interest in developing packaging materials that help reduce spoilage and extend the shelf life of food while reducing the environmental impact of conventional plastic-based packaging. The use of biodegradable materials as a total or partial replacement of plastics used for packaging has allowed the development of products with specific characteristics (barrier, thermal, and mechanical properties) in certain packaging such as films, protectors, wrappings, foams, cups, trays, bags, etc. (Villada et al., 2007). The joint use of biodegradable packaging and extracts with antimicrobial activity constitutes an alternative to counteract pathogenic microorganisms that affect food in post-harvest stages, among which fungi stand out (Suárez-Quiroz et al., 2013). Biotechnological processes such as fermentations with the use of co-cultures are highly efficient tools for the production of antimicrobial metabolites with potential use for biological control (De la Cruz-Quiroz et al., 2019), which can be used to extend the pre- and postharvest life of fruits and vegetables by reducing their losses through their incorporation in the formulation of biodegradable packaging.

9.2 BIOLOGICAL CONTROL

The increasing demand for agricultural products has prompted the search for non-synthetic and environmentally friendly alternatives to minimize losses caused by pathogens (Mhatre et al., 2019). These mainly rely on effective and sustainable strategies to control crop diseases (Sabri et al., 2021). Thus, biological control is presented as a practical solution with ecological approaches to these problems.

Biological control or biocontrol is defined as the use of different organisms or compounds obtained from them, which alone or in combination have the ability to reduce the harmful effects caused by a population on the growth and/or productivity of a crop (Vinchira & Moreno, 2019). Stenberg et al. (2021) define biological control as the use of living agents including viruses to combat undesirable organisms (pests, fungi, bacteria, weeds, etc.) to provide benefits to humans.

In biological control, various organisms are used, including fungi, bacteria, viruses, and parasitoids (Viera & Jackson, 2018) which have a great capacity to generate multiple interactions between themselves and their environment and are used to contribute greatly to the effectiveness of biological control (Rolim et al., 2020). This type of sustainable and economically important strategy seeks to reestablish the disturbed ecological balance through the use of living organisms or their metabolites in order to reduce or eliminate the damage caused by pathogenic organisms (Guédez et al., 2008).

9.2.1 TYPES OF BIOLOGICAL CONTROL

Biological control mimics the ecosystemic harmony evolved by natural enemies controlling pests and pathogens in the absence of human intervention (Stenberg et al., 2021). The three types of biological control can be used individually or combined as shown in Figure 9.1. These are classified as augmentative or inundative, which requires an initial or repeated release of natural enemies, classical, in which exotic natural enemies are introduced for the control and management of invasive pests (Romeis et al., 2019), and conservative, which makes use of the resident natural enemies of the pests present through habitat management.

9.3 BIOLOGICAL CONTROL AGENTS (BCA)

The use of organisms for pathogen control is an alternative method that can reduce costs, decrease dependence on synthetic products, and minimize environmental impact. Biopesticides can be more selective than synthetic pesticides and only affect the desired species through various mechanisms, reducing the likelihood of inducing resistance in the pathogenic species (Omer et al., 2010).

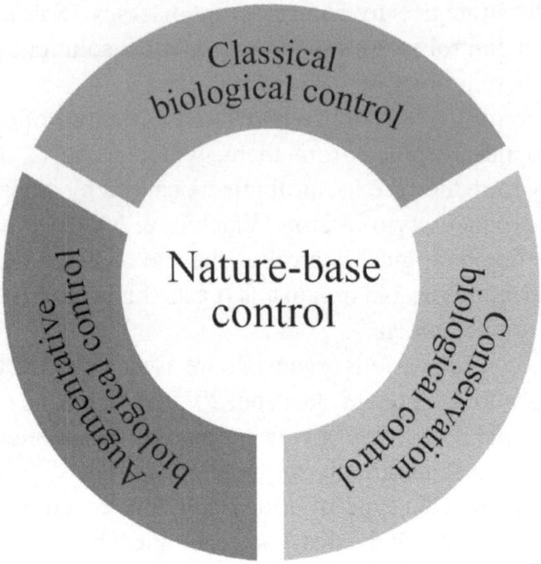

FIGURE 9.1 Types of biological control.

Biological control agents (BCAs) are utilized in crops to manage or eradicate pathogens responsible for plant diseases by employing a variety of action mechanisms. Some of the mechanisms used by BCAs include inducing a systemic immune response in plants, competing for nutrients and space, hyper-parasitism, and antibiosis (Köhl et al., 2019).

The continuous research and development of BCAs in crop management provide benefits and enhance the utilization of their diverse characteristics, such as reducing causal agents, safeguarding crops, preventing soil, water, and plant contamination, and eliminating waste management issues, among others (Tariq et al., 2020).

9.3.1 TYPES OF BIOLOGICAL CONTROL AGENTS

In most cases, the organism applied as a biological control agent belongs to a different taxonomic group than the target pest (King et al., 2011). A BCA must essentially have sufficient levels of antagonistic activity against a wide variety of pathogens or specific pathogens causing crop problems. There are many examples where biological control is achieved by a wide range of microorganisms, the main ones being bacteria and fungi of different genera (Jung et al., 2018) as shown in (Table 9.1).

TABLE 9.1 Examples and Applications of Biological Control Agents

Type	Gender or Species	Application	References
Bacteria	*Streptomyces atratus* *Streptomyces tsukiyonensis* *Streptomyces* sp.	Biocontrol against wood decay fungus *Gloeophyllum trabeum*	Jung et al. (2018)
Bacteria	*Bacillus firmus*	Biological control agent of *Phytophthora capsici* in tomato.	Lagunas-Lagunas et al. (2001)
Bacteria	*Pasteuria thornei*	Biological control in maize (seeds) against *Pratylenchus zeae.*	Confort & Inomoto (2023)
Bacteria	*Bacillus* sp.	Mitigation of fish pathogens in aquaculture.	Kuebutornye et al. (2020)
Bacteria Fungus	*Bacillus megaterium* *Trichoderma viride*	Biological control of root rot/rot in broad bean plants.	Abdel-Monaim (2013)
Fungus	*Trichoderma harzianum*	Biofungicide for the biocontrol of *Phytophthora capsici* in bell pepper plants.	Ezziyyani & Sánchez (2004)
Fungus	*Trichoderma harzianum*	Biological control of wheat weevil (*Sitophilus granarius*).	Andrade-Bustamante et al. (2022)
Fungus	*Trichoderma harzianum*	Biological control of avocado (*Neofusicoccum parvum*, *Colletotrichum gloeosporioides*, *Diaporthe* sp. *Phomopsis perseae*)	López-lópez et al. (2022)
Fungus Nematode	*Trichoderma harzianum* *Steinernema feltiae* *Fitonematocida Nemarioc-AL*	Suppression of agglutinating nematodes *Meloidogyne* spp.	Madaure et al. (2019)
Arachnids	*Crytophora citricola*	Biological control in tomato crops against tomato leafminer *Tuta absoluta.*	Roberts-McEwen et al. (2022)

9.3.2 TRICHODERMA

Trichoderma is a BCA and plant growth promoter. Its applications in agriculture range from biopesticides, biofertilizers, yield promoters, nutrient solubilizers, and organic matter decomposers (Eduardo et al., 2023). Trichoderma is a cosmopolitan fungus, and frequently its species are dominant components of the soil microbiota in very diverse habitats (López et al., 2020). This genus is of great biotechnological and environmental interest due to the large amount of metabolites it produces, among which enzymes, plant growth-promoting compounds, and volatile compounds can be mentioned (Hernández Melchor et al., 2019).

Trichoderma species grow rapidly and produce abundant conidia, as well as a wide range of enzymes (Ramírez-Guzman, 2016). The colonies of this biocontrol fungus are very easy to recognize by their characteristic green color. First, it presents a white mycelium, and days later it takes on a green color due to the appearance of reproductive structures (Sanabria, 2020).

The versatility, adaptability, and easy manipulation of this genus have allowed its study and use as BCA due to its mechanisms of action against pathogens such as parasitism, antibiosis, paralysis, and the production of lytic enzymes. They can promote greater absorption of nutrients and water to the plant, modify the morphology of the root, and make interactions with the rhizosphere easier (Poveda et al., 2020). They are responsible for the promotion of plant growth, the elimination and/or control of plant pathogens, and the induction of defense mechanisms, making them promising biological control agents (Asad, 2022).

9.3.3 METABOLITES

9.3.3.1 PRIMARY METABOLITES

Primary metabolites are ubiquitous and essential biomolecules involved as plant growth factors or as lytic enzymes, siderophores, antibiotics, carbon and nitrogen permeases, and their concentrations in cells are controlled (Mesa Vanegas et al., 2020).

9.3.3.2 SECONDARY METABOLITES

Microbial secondary metabolites are low molecular weight substances produced during the late phase of the growth cycle of microorganisms. Because they are not essential for microbial growth, these metabolites can be used as a source for the development of biopesticides, biostimulants, and drugs (Li et al., 2021). They are broadly divisible into several characteristic groupings such as polyketides, terpenes, terpenoids, pyrogens, phenols, and alkaloids derived from indole compounds and non-ribosomal peptides, which give their origin and biosynthesis (Mesa Vanegas et al., 2020).

9.3.3.3 SECONDARY METABOLITES OF TRICHODERMA

As mentioned above, *Trichoderma* has the ability to synthesize and release hydrolytic enzymes (cellulases, glucanases, xylanases, proteases, and laminarinases) associated with the degradation of elemental structures that make up the cell walls of phytopathogens, as well as a wide range of volatile and non-volatile metabolites that act as fungicides (Cubilla-Ríos et al., 2019; Hernández Melchor et al., 2019) (Table 9.2). The expression of secondary metabolites depends on the genetic and environmental characteristics to which the strains are subjected (Mesa Vanegas et al., 2020).

One of the most important and best-documented antifungal compounds of this genus is 6-n-pentyl-2H-pyran-2-one (6-PP). The metabolite is responsible for its characteristic coconut odor and plays an important role in the regulation of plant growth and activation of plant defense responses (El-Benawy et al., 2020).

TABLE 9.2 Secondary Metabolites of the Gender Trichoderma

Metabolite	Gender or Species	Application	References
Viridin	*Trichoderma* spp.	Nematicide	Tariq et al. (2020)
Acetic acid	*T. longibrachiatum*		Tariq et al. (2020)
Gliotoxin	*T. virens*		Khan et al. (2022).
Cyclosporine peptide	*T. polysporum*		Tariq et al. (2020)
Trichosetin peptaibiotics	*T. viride*	Antimicrobial and anticancer drugs	Awad et al. (2018)
Furfural and 2-furamethanol	*T. atroviride*	Natural colorant	El-Benawy et al. (2020).
Glacial acetic acid	*T. harzianum*	Antifungal	Meena et al. (2017)
17-Octadecinoic acid	*T. viride*		
6-pentyl-α-pyrone	*T. harzianum*	Antifungal	Yassin et al. (2022)
Cyclooctanol	*T. viride*		
1,2-benzene dicarboxylic acid, 2-butoxy-2-oxoethyl butylic acid	*T. asperellum*	Pesticide activity	Bhardwaj & Kumar (2017)
Viridine and glovirine	*T. viride*	Conidia germination and mycelial growth.	Pakora et al. (2018)

9.4 CO-CULTURES

Microorganisms do not exist in isolation in the environment; instead, they form complex communities among themselves (Tan et al., 2015). In nature, microorganisms live in communities where multiple species interact. These interactions are lost when establishing a pure culture, increasing the metabolic load but limiting the metabolic potential of the isolated microorganism (Diender et al., 2021).

Systems consisting of the cultivation of two or more microorganisms in the same space are called microbial communities or consortiums, and mixed cultures or co-cultures. These types of systems allow for obtaining metabolites that in monoculture have no or low production (Canon et al., 2020). Mhatre et al. (2019) define at a basic level that a co-culture is a configuration in a specific medium where two or more different populations of microorganisms are cultured with some degree of contact between them. There are three types of co-cultures: co-culture of different fungi, co-culture of fungi with bacteria, and co-cultures of different bacteria (Peng et al., 2021). In a co-culture, substances get broken down because of the mix of 2 or more strains in the same substance (Rodriguez Carreón, 2016). It has a big effect on how the strains work, can produce more substances, initiate new processes, and alter how the organisms' function (Divya Arora et al., 2020). When you conduct co-cultures, you need to consider various factors such as the number of organisms, their proximity, their diet, consumption rate, population size, and duration (Martinez, 2019).

Co-cultures are beneficial and are utilized in the production of vitamins, acids, and other substances. They promote faster growth of organisms, increased production, rapid changes, enhanced yield, and decomposition of substances (Canon et al., 2020) (Figure 9.2).

9.4.1 APPLICATION

The use of co-cultures has been reported since the 20th century as an efficient strategy in which complex substrates are used for the production of high-value-added compounds, taking advantage of the activation of hidden bioactive pathways that cannot be obtained in monoculture (Rosero-Chasoy et al., 2021).

Microbial communities are found in all ecosystems of the Earth, which makes it possible to have access to a large number of metabolites of interest through the use of multiple species within the same culture. These systems have been considered a promising platform for the production of various metabolites (Martinez et al., 2022). For example, we can find them as a key point in the human intestine for the digestion of food. They are also important in various environmental processes such as the decomposition of organic matter and are essential producer organisms for biotechnological applications (Burmeister & Grünberger, 2020). The most important applications using microbial co-cultures are discussed in subsections.

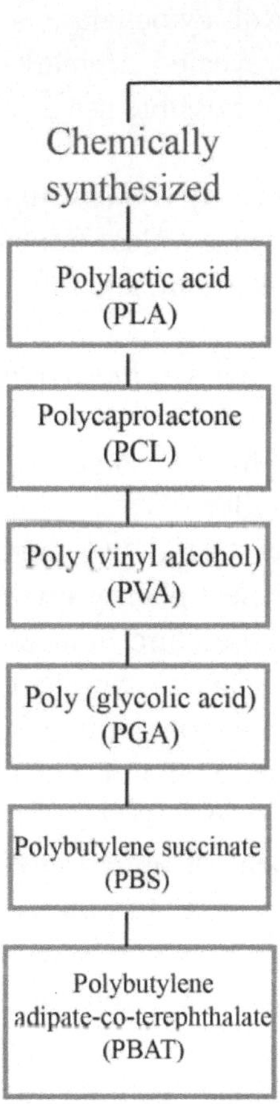

FIGURE 9.2 Types and areas of application of co-cultures.

9.4.1.1 BIOLOGICAL CONTROL

The use of fungi in co-cultures presents a great advantage, thanks to the immense diversity of the fungal kingdom, and since these microorganisms generally coexist symbiotically on substrates in nature, the use of co-cultures in fermentations is beneficial for the production of substances of biotechnological interest and applications such as biological control (Lodha et al., 2020).

Aspergillus is a genus capable of synthesizing a chemical diversity of compounds and using co-cultures of different strains, and even with other microorganisms including bacteria, plants, and fungi, the co-cultures of this genus produced or increased compounds such as taxol, cytochalasans, notamides, pentapeptides, etc. Most of these co-cultures revealed a remarkable improvement in their antimicrobial behavior and their possible use as a biocontrol agent (Alanzi et al., 2023).

Senthilraja et al. (2010) reported that co-culture formed by *B. bassiana* and *P. fluorescens* is able to significantly reduce the incidence of leaf miner disease and peanut rot when talc base is applied through seed, soil, and foliar application.

Mitsuboshi et al. (2022) investigated suppression against *Fusarium oxysporum*-borne diseases in spinach using a co-culture consisting of *F. oxysporum f.* sp. *spinaciae* – soil microbiota, demonstrating that the co-culture obtained higher growth rates in the incidence of spinach wilt.

The co-culture formed by *Trichoderma asperellum* GDFS1009 and Bacillus amyloliquefaciens 1841 showed an impact on wheat plant growth and biocontrol potential, observing that the use of the co-culture induced the expression of genes of *T. asperellum* related to secondary metabolism, myco-parasitism, antioxidants, and plant growth. Likewise, competition in the system induced compounds not detected in monoculture, and provided protection against *Fusarium graminearum* (Karuppiah et al., 2019). This same co-cultivation can be used for the proper treatment of agricultural biomass residues, obtaining benefits such as enrichment of lignocellulose biodegradation, corn plant growth, defense potential, and disease resistance (*Fusarium verticillioides* and *Cochliobolus heterostrophus*) (Karuppiah et al., 2021).

Co-culture of microbial strains can activate silent biosynthetic gene clusters. Shakeri Moghaddam et al. (2021) reported an increase in antibacterial activity using a co-culture composed of *Streptomyces* sp. and the pathogen *Staphylococcus aureus* inactivated by heat, which was efficient against Bacillus subtilis and *S. aureus*, obtaining the activation of silent genes and obtaining secondary metabolites hidden in the genome of the pathogen (anthraquinone derivatives).

9.4.1.2 PHARMACEUTICAL PRODUCTS

Co-cultures are an option for drug generation and discovery because these systems are capable of inducing the production of a wide range of metabolites (polyketides, oxylipins, peptides, and alkaloids) that exhibit physicochemical properties targeted to treat specific diseases (Arora et al., 2020).

9.4.1.3 SOIL AND WATER BIOREMEDIATION

The use of microbial communities or co-cultures isolated from the contaminated environment for soil and water bioremediation is more efficient for the degradation of harmful compounds due to unique interactions, production of degradative enzymes, production of metabolites, and transport of bacteria through fungal mycelia (Angeles-de Paz et al., 2023).

Co-cultures can be used for the elimination of toxic compounds present in the environment, to create ideal conditions that make the processes more favorable for other microorganisms. An example of this is the growth of strict aerobes in co-culture with aerobes, making use of the high energy yields of aerobic metabolism while simultaneously taking advantage of the fermentative aspects of strict anaerobes, as in the case of a co-culture formed by the aerobic fungus Trichoderma reesei and lactic acid bacteria used to convert lignocellulose into different end products such as propionate, butyrate, caproate (Diender et al., 2021).

Several recent studies have shown that the use of co-culture systems between microalgae and bacteria increases the productivity of microalgal biomass and greater efficiency in wastewater treatment (Arroyo Cevallos, 2023), for example, the microalgae-bacteria consortium (two microalgae Chlorella sorokiniana DBWC2 and Chlorella sp. DBWC7 and two bacteria Klebsiella pneumoniae ORWB1 and Acinetobacter calcoaceticue ORWB3) showed an improvement in the treatment of wastewater from raw dairy products, additionally this co-culture produced higher intracellular levels of lipids, carbohydrates , proteins and color pigments, in addition to increasing biomass production (Tong et al., 2023).

9.4.1.4 FOODS

The co-cultivation system within the food area can be used to ferment mixtures of plant and animal resources, allowing the diversification of the products obtained. Such is the case of lactic acid bacteria (LAB) commonly used as starters in multiple fermented foods. As an example of this, Canon et al. (2020) reported the application of LAB co-cultures for carbohydrate degradation and protein hydrolysis of milk and lupine during the production of food made by fermenting bovine milk and lupine flour (Canon et al., 2020).

Smoke-cured tobacco presents an irritating and undesirable flavor, and to eliminate this flavor, it must be aged. However, it usually takes a long time. The co-culture of *B. amyloliquefaciens-B. kochii* in fermentation showed an improvement in quality and higher functional enzyme activity, interpreted in the presentation of the Maillard reaction. The terpenoid metabolites present promoted an increase in aroma, smoothness, and decreased irritation, thus shortening the curing time or making this step null (Wu et al., 2021).

9.4.1.5 BIOPOLYMERS

Jin et al. (2022) reported obtaining bacterial cellulose (BC) by using a microbial co-culture consisting of Gram-negative (*Acetobacter, Salmonella*) and Gram-positive (Sarcina) bacteria, in which the microorganism in charge of microbial cellulase production manufactures the networks while the other species secretes components (metabolites) into the matrix. Because BC lacks antimicrobial, antioxidant, electrical conductivity, magnetic, and optical properties, this difficulty is solved by adding reinforcing materials (synthetic polymers or natural polymers) to improve its performance, making the biopolymer more stable and easier to use.

9.4.1.6 METABOLIC ENGINEERING

In metabolic engineering, microbial production of a wide range of chemicals has been demonstrated, which are obtained from the biosynthesis of complex natural products using cheap, available, and renewable carbon sources (Jones & Wang, 2018).

In biotechnology, a system that can convert a cheap and available energy source/substrate (CO_2, complex organic waste, light, etc.) into metabolites that can subsequently be converted into higher-value products by sequential processes is preferred. An example of this is the co-culture of the genetically modified cyanobacterium *Synechococcus elongatus* in conjunction with associated bacteria to obtain polyhydroxybutyrate or α-amylase (Diender et al., 2021).

Low-cost biofuel production employing co-cultures is reported in a recent case where two *E. coli* strains were assembled that cooperate in the transformation of xylan into ethanol, where one strain secretes two hemicelluloses while the other uses the released sugars to produce ethanol (Shong et al., 2012).

9.5 BIOTECHNOLOGICAL PROCESSES FOR OBTAINING EXTRACTS

Biotechnology is a multidisciplinary branch that is defined as the use of living organisms or products obtained for the benefit of humans and their environment, in order to solve a problem or develop a product (Thieman & Palladino, 2010). Le Borgne & Quintero (2003) describes biotechnology as the use of living organisms or parts thereof to manufacture or modify products, plants, animals, or develop microorganisms for specific uses as well as their enzymes. It is also defined as any technological application that uses biological systems and living organisms or their derivatives to create or modify products or processes for specific uses (Saval et al., 2012).

Due to these unique characteristics of biotechnology, it is considered an alternative for obtaining extracts that could be used in the biological control of pathogens in pre- and post-harvest of perishable foods.

Microbial fermentation is a biotechnological process through which various products (metabolites, enzymes, peptides, etc.) can be obtained. The microorganisms generally used are bacteria, fungi, or yeasts, which can be present in the substrate in an autochthonous manner or added as a starter culture (Cruz-Casas et al., 2021).

In the fermentation process, submerged fermentation and solid-state fermentation are the most commonly used. Submerged fermentation consists of a culture of microbial cells in a liquid medium, consisting essentially of a series of nutrients dissolved in water, under controlled conditions. In this type of fermentation, it is possible to control parameters of importance in microbial processes such as the composition of the culture medium, inoculum, aeration, agitation, pH, oxygen, and temperature (Martinez Medina, 2018). The bioactive compounds of interest are secreted in the broth, and it is necessary to replace the substrates or supplement them constantly because they are used quite quickly, presenting this technique as a great advantage when purifying the products obtained (Ravichandran, 2017). On the other hand, solid-state fermentation consists of a microbiological process that commonly occurs on the surface of solid materials that have the property of absorbing and containing water, with or without soluble nutrients, and occurs in an aseptic and natural state (Julián Ricardo & Sánchez Ramos, 2007; Robinson & Singh, 2002).

This type of fermentation has emerged as a potential technology for obtaining microbiological products, which are very useful in the food, chemical, and pharmaceutical industries (Chávez González et al., 2009).

Biotechnological processes, such as both solid and submerged fermentations, can be used to improve and obtain the production of compounds in extracts through the breaking of bonds and the cell wall of the substrate, increasing the concentration of compounds and consequently their functional properties, and diversifying their applications (Santos da Silveira et al., 2019). Biological activities such as antioxidant, antimicrobial, antiproliferative, and anti-inflammatory activity, to mention a few, can be obtained by biotechnological processes such as fermentation and can be improved using co-cultures (Villarreal-Soto et al., 2019).

9.6 THE IMPORTANCE OF BIODEGRADABLE POLYMERS IN THE ENVIRONMENT AND AGRO-INDUSTRY

A polymer is a large molecule (macromolecule) composed of repeated structural units; these subunits are usually linked by covalent chemical bonds (Kulkarni et al., 2012). In general, polymers can be classified according to their origin into two large groups: natural polymers, which come from natural resources, and synthetic polymers, obtained from fossil fuels such as petroleum (Sánchez & Muñoz, 2021).

Based on this classification, new strategies are focused on the production of natural polymers obtained from renewable sources, as these can be a viable alternative to the problem of conventional plastics, driving the development of sustainable, environmentally friendly, and economically attractive technologies (Khan, 2017).

Natural biopolymers come from four major sources:

- Animal origin (collagen/gelatin) 2;
- Marine origin (chitin/chitosan);
- Agricultural origin (lipids and fats and hydrocolloids: polysaccharides and proteins);
- Microbial origin (polylactic acid (PLA) and polyhydroxyalkanoates (PHA) (Villada et al., 2007).

Biopolymers can play a key role and are available in nature; each biopolymer has different types of unique characteristics, chain arrangements, and physicochemical properties that make them potential raw materials in the fields of tissue engineering, pharmaceuticals, cosmetics, food packaging, bioadsorbents, energy, and environmental sciences (Shekh et al., 2023). Likewise, biodegradable biopolymers are generally classified according to their source as shown in Figure 9.3: chemically synthesized polymers derived from chemicals (based on fossil or natural materials), polymers derived from biomass, and biosynthesized polymers obtained from microbial fermentation (Babaremu et al., 2023).

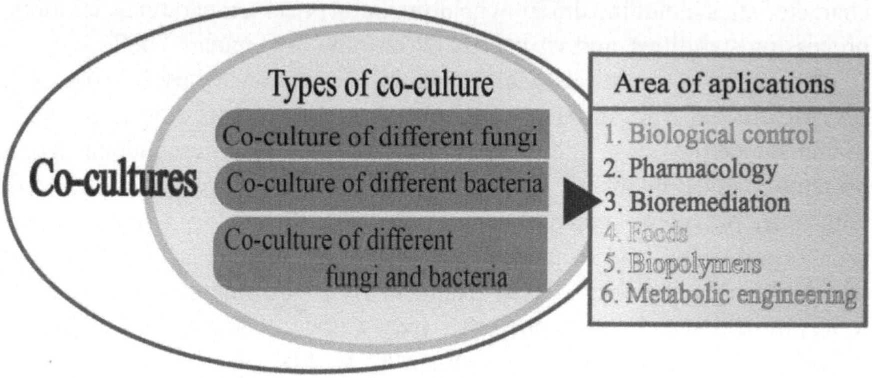

FIGURE 9.3 Types of polymers.

9.6.1 *BIOPOLYMERS AS PACKAGING*

Preservation and quality assurance of foods, especially perishable foods, are necessary to meet the demand for food today; food protection is important at every step, including pre-harvest, post-harvest, transportation, storage, and market shelves for longer shelf life (Ullah et al., 2022). Bringing with it one of the trends with greater influence on this type of food is the growth of environmental awareness, which generates expectations and demands towards companies making their products and processes not negatively impact the environment. From this need arises biodegradable packaging as a response to consumer demand (Rivera et al., 2019).

In 2020, 2.1 million tons of bioplastic were produced worldwide and 47% was destined for packaging processing (Babaremu et al., 2023).

The scientific community is currently searching for new biodegradable compositions to replace traditional materials such as plastics from oil and gas. Following this need, there are natural sources such as starch, cellulose, and chitin due to their availability and low cost of raw materials (Kunarbekova et al., 2023). Chitosan is a linear macromolecular polysaccharide derived from the abundant chitin found in the shells of crabs and shrimp and is used due to its natural availability, biodegradability, film-forming capacity, and antimicrobial activity (Wang et al., 2020). Starch is a polysaccharide and the main sources of its production are potatoes, rice, wheat, and corn (Kunarbekova et al., 2023).

Rice starch and its main components (amylose and amylopectin) are attractive raw materials for use as biopolymers due to their low cost, renewability, and especially due to the wide variety of rice. Rice starch has different characteristics including different gelatinization onset temperatures, textures, processing stabilities, and viscosities (Bourtoom & Chinnan, 2008).

Among the raw materials available for the elaboration of these biopolymers, starch extracted from different renewable sources presents advantages such as abundance, low cost, and non-toxicity. However, when used alone, starch presents many limitations, which can be avoided by developing a mixture with other polymers, preferably from renewable and biodegradable sources such as polylactic acid (Carneiro da Silva et al., 2023). PLA is considered one of the most promising and environmentally friendly natural polymers due to its physical and chemical properties (renewable, biodegradability, biocompatibility, GRAS) (Gumienna & Górna, 2021). Likewise, natural fibers are candidates to be used as a complement and/or reinforcement for the elaboration of biopolymers, low cost, biodegradability, and availability due to their low utilization or even they are often considered waste (Alharbi et al., 2020). The incorporation of nanomaterials and natural plant extracts in biopolymer formulations are strategy used to improve physicochemical functional properties, such as ultraviolet light, gas barrier, and antioxidant, and antibacterial properties to preserve food quality and prolong shelf life (Ezati et al., 2023).

Properties such as water adsorption, mechanical resistance, water vapor permeability, and gas permeability are of great importance since they determine the functional behavior of the packaging obtained, but they can also be affected not only by the origin of the raw materials used but also by the processing (temperature and compression pressure) and storage (temperature and relative humidity) conditions (Navia & Villada, 2013).

Antibacterial, antifungal, and antioxidant activities can be induced by the main polymer used for packaging or by the addition of various components such as natural agents (bacteriocins, essential oils, natural extracts, etc.) to synthetic agents both organic and inorganic (Motelica et al., 2020). Packaging modified with antimicrobial agents that can be obtained through biotechnological processes such as submerged fermentation is a promising and rapidly growing technology that can provide safe, high-quality packaged foods and prolong their shelf life, mainly the growth of pathogenic microorganisms such as fungi, and can be used to prevent the growth of pathogenic microorganisms such as fungi (Perumal et al., 2021).

9.7 POST-HARVEST PACKAGING

All fresh horticultural crops are metabolically active organisms, even from harvest until they are processed or consumed; the most frequent approach to preserve fruit quality and reduce postharvest product loss is to delay ripening and senescence (Ziv & Fallik, 2021).

Postharvest losses are described as accidental losses resulting from events that occur from crop to table, such as physical damage, internal bruising, premature spoilage, damage by microorganisms, insects, etc. (Shipman et al., 2021). Derived from this problem, it is estimated that 7% of postharvest losses of fruits and vegetables occur in crops, while 17% and 18% are wasted in consumer-oriented businesses and households (ReFED, 2022).

Approximately 33% of the produce that is harvested is never consumed because these products naturally have a short shelf life, leading to postharvest losses and waste. Worldwide estimates of the amount of postharvest losses of fruits and vegetables range from 37% to 55% (Opara et al., 2021).

Postharvest losses of fruits and vegetables are difficult to predict, as they can occur at any step of the chain, mainly impacting economically. The main spoilage-producing agents are those attributed to fungal pathogens such as those of the genera Penicillium, Botrytis, Monilinia, Rhizopus, Alternatia, Fusarium, Aspergillus, Mucor, Colletotrichum, or phytosociological disorders, especially cold damage (Mari et al., 2018).

Packaging is the most important barrier against contamination by microorganisms. Its main advantages are protection against physical damage, maintaining food safety, facilitating product presentation, including product identification, and prolonging the shelf life of the product (Porat et al., 2018).

9.7.1 POST-HARVEST USE OF BIOPOLYMERS

Biopolymers maintain fruit quality by reducing desiccation that occurs during storage, and regulating the transfer of water vapor, oxygen, and carbon dioxide into or out of the product; when used properly, biodegradable polymers can bring advantages such as delaying fruit ripening, suppressing the breakdown of essential fruit compounds, reducing weight loss, improving fruit appearance, and prolonging shelf life (Munhuweyi et al., 2020). In particular, the addition of natural antimicrobial agents and antioxidants to bioplastic materials provides a more sustainable alternative to conventional synthetic food packaging (Bof et al., 2021).

Polylactic acid (PLA) and starch with compounds that are used in the manufacture of packaging to replace petroleum derivatives as biodegradable and environmentally friendly materials (del Rosario Salazar-Sánchez et al., 2019). TPS/MA/PLA is a blend of thermoplastic starch (TPS) and polylactic acid (PLA) compatibilized with maleic anhydride that has the potential to be a substitute for petroleum-based plastics (Camacho-Muñoz et al., 2020).

9.8 CONCLUSION

The use of biopolymers is an environmentally friendly strategy, and through the use of extracts with antimicrobial activity, they present an enormous potential against postharvest losses. However, there is not much reported in the literature on the joint use of these technologies, especially in conjunction with co-cultures, which makes it a new field to be explored and an alternative to reduce the consumption of petroleum-based plastic in the packaging area. Last but not least, it should be noted that every measure taken to reduce postharvest losses has a direct impact on the nutritional and sensory quality of fresh produce. All pre- and post-harvest practices directly affect product quality and should be analyzed from different points of view (physiology, pathology, chemistry, and nutrients).

KEYWORDS

- **biodegradable packaging**
- **biological control**
- **biopolymers**
- **co-cultures**
- **postharvest**
- **postharvest losses**
- **sensory quality**
- **Trichoderma**

REFERENCES

Abdel-Monaim, M. F. (2013). Improvement of biocontrol of damping-off and root rot/wilt of faba bean by salicylicacid and hydrogen peroxide. *Mycobiology, 41*(1), 47–55. https://doi.org/10.5941/MYCO.2013.41.1.47.

Alanzi, A., Elhawary, E. A., Ashour, M. L., & Moussa, A. Y. (2023). Aspergillus co-cultures: A recent insight into their secondary metabolites and microbial interactions. *Archives of Pharmacal Research*. https://doi.org/10.1007/s12272-023-01442-5.

Alharbi, M. A. H., Hirai, S., Tuan, H. A., Akioka, S., & Shoji, W. (2020). Effects of chemical composition, mild alkaline pretreatment and particle size on mechanical, thermal, and structural properties of binderless lignocellulosic biopolymers prepared by hot-pressing raw microfibrillated Phoenix dactylifera and Cocos nucifera. *Polymer Testing, 84*, 106384. https://doi.org/10.1016/j.polymertesting.2020.106384.

Anaya-Esparza, L. M., Pérez-Larios, A., Ruvalcaba-Gómez, J. M., Sánchez-Burgos, J. A., Romero-Toledo, R., & Montalvo-González, E. (2020). Funcionalización de los recubrimientos a base de quitosano para la conservación postcosecha de frutas y hortalizas. *TIP Revista Especializada En Ciencias Químico-Biológicas, 23*, 1–14. https://doi.org/10.22201/fesz.23958723e.2020.0.241.

Andrade-Bustamante, G., Suárez Hernández, A. M., Aispuro-Hernández, E., & Martínez-Ruiz, F. E. (2022). Trichoderma harzianum y espinosina en el control de gorgojo del trigo *Sitophilus granarius* (L. 1758). *Biotecnia, 25*(1), 94–99. https://doi.org/10.18633/biotecnia.v25i1.1819.

Angeles-de Paz, G., Ledezma-Villanueva, A., Robledo-Mahón, T., Pozo, C., Calvo, C., Aranda, E., & Purswani, J. (2023). Assembled mixed co-cultures for emerging pollutant removal using native microorganisms from sewage sludge. *Chemosphere, 313*, 137472. https://doi.org/10.1016/j.chemosphere.2022.137472.

Arora, D., Gupta, P., Jaglan, S., Roullier, C., Grovel, O., & Bertrand, S. (2020). Expanding the chemical diversity through microorganisms co-culture: Current status and outlook. *Biotechnology Advances, 40*(January), 107521. https://doi.org/10.1016/j.biotechadv.2020.107521.

Arroyo Cevallos, S. A. (2023). Obtención de lípidos a partir de biomasa de consorcio microalgal cultivado en agua residual de la industria cervecera.

Asad, S. A. (2022). Mechanisms of action and biocontrol potential of *Trichoderma* against fungal plant diseases – A review. *Ecological Complexity, 49*. https://doi.org/https://doi.org/10.1016/j.ecocom.2021.100978.

Awad, N. E., Kassem, H. A., Hamed, M. A., El-Feky, A. M., Elnaggar, M. A. A., Mahmoud, K., & Ali, M. A. (2018). Isolation and characterization of the bioactive metabolites from the soil derived fungus *Trichoderma viride*. *Mycology, 9*(1), 70–80. https://doi.org/10.1080/21501203.2017.1423126.

Babaremu, K., Oladijo, O. P., & Akinlabi, E. (2023). Biopolymers: A suitable REPLACEMENT for plastics in product packaging. *Advanced Industrial and Engineering Polymer Research, 6, 333–340*. https://doi.org/10.1016/j.aiepr.2023.01.001.

Bhardwaj, N. R., & Kumar, J. (2017). Characterization of volatile secondary metabolites from *Trichoderma asperellum*. *Journal of Applied and Natural Science, 9*(2), 954–959. https://doi.org/10.31018/jans.v9i2.1303.

Bof, M. J., Laurent, F. E., Massolo, F., Locaso, D. E., Versino, F., & García, M. A. (2021). Bio-packaging material impact on blueberries quality attributes under transport and marketing conditions. *Polymers, 13*(4), 1–20. https://doi.org/10.3390/polym13040481.

Bourtoom, T., & Chinnan, M. S. (2008). Preparation and properties of rice starch–chitosan blend biodegradable film. *LWT – Food Science and Technology, 41*(9), 1633–1641. https://doi.org/10.1016/j.lwt.2007.10.014.

Burmeister, A., & Grünberger, A. (2020). Microfluidic cultivation and analysis tools for interaction studies of microbial co-cultures. *Current Opinion in Biotechnology, 62*, 106–115. https://doi.org/10.1016/j.copbio.2019.09.001.

Camacho-Muñoz, R., Villada-Castillo, H. S., & Solanilla-Duque, J. F. (2020). Anaerobic biodegradation under slurry thermophilic conditions of poly(lactic acid)/starch blend compatibilized by maleic anhydride. *International Journal of Biological Macromolecules, 163*, 1859–1865. https://doi.org/10.1016/j.ijbiomac.2020.09.183.

Canon, F., Mariadassou, M., Maillard, M. B., Falentin, H., Parayre, S., Madec, M. N., Valence, F., Henry, G., Laroute, V., Daveran-Mingot, M. L., Cocaign-Bousquet, M., Thierry, A., & Gagnaire, V. (2020). Function-Driven Design of Lactic Acid Bacteria Co-cultures to Produce New Fermented Food Associating Milk and Lupin. *Frontiers in Microbiology, 11*(November). https://doi.org/10.3389/fmicb.2020.584163.

Carneiro da Silva, L. R., de Oliveira Rios, A., & Campomanes Santana, R. M. (2023). Polymer blends of poly(lactic acid) and starch for the production of films applied in food packaging: A brief review. *Polymers from Renewable Resources.*

Chávez González, M. L., Rodríguez Durán, L. V., Rodríguez Herrera, R., & Aguilar, C. N. (2009). Aspectos básicos de la fermentación en medio sólido. *CIENCIACIERTA, No. 20.*

Confort, P. M. S., & Inomoto, M. M. (2023). Effect of seed treatment with the biological agent *Pasteuria thornei* on the control of *Pratylenchus zeae* in Maize. *European Journal of Plant Pathology*. https://doi.org/10.1007/s10658-023-02659-2.

Cruz-Casas, D. E., Aguilar, C. N., Ascacio-Valdés, J. A., Rodríguez-Herrera, R., Chávez-González, M. L., & Flores-Gallegos, A. C. (2021). Enzymatic hydrolysis and microbial fermentation: The most favorable biotechnological methods for the release of bioactive peptides. *Food Chemistry: Molecular Sciences, 3.* https://doi.org/10.1016/j.fochms.2021.100047.

De la Cruz-Quiroz, R., Ascacio-Valdés, J. A., Rodríguez-Herrera, R., Roussos, S., & Aguilar, C. N. (2019). Phytopathogen Biomass as Inducer of Antifungal Compounds by *Trichoderma asperellum* Under Solid-State Fermentation. *Secondary Metabolites of Plant Growth Promoting Rhizomicroorganisms*, 113–124. https://doi.org/10.1007/978-981-13-5862-3_6.

Del Rosario Salazar-Sánchez, M., Campo-Erazo, S. D., Villada-Castillo, H. S., & Solanilla-Duque, J. F. (2019). Structural changes of cassava starch and polylactic acid films submitted to biodegradation process. *International Journal of Biological Macromolecules, 129*, 442–447. https://doi.org/10.1016/j.ijbiomac.2019.01.187.

Diender, M., Parera Olm, I., & Sousa, D. Z. (2021). Synthetic co-cultures: novel avenues for bio-based processes. *Current Opinion in Biotechnology, 67*, 72–79. https://doi.org/10.1016/j.copbio.2021.01.006.

Divya Arora, Gupta, P., Jaglan, S., Roullier, C., Grovel, O., & Samuel, B. (2020). Expanding the chemical diversity through microorganisms co-culture: Current status and outlook. *Biotechnology Advances*.

Eduardo, J., Sánchez, B., De, I., Estatal, U., Rafael, A., & Bacusoy, F. (2023). *Trichoderma harzianum* como biofertilizante en el cultivo de arroz (*Oryza sativa L.*) para una producción ecosostenible, *7*, 9762–9776.

El-Benawy, N. M., Abdel-Fattah, G. M., Ghoneem, K. M., & Shabana, Y. M. (2020). Antimicrobial activities of *Trichoderma atroviride* against common bean seed-borne *Macrophomina phaseolina* and *Rhizoctonia solani*. *Egyptian Journal of Basic and Applied Sciences, 7*(1), 267–280. https://doi.org/10.1080/2314808X.2020.1809849.

Ezati, P., Khan, A., Priyadarshi, R., Bhattacharya, T., Tammina, S. K., & Rhim, J.-W. (2023). Biopolymer-based UV protection functional films for food packaging. *Food Hydrocolloids, 142*, 108771. https://doi.org/10.1016/j.foodhyd.2023.108771.

Ezziyyani, M., & Sánchez, C. P. (2004). *Trichoderma harzianum* como biofungicida para el biocontrol de *Phytophthora capsici* en plantas de pimiento (*Capsicum annuum* L.). *Anales de Biología, 26*, 35–45. http://dialnet.unirioja.es/servlet/articulo?codigo=2198154 (accessed on 28 May 2024).

Fernández Valdés, D., Bautista Baños, S., Fernández Valdés, D., Ocampo Ramírez, A., García Pereira, A., & Falcón Rodríguez, A. (2015). Películas y recubrimientos comestibles: una alternativa favorable en la conservación poscosecha de frutas y hortalizas. *Revista Ciencias Técnicas Agropecuarias, 24*(3), 52–57.

Guédez, C., Castillo, C., Cañizales, L., & Olivar, R. (2008). Control Biológico: Una Herramienta Para El Desarrollo Sustentable Y Sostenible. *Academia, 7*(November), 50–74.

Gumienna, M., & Górna, B. (2021). Antimicrobial food packaging with biodegradable polymers and bacteriocins. *Molecules, 26*(12). https://doi.org/10.3390/molecules26123735.

Hernández Melchor, D. J., Ferrera Cerrato, R., & Alarcón, A. (2019). *Trichoderma*: importancia agrícola, biotecnológica, y sistemas de fermentación para producir biomasa y enzimas de interés industrial. *Chilean Journal of Agricultural & Animal Sciences, ahead*, 0–0. https://doi.org/10.4067/s0719-38902019005000205.

Herrera-Cebreros, J. M., Preciado-Rodríguez, J. M., & Robles-Parra, J. M. (2022). Economic impact of postharvest losses in farming systems: The table grape system. *Revista Iberoamericana de Tecnología Postcosecha, 23–1*, 2–17.

Jin, K., Jin, C., & Wu, Y. (2022). Synthetic biology-powered microbial co-culture strategy and application of bacterial cellulose-based composite materials. *Carbohydrate Polymers, 283*, 119171. https://doi.org/10.1016/j.carbpol.2022.119171.

Jones, J. A., & Wang, X. (2018). Use of bacterial co-cultures for the efficient production of chemicals. *Current Opinion in Biotechnology, 53*, 33–38. https://doi.org/10.1016/j.copbio.2017.11.012.

Julián Ricardo, M. C., & Sánchez Ramos, L. B. (2007). Fermentacion en estado solido. *Tecnología Química, 17–22.* http://www.redalyc.org/articulo.oa?id=445543754003 (accessed on 28 May 2024).

Jung, S. J., Kim, N. K., Lee, D. H., Hong, S. Il, & Lee, J. K. (2018). Screening and evaluation of *Streptomyces* species as a potential biocontrol agent against a wood decay fungus, *Gloeophyllum trabeum*. *Mycobiology, 46*(2), 138–146. https://doi.org/10.1080/12298093.2018.1468056.

Karuppiah, V., Sun, J., Li, T., Vallikkannu, M., & Chen, J. (2019). Co-cultivation of Trichoderma asperellum GDFS1009 and Bacillus amyloliquefaciens 1841 Causes Differential Gene Expression and Improvement in the Wheat Growth and Biocontrol Activity. *Frontiers in Microbiology, 10*(MAY), 1–16. https://doi.org/10.3389/fmicb.2019.01068.

Karuppiah, V., Zhixiang, L., Liu, H., Vallikkannu, M., & Chen, J. (2021). Co culture of Vel1-overexpressed Trichoderma asperellum and Bacillus amyloliquefaciens: An eco-friendly strategy to hydrolyze the lignocellulose biomass in soil to enrich the soil fertility, plant growth and disease resistance. *Microbial Cell Factories, 20*(1), 1–14. https://doi.org/10.1186/s12934-021-01540-3.

Khan, A., Bani Mfarrej, M. F., Nadeem, H., Ahamad, L., Hashem, M., Alamri, S., Gupta, R., & Ahmad, F. (2022). Trichoderma virens mitigates the root-knot disease progression in the chickpea plant. *Acta Agriculturae Scandinavica Section B: Soil and Plant Science, 72*(1), 775–787. https://doi.org/10.1080/09064710.2022.2080107.

Khan, B. A. (2017). Development of antibacterial hemp hurd/poly(lactic acid) biocomposite for food packaging.

King, E. D., Bassi, A. B., Ross, D. C., & Druebbisch, B. (2011). An industry perspective on the use of "atoxigenic" strains of Aspergillus flavus as biological control agents and the significance of cyclopiazonic acid. *Toxin Reviews, 30*(2–3), 33–41. https://doi.org/10.3109/15569543.2011.588818.

Köhl, J., Kolnaar, R., & Ravensberg, W. J. (2019). Mode of action of microbial biological control agents against plant diseases: Relevance beyond efficacy. *Frontiers in Plant Science, 10*(July), 1–19. https://doi.org/10.3389/fpls.2019.00845.

Kuebutornye, F. K. A., Abarike, E. D., Lu, Y., Hlordzi, V., Sakyi, M. E., Afriyie, G., Wang, Z., Li, Y., & Xie, C. X. (2020). Mechanisms and the role of probiotic Bacillus in mitigating fish pathogens in aquaculture. *Fish Physiology and Biochemistry, 46*, 819–841.

Kulkarni, V., Butte, K., & Rathod, S. (2012). Natural Polymers – A Comprehensive Review. *International Journal of Research in Pharmaceutical and Biomedical Sciences, 3*(4), 1597–1613.

Kunarbekova, M., Shynzhyrbai, K., Mataev, M., Bexeitova, K., Kudaibergenov, K., Sailaukhanuly, Y., Azat, S., Askaruly, K., Tuleshov, Y., Zhantikeyev, S. U., & Ybyraiymkul, D. (2023). Biopolymers synthesis and application. *Materials Today: Proceedings*. https://doi.org/https://doi.org/10.1016/j.matpr.2023.02.367.

Lagunas-Lagunas, J., Zavaleta-Mejía, E., Osada-Kawasoe, S., Aranda-Ocampo, S., Luna-Romero, I. J., & Vaquera-Huerta, H. (2001). Bacillus firmus como agente de control biológico de Phytophthora capsici Leo. en jitomate (Lycopersicon esculentum Mill.). *Revista Mexicana de Fitopatología, 19*(1), 57–65.

Le Borgne, S., & Quintero, R. (2003). Biotechnological processes for the refining of petroleum. *Fuel Processing Technology, 81*(2), 155–169. https://doi.org/10.1016/S0378-3820(03)00007-9.

Li, D., Liu, R., Cui, X., He, M., Zheng, S., Du, W., Gao, M., & Wang, C. (2021). Co-culture of bacteria and microalgae for treatment of high concentration biogas slurry. *Journal of Water Process Engineering, 41*(March), 102014. https://doi.org/10.1016/j.jwpe.2021.102014.

Lodha, A., Pawar, S., & Rathod, V. (2020). Optimised cellulase production from fungal co-culture of Trichoderma reesei NCIM 1186 and Penicillium citrinum NCIM 768 under solid state fermentation. *Journal of Environmental Chemical Engineering, 8*(5), 103958. https://doi.org/10.1016/j.jece.2020.103958.

López, A. C., Alvarenga, A. E., Vereschuk, M. L., Barua, R. C., Zapata, P. D., Luna, M. F., & Villaba, L. L. (2020). Trichoderma strains isolated from Ilex paraguariensis ST. HIL: promising biocontrol agents with chitinolytic activity and plant growth promoter on Lycopersicum esculentum. *Arab Journal of Basic and Applied Sciences, 27*(1), 105–113. https://doi.org/10.1080/25765299.2020.1732033.

López-lópez, M. E., Del-toro-sánchez, C. L., Ochoa-ascencio, S., Aguilar-, J. A., Martínez-cruz, O., Madrigal-pulido, J. A., Robles-garcía, M. A., Bernal-, A. T., Ávila-novoa, M. G., Guerrero-medina, P. J., & Gutiérrez-lomelí, M. (2022). Antagonismo de cepas de Trichoderma aisladas en Tanaxuri, Michoacán, México contra patógenos postcosecha del fruto de aguacate (Persea americana Mill). *Biotecnia, 25*(1), 24–33. https://doi.org/10.18633/biotecnia.v25i1.1726.

Madaure, J. T., Mashela, P. W., & De Waele, D. (2019). Interactive effects of Nemarioc-AL phytonematicide, Steinernema feltiae and Trichoderma harzianum on the reproduction of Meloidogyne incognita race 2 under greenhouse conditions. *Acta Agriculturae Scandinavica Section B: Soil and Plant Science, 69*(3), 235–240. https://doi.org/10.1080/09064710.2018.1539190.

Mari, M., Mari, M., Francesco, A. Di, & Bertolini, P. (2018). Control of fruit postharvest diseases: Old issues and innovative approaches Control of fruit postharvest diseases : old issues and innovative approaches. December. https://doi.org/10.2212/spr.2014.1.1.

Martinez Medina, G. A. (2018). Desarrollo de un bioproceso para la producción de proteasas con potencial en el área de alimentos. Universidad Autonoma de Coahuila.

Martinez, J. A., Delvenne, M., Henrion, L., Moreno, F., Telek, S., Dusny, C., & Delvigne, F. (2022). Controlling microbial co-culture based on substrate pulsing can lead to stability through differential fitness advantages. *PLoS Computational Biology, 18*(10), 1–27. https://doi.org/10.1371/journal.pcbi.1010674.

Martinez, P. (2019). Co-cultivo de microorganismos de origen marino como estrategia para la producción diferencial de metabolitos especializados. Universidad Nacional de Colombia, 6–19.

Meena, M., Swapnil, P., Zehra, A., Dubey, M. K., & Upadhyay, R. S. (2017). Antagonistic assessment of Trichoderma spp. by producing volatile and non-volatile compounds against different fungal pathogens. *Archives of Phytopathology and Plant Protection, 50*(13–14), 629–648. https://doi.org/10.1080/03235408.2017.1357360.

Mesa Vanegas, A. M., Calle Osorno, J., & Marín Pavas, D. A. (2020). Metabolitos secundarios en Trichoderma spp. y sus aplicaciones biotecnológicas agrícolas. *Actualidades Biológicas, 41*(111), 32–44. https://doi.org/10.17533/udea.acbi.v41n111a02.

Mhatre, P. H., Karthik, C., Kadirvelu, K., Divya, K. L., Venkatasalam, E. P., Srinivasan, S., Ramkumar, G., Saranya, C., & Shanmuganathan, R. (2019). Plant growth promoting rhizobacteria (PGPR): A potential alternative tool for nematodes bio-control. *Biocatalysis and Agricultural Biotechnology, 17*, 119–128. https://doi.org/10.1016/J.BCAB.2018.11.009.

Mitsuboshi, M., Kioka, Y., Noguchi, K., & Asakawa, S. (2022). Evaluation of Disease Suppressiveness of Soils in Croplands by Co-Cultivation of Pathogenic Fusarium oxysporum and Indigenous Soil Microorganisms. *Microbes and Environments, 37*(4), 1–10. https://doi.org/10.1264/jsme2.ME21063.

Moncayo Martínez, D. C. (2013). Desarrollo de un recubrimiento comestible a partir de un biopolímero para prolongar la vida útil de frutas frescas [Universidad Nacional de Colombia sede Bogotá]. https://repositorio.unal.edu.co/handle/unal/50858 (accessed on 28 May 2024).

Motelica, L., Ficai, D., Ficai, A., Oprea, O. C., Kaya, D. A., & Andronescu, E. (2020). Biodegradable Antimicrobial Food Packaging: Trends and Perspectives. *Foods, 9*.

Munhuweyi, K., Mpai, S., & Sivakumar, D. (2020). Extension of avocado fruit postharvest quality using non-chemical treatments. *Agronomy, 10*(2). https://doi.org/10.3390/agronomy10020212.

Navia, D. P., & Villada C, H. S. (2013). Impacto da embalagem biodegradável pesquisa de ciência, tecnologia e inovação Artículos de Reflexión. *Biotecnología En El Sector Agropecuario y Agroindustrial, 11*(2), 173–180.

Omer, Z. S., Karin, J., Eberhard, T. H., & Johansson, L. K. H. (2010). Bacteria considered as biocontrol agents to control growth of white clover on golf courses. *Acta Agriculturae Scandinavica Section B: Soil and Plant Science, 60*(3), 193–198. https://doi.org/10.1080/09064710902773637.

Opara, I. K., Fawole, O. A., Kelly, C., & Opara, U. L. (2021). *Quantification of on-Farm Pomegranate Fruit Postharvest Losses and Waste, and Implications on Sustainability Indicators: South African Case Study*, Stewart Postharvest Review 2014, *1*, 1–20. www.stewartpostharvest.com, ISSN: 1745-9656.

Pakora, G.-A., Mpika, J., Kone, D., Ducamp, M., Kebe, I., Nay, B., & Buisson, D. (2018). Inhibition of Phytophthora species, agents of cocoa black pod disease, by secondary metabolites of Trichoderma species. *Environmental Science and Pollution Research, 25*(30), 29901–29909. https://doi.org/10.1007/s11356-017-0283-9.

Perumal, A. B., Nambiar, R. B., Sellamuthu, P. S., & Emmanuel, R. S. (2021). Use of modified atmosphere packaging combined with essential oils for prolonging post-harvest shelf life of mango (cv. Banganapalli and cv. Totapuri). *Lwt., 148*, 111662. https://doi.org/10.1016/j.lwt.2021.111662.

Porat, R., Lichter, A., Terry, L. A., Harker, R., & Buzby, J. (2018). Postharvest Biology and Technology Postharvest losses of fruit and vegetables during retail and in consumers' homes: Quantifications, causes, and means of prevention. *Postharvest Biology and Technology, 139*(February), 135–149. https://doi.org/10.1016/j.postharvbio.2017.11.019.

Poveda, J., Abril-Urias, P., & Escobar, C. (2020). Biological Control of Plant-Parasitic Nematodes by Filamentous Fungi Inducers of Resistance: Trichoderma, Mycorrhizal and Endophytic Fungi. *Frontiers in Microbiology, 11*(May), 1–14. https://doi.org/10.3389/fmicb.2020.00992.

Ravichandran, S. (2017). Solid state and submerged fermentation for the production of bioactive substances: a comparative study.

ReFED, R. F. W. T. E. D. (2022). A roadmap to reduce U.S. food waste by 20%.

Rivera, C., Contreras, F., Ariza, W., Bonilla, S., & Cruz, A. (2019). Los empaques biodegradables, una respuesta a la consciencia ambiental de los Consumidores. *Realidad Empresarial, 7*, 2–8.

Roberts-McEwen, T. A., Deutsch, E. K., Mowery, M. A., & Grinsted, L. (2022). Group-Living Spider Cyrtophora citricola as a Potential Novel Biological Control Agent of the Tomato Pest Tuta absoluta. *Insects, 14*(1), 34. https://doi.org/10.3390/insects14010034.

Robinson, Singh, N. (2002). Fermentación En Estado Sólido: Una Tecnología Microbiana Promisoria Para La Producción De Metabolitos Secundarios. *Vitae, 9*(2), 27–36.

Robinson, T., Shingh, D., & Nigman, P. (2002). Fermentación en estado sólido: una tecnología microbiana promisora para la producción de metabolitos secundarios. *Vitae, 9*, 27–36.

Rodriguez Carreón, A. (2016). Degradación de endosulfán en aguas contaminadas mediante procesos enzimáticos pormovidos por co-cultivos fúngicos en un reactor tipo Airlift. *Centro de Investigación y de estudios avanzados del Instituto Politécnico Nacional.*

Rolim, L. A., Santos, F. C. M. dos, Chaves, L. L., Gonçalves, M. L. C. M., Freitas-Neto, J. L., Nascimento, A. L. da S. do, Soares-Sobrinho, J. L., Albuquerque, M. M. de, Lima, M. do C. A. de, & Rolim-Neto, P. J. (2020). Enhanced Reader.pdf. *Nature, 388*, 539–547.

Romeis, J., Naranjo, S. E., Meissle, M., & Shelton, A. M. (2019). Genetically engineered crops help support conservation biological control. *Biological Control, 130*(July 2018), 136–154. https://doi.org/10.1016/j.biocontrol.2018.10.001.

Rosero-Chasoy, G., Rodríguez-Jasso, R. M., Aguilar, C. N., Buitrón, G., Chairez, I., & Ruiz, H. A. (2021). Microbial co-culturing strategies for the production high value compounds, a reliable framework towards sustainable biorefinery implementation – an overview. *Bioresource Technology, 321*(October 2020). https://doi.org/10.1016/j.biortech.2020.124458.

Ruelas, X., De la Luz Reyes, M., Valdivia, B., Contreras, J., Montañez, J., Aguilera, A., & Peralta, R. (2013). Conservación de Frutas y Hortalizas Frescas y Mínimamente Procesadas con Recubrimientos Comestibles. *Revista Científica de la Universidad Autónoma de Coahuila, 5*(9), 7. https://refed.org/downloads/Executive-Summary.pdf (accessed on 28 May 2024).

Sabri, M., Benkirane, R., Habbadi, K., Sadik, S., Ou-Zine, M., Diouri, M., & Achbani, E. H. (2021). Phages as a potential biocontrol of phytobacteria. *Archives of Phytopathology and Plant Protection, 54*(17–18), 1277–1291. https://doi.org/10.1080/03235408.2021.1902033.

Sánchez Aguilar, R. F., & Muñoz Chaves, J. A. (2021). Natural polymers based on agro-industrial wastes: an alternative for biodegradable packaging in Colombia, 233–245.

Santos da Silveira, J., Durand, N., Lacour, S., Belleville, M. P., Perez, A., Loiseau, G., & Dornier, M. (2019). Solid-state fermentation as a sustainable method for coffee pulp treatment and production of an extract rich in chlorogenic acids. *Food and Bioproducts Processing, 115*, 175–184. https://doi.org/10.1016/j.fbp.2019.04.001.

Saval, S., Ingeniería, I. De, & Universitaria, C. (2012). Aprovechamiento de Residuos Agroindustriales: Pasado, Presente y Futuro. *BioTecnologia, 16*(2), 14–46.

Senthilraja, G., Anand, T., Durairaj, C., Kennedy, J. S., Suresh, S., Raguchander, T., & Samiyappan, R. (2010). A new microbial consortia containing entomopathogenic fungus, Beauveria bassiana and plant growth promoting rhizobacteria, Pseudomonas fluorescens for simultaneous management of leafminers and collar rot disease in groundnut. *Biocontrol Science and Technology, 20*(5), 449–464. https://doi.org/10.1080/09583150903576949.

Shipman, E. N., Yu, J., Zhou, J., Albornoz, K., & Beckles, D. M. (2021). Can gene editing reduce postharvest waste and loss of fruit, vegetables, and ornamentals? *Horticulture Research*, 1–21. https://doi.org/10.1038/s41438-020-00428-4.

Shong, J., Jimenez Diaz, M. R., & Collins, C. H. (2012). Towards synthetic microbial consortia for bioprocessing. *Current Opinion in Biotechnology, 23*(5), 798–802. https://doi.org/10.1016/j.copbio.2012.02.001.

Stenberg, J. A., Sundh, I., Becher, P. G., Björkman, C., Dubey, M., Egan, P. A., Friberg, H., Gil, J. F., Jensen, D. F., Jonsson, M., Karlsson, M., Khalil, S., Ninkovic, V., Rehermann, G., Vetukuri, R. R., & Viketoft, M. (2021). When is it biological control? A framework of definitions, mechanisms, and classifications. *Journal of Pest Science, 94*(3), 665–676. https://doi.org/10.1007/s10340-021-01354-7.

Suárez-Quiroz, M. L., Mendoza-Bautista, I., Monroy-Rivera, J. A., De la Cruz-Medina, J., Angulo-Guerrero, O., & González-Ríos, O. (2013). Aislamiento, identificación y sensibilidad a antifúngicos de hongos fitopatógenos de papaya cv. maradol (Carica papaya L.). *Revista Iberoamericana de Tecnología Postcosecha, 14*(2), 115–124.

Tariq, M., Khan, A., Asif, M., Khan, F., Ansari, T., Shariq, M., & Siddiqui, M. A. (2020). Biological control: a sustainable and practical approach for plant disease management. *Acta Agriculturae Scandinavica Section B: Soil and Plant Science*, 507–524. https://doi.org/10.1080/09064710.2020.1784262.

Thieman, W., & Palladino, M. (2010). *Introducción a la Biotecnología*. Pearson Educacion.

Tong, C. Y., Honda, K., & Derek, C. J. C. (2023). A review on microalgal-bacterial Co-culture: The multifaceted role of beneficial bacteria towards enhancement of microalgal metabolite production. *Environmental Research, 115872*. https://doi.org/10.1016/j.envres.2023.115872.

Ullah, S., Hashmi, M., Lee, J., Youk, J. H., & Kim, I. S. (2022). Recent Advances in Pre-harvest, Post-harvest, Intelligent, Smart, Active, and Multifunctional Food Packaging. *Fibers and Polymers, 23*(8), 2063–2074. https://doi.org/10.1007/s12221-022-4192-2.

Viera, W., & Jackson, T. (2018). Biocontrol para sistemas de agricultura sustentable, Ecuador. *Memorias Del Primer Congreso de Control Biológico Aplicado, 1*.

Villada, H. S., Acosta, H., & Velasco, R. (2007). Biopolímeros naturales usados en empaques biodegradables. *Temas Agrarios, 12*(2), 5–13. https://doi.org/10.21897/rta.v12i2.652.

Villarreal-Soto, S. A., Beaufort, S., Bouajila, J., Souchard, J.-P., Renard, T., Rollan, S., & Taillandier, P. (2019). Impact of fermentation conditions on the production of bioactive compounds with anticancer, anti-inflammatory and antioxidant properties in kombucha tea extracts. *Process Biochemistry, 83*, 44–54. https://doi.org/10.1016/j.procbio.2019.05.004.

Villarroel, D., Romero, L., Brito, M., & Yndira, A. (2014). Luz Ultravioleta: Inactivación Microbiana En Frutas. *Universidad de Oriente, Venezuela, 27*(3), 454–469. https://doi.org/ISSN 1315-0162.

Vinchira Villarraga, D. M., & Moreno Sarmiento, N. (2019). Control biológico: Camino a la agricultura moderna. *Revista Colombiana de Biotecnología, 21*(1), 2–5. https://doi.org/10.15446/rev.colomb.biote.v21n1.80860.

Wang, C., Chang, T., Dong, S., Zhang, D., Ma, C., Chen, S., & Li, H. (2020). Biopolymer films based on chitosan/potato protein/linseed oil/ZnO NPs to maintain the storage quality of raw meat. *Food Chemistry, 332*, 127375. https://doi.org/10.1016/j.foodchem.2020.127375.

Wu, X., Zhu, P., Li, D., Zheng, T. F., Cai, W., Li, J., Zhang, B., Zhu, B., Zhang, J., & Du, G. (2021). Bioaugmentation of *Bacillus amyloliquefaciens–Bacillus kochii* co-cultivation to improve sensory qualities of flue-cured tobacco, 1–17. https://doi.org/10.21203/rs.3.rs-626237/v1.

Yassin, M. T., Mostafa, A. A. F., & Al-Askar, A. A. (2022). In vitro antagonistic activity of Trichoderma spp. against fungal pathogens causing black point disease of wheat. *Journal of Taibah University for Science, 16*(1), 57–65. https://doi.org/10.1080/16583655.2022.2029327.

Ziv, C., & Fallik, E. (2021). Postharvest storage techniques and quality evaluation of fruits and vegetables for reducing food loss. *Agronomy, 11*(6), 1–5. https://doi.org/10.3390/agronomy11061133.

CHAPTER 10

Biodiesel Production from Residual Fats and Oils by Enzymatic Catalysis: An Environmentally Friendly Alternative

VEYMAR G. TACIAS-PASCACIO, GILBER VELA-GUTIÉRREZ, and ERIKA J. LÓPEZ-ZÚÑIGA

Faculty of Nutrition and Food Sciences, University of Sciences and Arts of Chiapas, Lib. Norte Pte., Tuxtla Gutierrez, Chiapas, Mexico

ABSTRACT

The high costs of extraction and the constant global reduction in oil reserves, as well as the environmental problems related to its use, have led to an intense and increasing search for renewable fuels that can reduce or replace the consumption of fossil fuels. Biodiesel represents a good option since it has properties similar to diesel, it is non-toxic, biodegradable, and renewable. Biodiesel is produced at an industrial level by alkaline transesterification of edible vegetable oils (mainly) and methanol; however, this process generates a product between 10 and 50% more expensive than petroleum diesel. To improve the economic viability of biodiesel production, the use of residual fats and oils generated in the food industry has been explored, which are produced in large quantities, are difficult to eliminate, and potential contaminants of water and land resources. However, the high content of free fatty acids in the residual oils negatively affects alkaline transesterification, considerably reducing the reaction yields. In this sense, the use of biocatalysts (lipases) in biodiesel production is a very encouraging alternative, which presents environmental and economic advantages over the conventional method. In this context, this work deals with the enzymatic production of biodiesel from residual fats and oils, as an alternative that contributes positively to the

Sustainable Agriculture and Global Environmental Health: Toward the Efficient Management of Resources. Mónica L. Chávez González, Cristóbal Noé Aguilar, Juan Carlos Contreras-Esquivel, and A. K. Haghi (Eds.)

economic viability of biodiesel production, avoids the uncontrolled disposal of residual oils into the environment, and allows the production of a renewable fuel through an environmentally friendly process.

10.1 INTRODUCTION

Currently, there is a strong dependence on fossil fuels, since these are widely used in industry, agriculture, and transport; however, the constant global reduction in oil reserves and the possible depletion of it, added to the high extraction costs and the increase in greenhouse gas emissions, acid rain, and other environmental problems, have stimulated the search for renewable fuels that can reduce consumption or completely replace fuels derived from petroleum (Palmer & Brigham, 2016).

Biodiesel represents a good alternative to replace part of the fossil fuels used, since it has similar properties to diesel, is non-toxic, biodegradable, renewable, and has fewer greenhouse gas emissions (Alptekin & Canakci, 2011; Dias et al., 2009).

Biodiesel is produced by transesterification of triglycerides with a low molecular weight alcohol (mainly methanol), in the presence of a catalyst (alkaline, acid, or enzymatic). Alkaline transesterification of vegetable oils is the most widely used method for obtaining biodiesel at an industrial level because it generates high yields in short reaction times (Poppe, Fernandez-Lafuente et al., 2015; Poppe, Matte et al., 2015).

Edible oils (such as rapeseed, soybean, and sunflower) represent more than 95% of the raw material used in the production of biodiesel since the high quality of these oils allows obtaining biodiesel with characteristics closer to those of petroleum diesel. However, one of the main disadvantages of biodiesel obtained from edible vegetable oils, which represents around 70–80% of production costs, is that the product is generally 10–50% more expensive than diesel obtained from oil (Ataya et al., 2007; Leung & Guo, 2006). In addition, the extensive use of edible vegetable oils as raw material for biodiesel production can lead to a crisis related to the food supply and induce an economic imbalance (Canesin et al., 2014).

The need to combat the scarcity of fuels without affecting another sector of vital importance (such as food) has encouraged the search for new alternatives of raw materials for biodiesel production, which also allows obtaining a product economically competitive with petroleum diesel. A possible solution to these drawbacks is the use of residual fats and oils generated in the food industry, which, given their easy availability and low cost, can significantly improve the economic viability of biodiesel production, allowing waste recycling and non-competition with the food industry (Chhetri et al., 2008; Predojević, 2008).

Residual fats and oils, as raw materials for the production of biodiesel, have as their main limitation the presence of free fatty acids (FFA) and water, which strongly affect conventional alkaline catalysis, since they react with the catalyst, leading to the formation of soap, with the consequent decrease in the yield of methyl esters (Poppe, Matte, et al., 2015). To overcome these drawbacks, the use of other types of catalysts has been explored, such as homogeneous acid catalysts, heterogeneous acids and alkaline catalysts, and biocatalysts.

Biodiesel can be produced by enzymatic transesterification using lipases. Lipases, which are mainly used immobilized (then known as biocatalysts), provide high yields and selectivity, facilitate glycerol removal and biodiesel purification, tolerate the presence of water in the oil, have low energy consumption (the reaction requires moderate conditions), generate small amounts of waste, carry out esterification and transesterification simultaneously, and do not carry out undesired reactions, such as saponification (the main drawback of alkaline catalysis) (Canesin et al., 2014; Zhang et al., 2003); for these reasons, they are highly desirable catalysts in obtaining biodiesel from residual fats and oils.

In this chapter, the authors aim to show the enzymatic production of biodiesel from residual fats and oils as an alternative that contributes positively to the economic viability of biodiesel production, avoids the uncontrolled disposal of residual oils into the environment, and allows the production of a renewable fuel through an environmentally friendly process.

10.2 BIODIESEL OVERVIEW

In these times in which oil reserves are diminishing, fossil fuel prices are rising (Maheswari et al., 2022; Singh et al., 2022) and the most serious global environmental problems caused by its use are now being increased by pollution, acid rain, and climate change (Alengebawy et al., 2022; Benti et al., 2022, 2023; Edrisi & Abhilash, 2016; Ogunkunle & Ahmed, 2021), the interest in promoting the development and use of renewable energy sources has been renewed (Neupane, 2022), and biodiesel has gained increasing attention for several reasons.

Biodiesel has been defined by the American Society for Testing and Materials (ASTM) as monoalkyl esters of long-chain fatty acids derived from renewable feedstocks, such as vegetable oils or animal fats, for use in compression ignition engines (Saleh et al., 2010). This highly biodegradable biofuel, which is produced mainly by transesterification of oils with short-chain alcohols, is a renewable energy source that belongs to the ecological fuels due to its qualitative composition (77% carbon, 12% hydrogen, 11% oxygen, traces of nitrogen, and sulfur) (Tomasevic & Siler-Marinkovic, 2003), and it is considered carbon neutral because the plants that provide the raw material for the production of biodiesel take up more carbon dioxide than it releases into the atmosphere when it is used as an energy source (Ferella et al., 2010; Gunvachai et al., 2007).

Biodiesel presents physicochemical properties similar to petroleum diesel, and it is miscible with it in any proportion, which allows its use alone or mixed with conventional diesel in any diesel engine without having to change the fuel injector or the ignition system (Robles-Medina et al., 2009). It has several technical advantages over diesel fuel, such as inherent lubricity (which can improve engine performance and extend engine life) (Vasudevan & Briggs, 2008), low toxicity, derivation from renewable and even domestic raw materials, higher flash point (about 150°C), biodegradability, negligible sulfur content, and lower overall exhaust gas emissions (CO, particles in suspension, polycyclic aromatic hydrocarbons, unburned hydrocarbons and sulfur, SOx) (Atadashi et al., 2010; Fukuda et al., 2001; Moser, 2009; Phan & Phan, 2008; Tayari et al., 2020; Yadav et al., 2017). These advantages allow biodiesel to be handled more safely than diesel and can be stored, transported, and generally managed with the same infrastructure as petroleum diesel (Ahmad et al., 2011; Al-Zuhair, 2007). This represents an important advantage in economic and operational terms since it is not necessary to build a new infrastructure to transport, store, and handle biodiesel.

For all the aforementioned, it can be considered that biodiesel offers an excellent alternative to diversify energy sources, which can reduce the global reliance on imported petroleum fuels (Neupane, 2022), in such a way that the supply is guaranteed, especially in countries where do not have or their sources of fossil fuels are scarce (Robles-Medina et al., 2009), making it possible to orient the current energy system towards security of supply and sustainability (Jegannathan et al., 2011). However, it is important to mention that, since approximately 95% of the biodiesel produced comes from expensive edible oils (sunflower, soybean, palm, linseed, and sesame oil), which represents around 70–80% of biodiesel production costs, this product is not competitive with fuels derived from petroleum (Yadav et al., 2016). Nevertheless, the constant increase in oil prices suggests that biofuels may be economically viable in the medium term (Robles-Medina et al., 2009). In fact, it has been reported that global biodiesel production increased from 3.9 billion liters in 2005 to 37 billion liters in 2020 and is projected to exceed 46 billion liters by 2025, a 24% increase over 2020 levels (Benti et al., 2023). In addition, there are factors that can contribute positively to the feasibility of biodiesel, such as tax regulations derived from international agreements for environmental conservation; the tax rules with differential rates according to the type and origin of the oilseeds used, the genetic modification of certain oilseed plants or microalgae to improve lipid content (Akoh et al., 2007; Chisti, 2007; Li et al., 2007; Robles-Medina et al., 2009), the use of low-cost raw materials such as residual fats and oils, as well as the development of more efficient processes in the production, recovery, and purification of biodiesel, especially those alternatives that are respectful of the environment, such as the production of biodiesel by enzymatic transesterification (Robles-Medina et al., 2009).

10.3 FEEDSTOCKS FOR BIODIESEL PRODUCTION

The feedstock used in biodiesel production will largely determine the quality of the biodiesel produced. For this reason, the selection of the raw material is a crucial stage in biodiesel production and is affected by several factors such as cost, purity, fatty acid composition, oil content, yield, and quality of the biodiesel obtained (Neupane, 2022). One of the main advantages of biodiesel is that it can be produced from a wide range of oily raw materials. Four main categories of feedstocks for biodiesel production have been identified and classified into three generations (Figure 10.1), which are edible and non-edible vegetable oils (more than 350 oilseed crops have been identified worldwide as potential sources of biodiesel) (Bart et al., 2010), algae oil, used cooking oil, and animal fats (Benti et al., 2023; Kim et al., 2018).

The main feedstocks used to produce biodiesel are vegetable oils extracted from oleaginous plants; these oils can be edible, such as cottonseed, peanut, corn, rapeseed, soybean, palm, sunflower, olive, coconut, etc., or inedible, such as *Jatropha curcas*, Pongamia, Neem, rubber seed, jojoba, and castor oil (Atadashi et al., 2010, 2012). The availability of raw materials for biodiesel production is influenced by factors such as geographic location, local soil characteristics, climate, and cultivation methods, among others (Benti et al.,

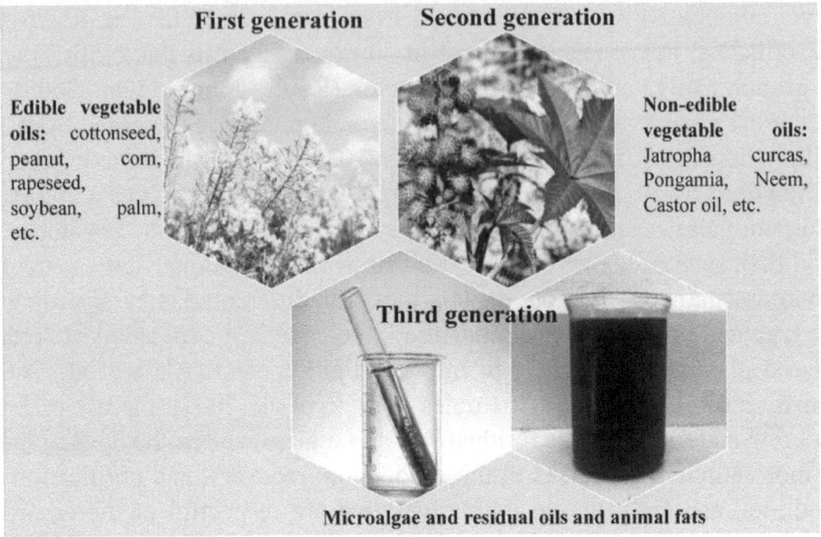

FIGURE 10.1 Categories of feedstocks for biodiesel production (Bart et al., 2010; Benti et al., 2023; Kim et al., 2018).

Source: Adapted from Bart et al., 2010; Benti et al., 2023; Kim et al., 2018

2023). Generally, the most abundant oils or fats in a region are the most common raw materials used for biodiesel obtaining. Therefore, rapeseed and sunflower oils are used mainly in Europe, palm oil predominates in tropical countries, and animal fats and soybean oil are more common in the United States (Demirbas, 2006, 2007).

Despite the wide range of possibilities, more than 95% of the raw material used in biodiesel production corresponds to edible vegetable oils (Balat, 2011; Balat & Balat, 2009). In 2020, approximately 71.7% of the biodiesel feedstock came from soybean, 10.7% from canola oil, and 13% from corn oil (Neupane, 2022). This is because the high quality of these oils allows obtaining a biodiesel with characteristics closer to those of petroleum diesel (Benti et al., 2023; Rodrigues et al., 2016).

The cost of oils used to obtain biodiesel represents around 70–80% of production costs (Figure 10.2) (Behzadi & Farid, 2007; Canesin et al., 2014); thus, the biodiesel produced from these high-quality raw materials, called first-generation biodiesel, is usually between 10 and 50% more expensive than petroleum diesel, which is an obstacle to the industrial development and commercialization of this biofuel (Canesin et al., 2014; Dias et al., 2009; Zhang et al., 2003). In addition, the large-scale consumption of edible oils for power generation creates a food supply crisis and can induce economic imbalance by promoting competition with the food market, which can also negatively affect the price of biodiesel, adding to significant environmental problems such as deforestation, severe loss of soil resources, and the use of most farmable land for oilseed cultivation (Balat, 2011; Benti et al., 2023; Rodrigues et al., 2016).

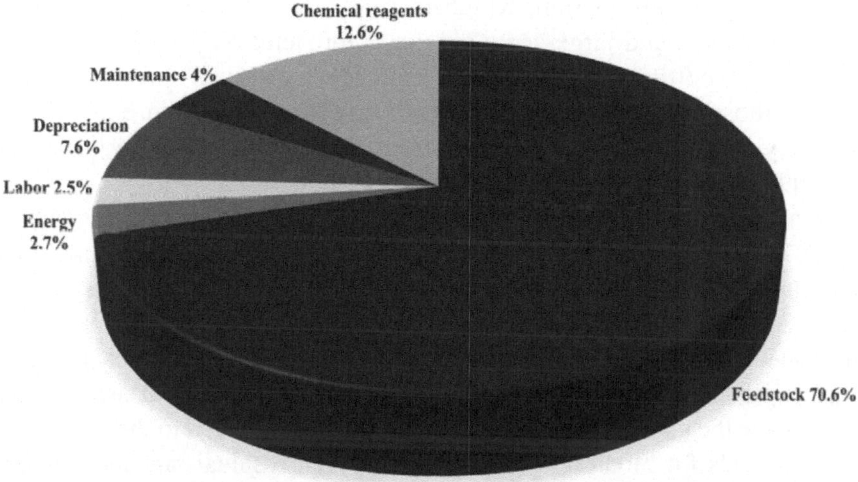

FIGURE 10.2 Distribution of factors involved in biodiesel production costs (%) (Kemp).

The use of non-edible vegetable oils for biodiesel production (second generation) arises as a strategy to reduce or replace the use of edible oils, and with it, overcome the aforementioned drawbacks. Non-edible vegetable oils, such as *Ricinus communis* and *Jatropha curcas*, have gained popularity because they are available around the world and grown carelessly in soils unsuitable for food crops (Benti et al., 2023). However, these oils have the disadvantage of a high concentration of free fatty acids that cause serious problems during the transesterification reaction and/or in the biodiesel separation and purification operations (Cao et al., 2008). In addition, to have these oils in large quantities, it is necessary to promote the cultivation of the plants from which they come, which must be carefully managed, evaluating the environmental and economic impacts that this may bring, since if it is not done in this way, it could incur again in promoting (unintentionally) deforestation, and the displacement of small farmers, generating food shortages and increasing poverty (Vasudevan & Briggs, 2008).

On the other hand, microalgae, a third-generation feedstock for the production of biodiesel, are photosynthetic microorganisms that convert sunlight, water, and CO_2 into algal biomass which contains about 20%–80% of oil that can be converted into biodiesel (Gissibl et al., 2019). Compared to any other raw material, microalgae grow easily and can generate 25 times more oil than palm oil and 250 times more than soybean when grown in a bioreactor or on a farm. However, the most significant impediment to the commercialization of microalgae is the high cost of production, which is related to the need for high-yield oil algae strains and large-scale efficient bioreactors; this is because the operating conditions that lead to a high level of oil in microalgae are usually those that provide a low growth rate (low temperature, low light intensity, and nitrogen deficiency) (Benti et al., 2023; Molina Grima et al., 1999; Robles-Medina et al., 2009).

Finally, in biodiesel production, the use of used cooking oils (Corsini et al., 2015), which are generally discarded after use, and residual animal fats and oils such as tallow (Öner & Altun, 2009), fish oil (Ching-Velasquez et al., 2020), chicken fat (Gürü et al., 2010), and lard (Lu et al., 2007), which are obtained mainly as by-products from meat processing facilities, has also been explored. These third-generation biodiesel feedstocks (Alaswad et al., 2015) have the main advantage of being low cost, compared to the rest of the raw materials for biodiesel production; however, residual fats and oils are characterized by their high content of impurities such as moisture and free fatty acids, which seriously affect conventional transesterification, generating by-products such as soap and severely complicating recovery processes and purification of the biodiesel produced, thus limiting their productivity at large scale (Talebian-Kiakalaieh et al., 2013). Nevertheless, even with the disadvantages that these materials present, in general, they are considered an economical, ecological, and sustainable alternative for biodiesel production, as will be discussed below.

10.4 RESIDUAL FATS AND OILS IN BIODIESEL PRODUCTION

The increase in the production of residual oils and fats is a growing problem in the world (Felizardo et al., 2006; Tacias Pascacio et al., 2016). In developed countries such as the United States, Canada, and the European Union, millions of tons of residual oils and animal fats are produced annually, mainly from the restaurant industry and meat processing facilities, respectively (Azócar et al., 2010; Singh et al., 2010; Tacias Pascacio et al., 2016). Animal fat wastes are mainly tallow from processing cattle meat, lard, and white fat from pigs, fish oils from fish processing, and poultry fat from processing chicken, turkey, or other birds (Toldrá-Reig et al., 2020). The management of this type of waste represents a significant challenge due to the problems related to its final disposal and the possible contamination of water and soil since in many countries these wastes are disposed of inappropriately, being discharged into sewers, rivers, and landfills, generating environmental contamination (Avhad & Marchetti, 2015; Chhetri et al., 2008; Fazal et al., 2011).

Residual fats and oils are not suitable for human consumption, and in addition, compared to "fresh" oils, they are up to 60% less expensive, depending on the source and its availability. In this context, and considering that to guarantee low biodiesel production costs, it is essential to choose the lowest-priced raw material (Pasha et al., 2021), the catalytic conversion of these residues into biodiesel results in outstanding benefits. For instance, it significantly improves the economic viability of biodiesel production, offering a solution to the growing problem of increasing production of residual fats and oils in the world (Predojević, 2008). In addition, since they are inedible products and do not need to be cultivated, but are obtained as wastes, their use in the production of biodiesel promotes food security because it prevents the use of food resources in the production of energy, thereby eliminating the controversial competition between biofuels and the food industry (Aniokete et al., 2022; Encinar et al., 2005; Varanda et al., 2011). Moreover, some characteristic properties of animal fats, such as lower unsaturated fatty acids, allow the production of biodiesel with good oxidative stability, elevated calorific value, and shorter ignition (Adewale et al., 2015). Due to these relevant advantages, countries like Brazil, Canada, the United States, the EU, and China, which generate large amounts of waste cooking oils, pork fat, chicken fats, lard, beef tallow, fish oils, etc., are already taking advantage of them for their conversion into biodiesel (Aniokete et al., 2022; Kirubakaran & Selvan, 2018).

Unfortunately, the production of biodiesel from residual fats and oils has important disadvantages. As can be seen in Table 10.1, the physicochemical characteristics of residual oils and fats are different from those of "fresh" oil. The deterioration of fats and the complex reactions of polymerization, oxidation, and hydrolysis that cooking oil undergoes when used for frying (Yaqoob et al., 2021) lead to the modification of important properties such as acidity and viscosity. This affects the yields and quality of the biodiesel obtained (Tacias Pascacio et al., 2016).

TABLE 10.1 Some Physicochemical Properties of Waste Cooking Oil, Residual Chicken Fat, and Fresh Edible Vegetable Oil

Properties	Fresh Edible Vegetable Oil (Tacias Pascacio et al., 2016)	Waste Cooking Oil (Tacias Pascacio et al., 2016)	Residual Chicken Fat (Odetoye et al., 2021)
Acid value (mg KOH/g of oil)	0.10	2.09	5.33
% Free fatty acid	0.05	1.05	2.68
Saponification value (mg KOH/g of oil)	196.23	198.54	109.40
Viscosity (mPa/s)	29.93	40.09	13.2

In operational terms, the main obstacle to the conversion of residual fats and oils into biodiesel is the presence of impurities, mainly free fatty acids and water (Avagyan & Singh, 2019; Avhad & Marchetti, 2015). As can be seen in Table 10.2, the content of free fatty acids in residual fats and oils is much higher than that found in fresh vegetable oils (refined and unrefined).

TABLE 10.2 Percentage of Free Fatty Acids (FFA) in Different Fats and Oils (Gerpen, 2005)

Feedstock	Free Fatty Acid Content (%)
Refined vegetable oils	<0.05%
Crude vegetable oils	0.3–0.7%
Used cooking oils	2–7%
Animal fats	5–30%
Trap grease	40–100%

Both water and free fatty acids seriously affect alkaline transesterification (a conventional method for obtaining biodiesel), decreasing the activity of the catalyst and favoring alternate reactions that lead to obtaining undesirable by-products (such as soap), which affect the yield of the reaction and make it difficult for the biodiesel recovery and purification processes. In the end, the cost of the process is higher than the price of conventional fossil fuels (Gerpen, 2005). However, the presence of impurities has not limited the use of these materials for the production of biodiesel. What is done is a two-stage process; the first stage consists of giving a pre-treatment to the raw material through an esterification reaction catalyzed by a homogeneous acid catalyst, such as sulfuric acid, which has proven to be tolerant to both the presence of FFA and water (Thomas Aniokete et al., 2019). This reaction converts free fatty acids directly to fatty acid methyl esters. Subsequently, in the second stage, the oil, which now has a low FFA content, can be subjected to the conventional homogeneous alkaline transesterification process without major problems. In addition to this widely known process, new alternative processes that are more tolerable to the presence of impurities are currently being explored, such as the use of heterogeneous catalysts (Aniokete et al., 2019; Degfie et al., 2019) and lipase biocatalysts (Tacias-Pascacio et al., 2017), as will be seen later.

10.5 CONVENTIONAL METHODS FOR BIODIESEL PRODUCTION

Interest in the use of renewable fuels began with the direct use of vegetable oils as a substitute for diesel. However, its direct use in compression ignition engines was restricted due to the high viscosity of the oil, which results in poor atomization, incomplete combustion, and carbon deposits in the injector and valves, causing serious engine fouling problems (Ramadhas et al., 2005). Subsequently, it was found that the products of the transesterification of these oils with methanol or ethanol were much better fuels due to their better physicochemical properties, mainly their lower viscosity (Alamu et al., 2008). The transesterification reaction of vegetable oils and animal fats, with alcohol (mainly methanol) in the presence of a catalyst, consists of three consecutive reversible reactions where triglycerides are converted to diglycerides, diglycerides are converted to monoglycerides, and the latter to glycerol. In each stage, an ester is produced, thus, for each triglyceride molecule, three methyl ester molecules are obtained, which are called biodiesel (Ma & Hanna, 1999; Sharma et al., 2008). The complete biodiesel production process includes the transesterification reaction, the separation of the ester-rich phase from the glycerol-rich phase, the purification of the ester-rich phase (biodiesel), and the recovery of glycerol from the glycerol-rich phase (Haas et al., 2006). Particularly, the purity of biodiesel is a matter of great importance, since the performance of biodiesel depends to a large extent on this; for this reason, the percentage of purity and other physicochemical parameters of biodiesel are specified in standards such as ASTM D6751 in North America and EN 14214 in Europe. The objective is to limit the presence of impurities that can affect the performance of the engine in the short or long term.

Stoichiometrically, three moles of alcohol are needed for every mole of triglyceride, but in general, higher molar ratios are used, depending on the type of feedstock, amount of catalyst, temperature, etc., to maximize ester production. Commonly used alcohols are methanol, ethanol, propanol, and butanol; however, since the performance of biodiesel is independent of the type of alcohol used, the selection of alcohol depends on costs and performance, which is why, due to its low cost and its physicochemical advantages (polar component and chain alcohol short), methanol is the most used alcohol (Berrios et al., 2010; Knothe, 2010; Ramadhas et al., 2005).

The most widely used method for the industrial production of biodiesel is alkaline transesterification. The reagents used in this reaction are methanol and catalysts such as potassium hydroxide (KOH) or sodium hydroxide (NaOH), due to their low cost and because they generate high yields in short reaction times (Darnoko & Cheryan, 2000; Ehsan & Chowdhury, 2015; Hameed et al., 2009). An important limitation of alkaline transesterification is that it requires high-quality raw materials since the presence of free fatty acids leads to the formation of soap. In order to make use of low-quality raw materials, it is necessary to carry out a two-stage process, which consists of an acid pretreatment to reduce the content of free fatty acids by esterification, and then continue with conventional transesterification (Gerpen, 2005). Due to these disadvantages, various types of catalysts have been explored, in such a way that, currently, biodiesel can be obtained by chemical catalysis with acid and alkaline homogeneous and acid, alkaline, or enzymatic heterogeneous catalysts (Figure 10.3) (Leung & Guo, 2006; Rodrigues et al., 2016). It is said that the catalysis is homogeneous when the catalyst used is in the same aggregation state as the rest of the reagents; while the catalysis will be heterogeneous when the catalyst used is in a different aggregation state than the rest of the reagents. Each of these types of catalysis has important advantages and disadvantages, which are shown in Table 10.3.

Due to the importance that alkaline homogeneous catalysis represents for biodiesel production, being the most widely used up to now, it is discussed in more detail in the next section.

10.5.1 ALKALINE HOMOGENEOUS CATALYSIS

Alkaline catalysis, due to its characteristics of high yields, low operating temperatures, and short reaction times, is the most widely used method in the industry for the production of biodiesel. Compared with acid catalysis, it is 4,000 times faster and requires a lower molar ratio of reactants (Kulkarni & Dalai, 2006).

The most commonly used alkaline catalysts are KOH and NaOH, which are characterized by having a high yield in the production of biodiesel, are

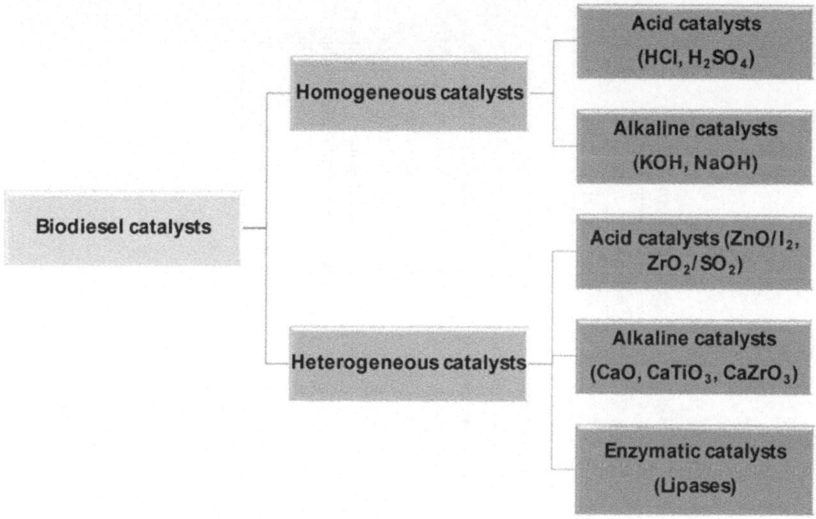

FIGURE 10.3 Types of catalysts used in biodiesel production.
Source: Adapted from Benti et al., 2023

easy to obtain, and generate a high conversion in a relatively short reaction time (Lotero et al., 2005). NaOH is cheaper and easier to obtain, so it is used more often, although the reaction catalyzed by KOH is faster (Atadashi et al., 2010). Methanol is the most widely used alcohol in the alkaline homogeneous transesterification reaction. Several studies agree that the optimum molar ratio of methanol:oil is 6:1; likewise, the reaction temperature considered optimal is in a range of 60°C to 70C (Ehsan & Chowdhury, 2015).

Some of the general advantages of alkaline catalysis are that it does not require special equipment, since the reaction occurs at moderate temperatures and atmospheric pressure, and the reaction time is short compared to other types of transesterifications; therefore, the capital and operating costs for biodiesel production facilities by the alkaline transesterification method are lower (Demirbas, 2006). The disadvantages of this method are the use of large amounts of energy, the glycerol produced is of low quality, so the separation and purification process is difficult and expensive, and it is a process that is very sensitive to the purity of the reagents, being affected by the content of free fatty acids and water (Rodrigues et al., 2016).

In alkaline homogeneous catalysis, the content of free fatty acids must not exceed certain limits, since their presence can promote saponification reactions (soap formation), which reduces the efficiency of the catalyst, decreases the yield of esters (biodiesel), and hinders subsequent separation

TABLE 10.3 Advantages and Disadvantages of the Types of Catalysis Used in the Production of Biodiesel (Leung & Guo, 2006)

	Homogeneous Catalysis		Heterogeneous Catalysis		
	Alkaline	**Acid**	**Alkaline**	**Acid**	**Enzymatic**
Advantages	High yield, short reaction time.	High yield; allows the use of low-quality raw materials; catalyzes esterification and transesterification simultaneously; avoids saponification.	Non-corrosive, recyclable; easy separation; high selectivity.	Catalyzes esterification and transesterification simultaneously; recyclable.	High yields and selectivity; catalyzes esterification and transesterification simultaneously; avoids the formation of soap; simplifies biodiesel purification; friendly to the environment; recyclable; allows the use of low-quality raw materials.
Disadvantages	It requires high-quality raw materials; saponification reactions can occur that make separation difficult and decrease biodiesel yield.	Equipment corrosion; excessive reaction times; high alcohol molar ratios and high amounts of catalyst.	Anhydrous conditions; high temperatures and pressures; diffusion limitations; catalyst poisoning.	Low microporosity; diffusion limitations.	Loss of activity due to pH, temperature, and solvents; high cost, long reaction times.

processes (Cao et al., 2008). It has been reported that an FFA content greater than 0.5% can affect the efficiency of the reaction when fresh oil is used as raw material (Freedman et al., 1986).

A successful transesterification by alkaline catalysis requires a raw material with a low content of free fatty acids, excess alcohol because it is the limiting reagent and has low solubility in oil, absence of moisture because the presence of water favors the saponification reaction, and a temperature that favors solubility and activation energy, although it must be kept below the boiling point of alcohol (Leung & Guo, 2006).

In the case of biodiesel production from feedstocks with a high content of free fatty acids (such as residual fats and oils and inedible oils), it is necessary to carry out alkaline transesterification to give a pretreatment to the raw material to reduce the free fatty acid content. This process is often referred to as the "integrated process" (Gerpen, 2005). In this sense, Alptekin et al. (2011) optimized the production of biodiesel from chicken fat with a high content of FFA through alkaline homogeneous catalysis. They used an integrated process, first performing the esterification using 20% sulfuric acid, a methanol/oil ratio of 40:1, 60°C, and 80 min; later, in transesterification, alkaline catalysts such as NaOMe, KOMe, KOH, and NaOH were used. The best yields found were 71.3% (4 h and 25°C) to 88.5% (1 h and 60°C) using NaOH and KOMe, respectively. In these results, it is possible to observe that, even when an optimized integrated process is used, the biodiesel yields do not reach the 99% yield, which has been widely reported for this type of catalysis when using high-purity edible vegetable oil. In addition, as has been mentioned in other sections of this work, the biodiesel produced from residual oils using this type of catalysis is biodiesel that, in addition to its low yield, will generate many difficulties during its recovery process, since in the middle reaction, impurities such as soap, catalyst residues, among others, will be found. That is why, in the case of low-quality raw materials, other methods that are more tolerant to the quality of the oil have been evaluated; among them, the enzymatic method has gained great importance due to the excellent results it has generated, as will be seen below.

10.6 ENZYMATIC TRANSESTERIFICATION OF RESIDUAL FATS AND OILS

Throughout this document, it has been mentioned that with the use of virgin vegetable oils, it is possible to obtain biodiesel in a simple way and with a high yield through homogeneous alkaline catalysis. This is because these oils are of high purity and have a low content of free fatty acids and moisture. However, the high cost of these oils means that biodiesel is not economically competitive against petroleum diesel. This has led to the need to use lower-cost raw materials characterized by their high content of free fatty acids. Base-catalyzed transesterification is not feasible for these materials since alkaline catalysts react with FFA to form soap and water, affecting yield and making biodiesel separation and purification difficult. In cases where it is necessary to process feedstocks with high FFA content, acid-catalyzed esterification is preferable to base-catalyzed transesterification because the former can convert FFA to biodiesel without producing soap. However, a large excess of alcohol and acid is needed to achieve a high yield of biodiesel. Additionally, this process generates highly toxic wastewater due to the presence of the catalyst and alcohol, which are difficult to recover at the end of the reaction (Robles-Medina et al., 2009). For this reason, when the oil is of low quality, biocatalysis is presented as an excellent alternative to carry out transesterification, since it overcomes the disadvantages of the conventional method.

Enzymatic catalysts have important advantages over conventional catalysts, such as greater compatibility with variations in the quality of the raw material, tolerating the use of cheap and low-quality raw materials, such as residual fats and oils and non-edible oils, with high FFA content; due to their high selectivity, they do not generate undesirable products such as the soap produced in alkaline catalysis; the recovery and purification of products is simpler; they require moderate reaction conditions and have low energy consumption. In addition, they considerably reduce the generation of waste, satisfying the demands of green chemistry, which, based on the current need to establish sustainable processes, makes the use of enzymes for the generation of biodiesel a viable alternative to chemical catalysts (Harding et al., 2008; Mandari & Devarai, 2021; Raza et al., 2022; Tongboriboon et al., 2010).

The enzymes mostly used in biodiesel production are lipases (triacylglycerol acylhydrolase EC 3.1.1.3) (Raza et al., 2022), which are carboxyl esterases that act on long-chain acyl glycerides. They are classified as serine hydrolases because they are inhibited by diethyl p-nitrophenol phosphate (Chapus et al., 1976). The active site of lipases is formed by a catalytic triad composed of serine, histidine, and aspartate or glutamate (Brady & Jordaan, 2009), and is protected by a cover or flap formed by an amphiphilic α-helical peptide sequence (Schmid & Verger, 1998). These enzymes are produced by microorganisms (fungi and bacteria), animals, and plants. However, commercial preparations are mainly derived from microbial sources due to their low cost of production and easy modification of their properties (by genetic or physicochemical methods) (Arpigny & Jaeger, 1999; Jaeger & Eggert, 2002). Some examples of lipase-producing microorganisms are *Candida antarctica, Rhizopus oryzae, Thermomyces lanuginosus, Mucor miehei, Pseudomonas alcaligenes, Bacillus subtilis, Pseudomonas mendocina, Pseudomonas cepacia,* and *Pseudomonas fluorescens* (Parawira, 2009).

Lipases can carry out not only hydrolytic but also synthetic reactions such as esterification, acidolysis, and alcoholysis, in aqueous and organic media; moreover, they have a broad substrate specificity and can catalyze a variety of chemo, regio, and enantioselective biotransformation's (Hasan et al., 2006).

In the case of biodiesel production, an "ideal" lipase should be non-stereospecific, in order to catalyze the conversion of all acylglycerides; it should be easy to produce, reusable, resistant to the presence of solvents, and variations of temperature and pH; present little inhibition by product, and achieve a high yield of biodiesel in a short reaction time. Until now, due to its excellent performance and characteristics, Novozyme 435 immobilized lipase (*Candida antarctica* lipase B) is considered the best commercial biocatalyst for biodiesel production. Much research has been carried out to develop and evaluate various lipase biocatalysts, with the objective of reaching or improving the performance of Novozyme 435 (Hirata et al., 2016; Palomo et al., 2002; Rodrigues et al., 2014; Tacias-Pascacio et al., 2017).

One of the main reasons that limits the industrial use of biocatalysis in biodiesel production is the high cost of the enzyme. This is partially overcome by the use of immobilized enzymes since immobilization allows reuse of the catalyst, and this has a positive effect on process costs (Narwal & Gupta, 2013); that is why, in biodiesel production, the use of immobilized enzymes is preferred instead of free enzymes. Immobilized enzymes are defined as "enzymes physically confined or localized in a certain defined region of space (support) with retention of their catalytic activities, and which can be used repeatedly and continuously" (Jegannathan et al., 2008). Enzyme immobilization produces reusable enzyme derivatives, which have better mechanical and thermal stability, and make the enzyme more tolerant to the presence of solvents such as ethanol or methanol; in addition, it simplifies the separation of the enzyme from the final product, avoiding biodiesel contamination by the presence of proteins (Tacias-Pascacio et al., 2016; Vargas et al., 2018). There are physical and chemical methods for enzyme immobilization; chemical methods include cross-linking and covalent binding, and physical methods include adsorption and entrapment (Zhang et al., 2012). An example of lipase immobilization by absorption on hydrophobic support is shown in Figure 10.4.

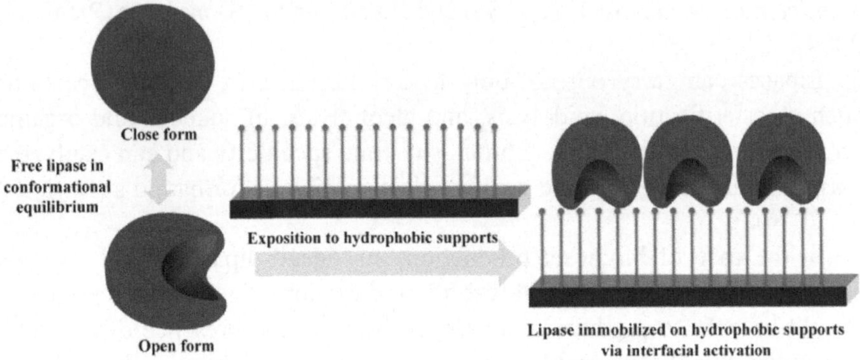

FIGURE 10.4 Lipase immobilization of hydrophobic supports by interfacial activation (Tacias-Pascacio et al., 2017).

There are several factors that affect the performance of enzyme catalysis; among them are the raw material used (an enzyme will respond differently depending on the nature of the triglycerides and fatty acids that are present); the presence of water; the temperature and pH used, since each enzyme has its optimum temperature and pH; the concentration and type of alcohol; the amount of enzyme, the stirring speed, among others (Vargas et al., 2018). That is why, in the different studies that have been carried out regarding biodiesel production by biocatalysis, these factors are studied, and in many cases optimized, in order to identify the conditions in which the biocatalyst generates the best biodiesel yields. Table 10.4 presents a compilation of works in which the enzymatic production of biodiesel from residual fats or oils is reported. It can be seen that the reaction conditions vary with respect to the raw material and the enzyme used, but in general, they are moderate conditions and the biodiesel yields obtained are high. In addition, Tacias et al. (2019), who made a comparison of four different types of catalysts (KOH, CaO, H_2SO_4, and lipase from *Thermomyces lanuginosus* immobilized on octadecyl methacrylate), mention that during their study they were able to corroborate that biodiesel obtained by enzymatic catalysis was cleaner, and the reaction required less energy than chemical catalysts (Tacias-Pascacio

TABLE 10.4 Studies Carried Out on the Biodiesel Production from Residual Fats and Oils by Enzymatic Catalysis with Lipases

Feedstock	Biocatalyst	Operational Conditions	Yield (%)	References
Waste cooking oil	Enzyme mixture (a 1:1 ratio of both *B. stearothermophilus* and *S. aureus* lipases) immobilized on $CaCO_3$.	Methanol to oil molar ratio: 1:6; catalysts content: 1% of the enzyme mixture; water content:1%; reaction time: 24 h; reaction temperature: 55°C.	97.66	Ben Bacha et al. (2022)
Waste palm cooking oil	Lipase AYS AMANO (immobilized *Candida rugosa* lipase).	Methanol to oil molar ratio: 10:1; catalysts content: 0.8 g; water content: 1 g; reaction time: 32 h; reaction temperature: 40°C.	85.72	Abdulla et al. (2022)
Chicken waste oil	A mixture of *C. rugosa* (1.5% wt.) and *R. miehei* (1.5% wt.) lipases immobilized on polyhydroxy butyrate particles.	Methanol to oil molar ratio: 6:1; catalysts content: 2.5%; water content: 5%; reaction time: 12 h; reaction temperature: 40°C; agitation rate: 200 rpm.	97.1	Sangkharak et al. (2020)
Waste cooking oil	Lipase B from *Candida Antarctica* immobilized on a magnetic hybrid sol-gel nanocomposite.	Methanol to oil molar ratio: 4:1; catalysts content: 1 g; reaction time: 30 h; reaction temperature: 40°C.	96	Jume et al. (2020)
Waste cooking oil	*Bacillus stratosphericus* PSP8 lipase immobilized on NH2-Pr-SBA-15.	Methanol to oil molar ratio: 3.12:1; catalysts content: 3.08%; reaction time: 3.19 h; reaction temperature: 48.6°C; agitation rate: 495.53 rpm.	97.01	Ismail et al. (2022)
Free fatty acids from residual chicken oil	The cocktail is composed of 67% lipase B from Candida antarctica and 33% lipase from *Rhizomucor miehei.*	Ethanol to FFA molar ratio: 5:1; catalysts content: 15%; reaction time: 3 h; reaction temperature: 30°C	89.95	Rocha et al. (2021)
Waste frying palm oil	Free lipase A from *Candida antarctica.*	Methanol to oil molar ratio: 7:1; catalysts content: 5.5%; water content: 16.6%; reaction time: 22 h; reaction temperature: 30°C.	94.6	Guo et al. (2020)
Waste cooking oil	Lipase from porcine pancreas immobilized on genipin cross-linked chitosan.	Methanol to oil molar ratio: 9:1; catalysts content: 7.5%; water content: 6%; reaction time: 10 h; reaction temperature: 40°C; agitation rate: 150 rpm.	92.3	Khan et al. (2020)
Fish waste oil	Lipase 1,3 regioselective from *Thermomyces lanuginosus* immobilized on octadecyl methacrylate.	Methanol to oil molar ratio: 3:1; catalysts content: 10%; water content: 1%; reaction time: 24 h; reaction temperature: 35°C; agitation rate: 216 rpm.	75.3	Ching-Velasquez et al. (2020)
Waste cooking oil	Lipase 1,3 regioselective from *Thermomyces lanuginosus* immobilized on octadecyl methacrylate.	Methanol to oil molar ratio: 9:1; catalysts content: 7%; water content: 10%; reaction time: 6 h; reaction temperature: 30°C; agitation rate: 350 rpm.	89	Tacias-Pascacio et al. (2019)

et al., 2019). Furthermore, many studies carried out show that the fuel and chemical properties of enzymatic biodiesel obtained from residual oils and fats can satisfy the international specifications that regulate the quality of biodiesel (ASTM D6751 and EN 14214) (Ben Bacha et al., 2022; Ching-Velasquez et al., 2020; Sangkharak et al., 2020).

Finally, the works presented in Table 10.4 show that agro-food waste can be successfully used as a raw material for biodiesel production through the use of lipases, which are environmentally friendly biocatalysts that avoid the technical disadvantages that conventional catalysts present when low-quality raw materials are used.

10.7 CONCLUSION

Residual fats and oils that are generated in large quantities around the world, if not reused or treated, can seriously contribute to environmental pollution. A solution to this growing problem is the conversion of these materials into biodiesel, which, in addition to reducing the environmental impact of these residues, reduces the cost of biodiesel production. However, the most widely used method for biodiesel production (alkaline transesterification) is not feasible for the conversion of this type of low-quality raw material, since the presence of free fatty acids and water leads to the formation of soap, which decreases the biodiesel yield, affects its quality, hinders the separation and purification processes, and generates large amounts of wastewater. In this case, enzymatic catalysis with lipases is presented as an excellent alternative to achieve the conversion of residual fats and oils, without the drawbacks of conventional catalysis, with the additional advantage of generating cleaner products with less energy consumption, since enzymatic transesterification is an environmentally friendly process. Until now, the main limitation for the industrial use of lipases in biodiesel production is the price of the enzyme; however, efforts continue to be made to reduce the cost of the catalyst, including the immobilization of the enzyme on suitable supports, which makes it more resistant to reaction conditions and allow its reuse in several cycles. Reuse is a key aspect to improve the economic feasibility of using enzymes in the production of this biofuel, which is why many investigations are carried out around enzymatic immobilization; in this way, through the contributions of current and future research, it is expected that there will soon be economically affordable, reusable biocatalysts that allow high biodiesel yields to be obtained in a short reaction time; in such a way that, the industrial use of residual fats and oils in the enzymatic production of biodiesel becomes a reality; which would bring enormous benefits to the environment and will contribute to the sustainability of biodiesel production, being consistent with the current paradigm of circular bioeconomy.

KEYWORDS

- **biocatalysts**
- **biodiesel**
- **circular bioeconomy**
- **enzymatic catalysis**
- **enzymatic production**
- **lipases**
- **residual fats**
- **residual fats and oils**
- **sustainability**

REFERENCES

Abdulla, R., Derman, E., K. Mathialagan, T., Yaser, A. Z., Abu Samah, M. A., Gansau, J. A., & Syed Najmuddin, S. U. F. (2022). Biodiesel Production from Waste Palm Cooking Oil Using Immobilized Candida rugosa Lipase. *Sustainability, 14*(20), 13632.

Adewale, P., Dumont, M. J., & Ngadi, M. (2015). Recent trends of biodiesel production from animal fat wastes and associated production techniques. *Renewable and Sustainable Energy Reviews, 45*, 574–588. https://doi.org/10.1016/j.rser.2015.02.039.

Ahmad, A. L., Yasin, N. M., Derek, C., & Lim, J. (2011). Microalgae as a sustainable energy source for biodiesel production: a review. *Renewable and Sustainable Energy Reviews, 15*(1), 584–593.

Akoh, C. C., Chang, S. W., Lee, G. C., & Shaw, J. F. (2007). Enzymatic approach to biodiesel production. *Journal of Agricultural and Food Chemistry, 55*(22), 8995–9005.

Alamu, O., Waheed, M., & Jekayinfa, S. (2008). Effect of ethanol–palm kernel oil ratio on alkali-catalyzed biodiesel yield. *Fuel, 87*(8–9), 1529–1533.

Alaswad, A., Dassisti, M., Prescott, T., & Olabi, A. G. (2015). Technologies and developments of third generation biofuel production. *Renewable and Sustainable Energy Reviews, 51*, 1446–1460.

Alengebawy, A., Mohamed, B. A., Ghimire, N., Jin, K., Liu, T., Samer, M., & Ai, P. (2022). Understanding the environmental impacts of biogas utilization for energy production through life cycle assessment: An action towards reducing emissions. *Environmental Research, 213*, 113632.

Alptekin, E., & Canakci, M. (2011). Optimization of transesterification for methyl ester production from chicken fat. *Fuel, 90*(8), 2630–2638.

Al-Zuhair, S. (2007). Production of biodiesel: possibilities and challenges. *Biofuels, Bioproducts and Biorefining: Innovation for a Sustainable Economy, 1*(1), 57–66.

Aniokete, T., Mbhele, S., Mdlalani, V., Ozonoh, M., & Daramola, M. (2019). Kinetic study of waste-derived solid hydroxy sodalite catalyst during transesterification of animal fat oil to biodiesel in a batch reactor. *Journal of Physics: Conference Series.*

Aniokete, T., Ozonoh, M., & Daramola, M. O. (2019). Synthesis of pure and high surface area sodalite catalyst from waste industrial brine and coal fly ash for conversion of waste cooking oil (WCO) to biodiesel. *International Journal of Renewable Energy Research (IJRER), 9*(4), 1924–1937.

Aniokete, T., Sadare, O., & Daramola, M. (2022). Prospects of biodiesel production from waste animal fats. In *Waste and Biodiesel* (pp. 17–44). Elsevier.

Arpigny, J. L., & JAEGER, K.-E. (1999). Bacterial lipolytic enzymes: classification and properties. *Biochemical Journal, 343*(1), 177–183.

Atadashi, I. M., Aroua, M. K., & Aziz, A. A. (2010). High quality biodiesel and its diesel engine application: A review. *Renewable and Sustainable Energy Reviews, 14*(7), 1999–2008. https://doi.org/10.1016/j.rser.2010.03.020.

Atadashi, I., Aroua, M. K., Aziz, A. A., & Sulaiman, N. (2012). Production of biodiesel using high free fatty acid feedstocks. *Renewable and Sustainable Energy Reviews, 16*(5), 3275–3285.

Ataya, F., Dubé, M. A., & Ternan, M. (2007). Acid-catalyzed transesterification of canola oil to biodiesel under single- and two-phase reaction conditions. *Energy & Fuels, 21*(4), 2450–2459. https://doi.org/10.1021/ef0701440.

Avagyan, A. B., & Singh, B. (2019). Biodiesel: feedstocks, technologies, economics and barriers. *Assessment of Environmental Impact in Producing and Using Chains.*

Avhad, M., & Marchetti, J. (2015). A review on recent advancement in catalytic materials for biodiesel production. *Renewable and Sustainable Energy Reviews, 50*, 696–718.

Azócar, L., Ciudad, G., Heipieper, H. J., & Navia, R. (2010). Biotechnological processes for biodiesel production using alternative oils. *Applied Microbiology and Biotechnology, 88*, 621–636.

Balat, M. (2011). Potential alternatives to edible oils for biodiesel production – A review of current work. *Energy Conversion and Management, 52*(2), 1479–1492.

Balat, M., & Balat, H. (2009). Recent trends in global production and utilization of bio-ethanol fuel. *Applied Energy, 86*(11), 2273–2282.

Bart, J. C., Palmeri, N., & Cavallaro, S. (2010). *Biodiesel Science and Technology: From Soil to Oil*. Elsevier.

Behzadi, S., & Farid, M. (2007). Examining the use of different feedstock for the production of biodiesel. *Asia-Pacific Journal of Chemical Engineering, 2*(5), 480–486.

Ben Bacha, A., Alonazi, M., Alharbi, M. G., Horchani, H., & Ben Abdelmalek, I. (2022). Biodiesel Production by Single and Mixed Immobilized Lipases Using Waste Cooking Oil. *Molecules, 27*(24), 8736.

Benti, N. E., Aneseyee, A. B., Geffe, C. A., Woldegiyorgis, T. A., Gurmesa, G. S., Bibiso, M., Asfaw, A. A., Milki, A, W., Mekonnen, Y. S. (2023). Biodiesel production in Ethiopia:

Current status and future prospects. *Scientific African, 19*, e01531. https://doi.org/10.1016/j.sciaf.2022.e01531.

Benti, N. E., Mekonnen, Y. S., Asfaw, A. A., Lerra, M. D., Woldegiyorgis, T. A., Gaffe, C. A., & Aneseyee, A. B. (2022). Techno-economic analysis of solar energy system for electrification of rural school in Southern Ethiopia. *Cogent Engineering, 9*(1), 2021838.

Berrios, M., Martín, M., Chica, A., & Martín, A. (2010). Study of esterification and transesterification in biodiesel production from used frying oils in a closed system. *Chemical Engineering Journal, 160*(2), 473–479.

Brady, D., & Jordaan, J. (2009). Advances in enzyme immobilisation. *Biotechnology Letters, 31*(11), 1639.

Canesin, E. A., de Oliveira, C. C., Matsushita, M., Dias, L. F., Pedrão, M. R., & de Souza, N. E. (2014). Characterization of residual oils for biodiesel production. *Electronic Journal of Biotechnology, 17*(1), 39–45. https://doi.org/10.1016/j.ejbt.2013.12.007.

Cao, F., Chen, Y., Zhai, F., Li, J., Wang, J., Wang, X., Wang, S., & Zhu, W. (2008). Biodiesel production from high acid value waste frying oil catalyzed by superacid heteropolyacid. *Biotechnology and Bioengineering, 101*(1), 93–100.

Chapus, C., Semeriva, M., Bovier-Lapierre, C., & Desnuelle, P. (1976). Mechanism of pancreatic lipase action. Interfacial activation of pancreatic lipase. *Biochemistry, 15*(23), 4980–4987.

Chhetri, A. B., Watts, K. C., & Islam, M. R. (2008). Waste cooking oil as an alternate feedstock for biodiesel production. *Energies, 1*(1), 3–18.

Ching-Velasquez, J., Fernández-Lafuente, R., Rodrigues, R. C., Plata, V., Rosales-Quintero, A., Torrestiana-Sánchez, B., & Tacias-Pascacio, V. G. (2020). Production and characterization of biodiesel from oil of fish waste by enzymatic catalysis. *Renewable Energy.*

Chisti, Y. (2007). Biodiesel from microalgae. *Biotechnology Advances, 25*(3), 294–306.

Corsini, A., Marchegiani, A., Rispoli, F., Sciulli, F., & Venturini, P. (2015). Vegetable oils as fuels in diesel engine. Engine performance and emissions. *Energy Procedia, 81*, 942–949.

Darnoko, D., & Cheryan, M. (2000). Kinetics of palm oil transesterification in a batch reactor. *Journal of the American Oil Chemists' Society, 77*(12), 1263–1267. https://doi.org/10.1007/s11746-000-0198-y.

Degfie, T. A., Mamo, T. T., & Mekonnen, Y. S. (2019). Optimized biodiesel production from waste cooking oil (WCO) using calcium oxide (CaO) nano-catalyst. *Scientific Reports, 9*(1), 18982.

Demirbas, A. (2006). Biodiesel production via non-catalytic SCF method and biodiesel fuel characteristics. *Energy Conversion and Management, 47*(15–16), 2271–2282.

Demirbas, A. (2007). Importance of biodiesel as transportation fuel. *Energy Policy, 35*(9), 4661–4670.

Dias, J. M., Alvim-Ferraz, M. C. M., & Almeida, M. F. (2009). Production of biodiesel from acid waste lard. *Bioresource Technology, 100*(24), 6355–6361. https://doi.org/10.1016/j.biortech.2009.07.025.

Edrisi, S. A., & Abhilash, P. (2016). Exploring marginal and degraded lands for biomass and bioenergy production: An Indian scenario. *Renewable and Sustainable Energy Reviews, 54*, 1537–1551.

Ehsan, M., & Chowdhury, M. T. H. (2015). Production of Biodiesel Using Alkaline Based Catalysts from Waste Cooking Oil: A Case Study. *Procedia Engineering, 105*, 638–645. https://doi.org/10.1016/j.proeng.2015.05.042.

Encinar, J. M., Gonzalez, J. F., & Rodríguez-Reinares, A. (2005). Biodiesel from used frying oil. Variables affecting the yields and characteristics of the biodiesel. *Industrial & Engineering Chemistry Research, 44*(15), 5491–5499.

Fazal, M., Haseeb, A., & Masjuki, H. (2011). Biodiesel feasibility study: an evaluation of material compatibility; performance; emission and engine durability. *Renewable and Sustainable Energy Reviews, 15*(2), 1314–1324.

Felizardo, P., Correia, M. J. N., Raposo, I., Mendes, J. F., Berkemeier, R., & Bordado, J. M. (2006). Production of biodiesel from waste frying oils. *Waste Management, 26*(5), 487–494.

Ferella, F., Di Celso, G. M., De Michelis, I., Stanisci, V., & Vegliò, F. (2010). Optimization of the transesterification reaction in biodiesel production. *Fuel, 89*(1), 36–42.

Freedman, B., Butterfield, R. O., & Pryde, E. H. (1986). Transesterification kinetics of soybean oil 1. *Journal of the American Oil Chemists' Society, 63*, 1375–1380.

Fukuda, H., Kondo, A., & Noda, H. (2001). Biodiesel fuel production by transesterification of oils. *Journal of Bioscience and Bioengineering, 92*(5), 405–416. https://doi.org/10.1016/S1389-1723(01)80288-7.

Gerpen, J. V. (2005). Biodiesel processing and production. *Fuel Processing Technology, 86*(10), 1097–1107. https://doi.org/10.1016/j.fuproc.2004.11.005.

Gissibl, A., Sun, A., Care, A., Nevalainen, H., & Sunna, A. (2019). Bioproducts from Euglena gracilis: synthesis and applications. *Frontiers in Bioengineering and Biotechnology, 7*, 108.

Gunvachai, K., Hassan, M., Shama, G., & Hellgardt, K. (2007). A new solubility model to describe biodiesel formation kinetics. *Process Safety and Environmental Protection, 85*(5), 383–389.

Guo, J., Sun, S., & Liu, J. (2020). Conversion of waste frying palm oil into biodiesel using free lipase A from Candida antarctica as a novel catalyst. *Fuel, 267*, 117323.

Gürü, M., Koca, A., Can, Ö., Çınar, C., & Şahin, F. (2010). Biodiesel production from waste chicken fat based sources and evaluation with Mg based additive in a diesel engine. *Renewable Energy, 35*(3), 637–643. https://doi.org/10.1016/j.renene.2009.08.011.

Haas, M. J., McAloon, A. J., Yee, W. C., & Foglia, T. A. (2006). A process model to estimate biodiesel production costs. *Bioresource Technology, 97*(4), 671–678.

Hameed, B. H., Lai, L. F., & Chin, L. H. (2009). Production of biodiesel from palm oil (*Elaeis guineensis*) using heterogeneous catalyst: An optimized process. *Fuel Processing Technology, 90*(4), 606–610. https://doi.org/10.1016/j.fuproc.2008.12.014.

Harding, K. G., Dennis, J. S., von Blottnitz, H., & Harrison, S. T. L. (2008). A life-cycle comparison between inorganic and biological catalysis for the production of biodiesel. *Journal of Cleaner Production, 16*(13), 1368–1378. https://doi.org/10.1016/j.jclepro.2007.07.003.

Hasan, F., Shah, A. A., & Hameed, A. (2006). Industrial applications of microbial lipases. *Enzyme and Microbial Technology, 39*(2), 235–251. https://doi.org/10.1016/j.enzmictec.2005.10.016.

Hirata, D. B., Albuquerque, T. L., Rueda, N., Virgen-Ortíz, J. J., Tacias-Pascacio, V. G., & Fernandez-Lafuente, R. (2016). Evaluation of different immobilized lipases in transesterification reactions using tributyrin: Advantages of the heterofunctional octyl agarose beads. *Journal of Molecular Catalysis B: Enzymatic, 133*, 117–123.

Ismail, A. R., Kashtoh, H., Betiha, M. A., Abu Amr, S. A., Baek, K. H., & El-Gendy, N. S. (2022). Valorization of Waste Cooking Oil into Biodiesel via *Bacillus stratosphericus* Lipase Amine-Functionalized Mesoporous SBA-15 Nanobiocatalyst. *International Journal of Chemical Engineering, 2022.*

Jaeger, K.-E., & Eggert, T. (2002). Lipases for biotechnology. *Current Opinion in Biotechnology, 13*(4), 390–397.

Jegannathan, K. R., Abang, S., Poncelet, D., Chan, E. S., & Ravindra, P. (2008). Production of biodiesel using immobilized lipase – A critical review. *Critical Reviews in Biotechnology, 28*(4), 253–264. https://doi.org/10.1080/07388550802428392.

Jegannathan, K. R., Eng-Seng, C., & Ravindra, P. (2011). Economic assessment of biodiesel production: Comparison of alkali and biocatalyst processes. *Renewable and Sustainable Energy Reviews, 15*(1), 745–751.

Jume, B. H., Gabris, M. A., Nodeh, H. R., Rezania, S., & Cho, J. (2020). Biodiesel production from waste cooking oil using a novel heterogeneous catalyst based on graphene oxide doped metal oxide nanoparticles. *Renewable Energy, 162*, 2182–2189.

Kemp, W. H. (2006). *Biodiesel: Basics and Beyond: A Comprehensive Guide to Production and Use for the Home and Farm.* Tamworth, Ont.: Aztext Press.

Khan, N., Maseet, M., & Basir, S. F. (2020). Synthesis and characterization of biodiesel from waste cooking oil by lipase immobilized on genipin cross-linked chitosan beads: A green approach. *International Journal of Green Energy, 17*(1), 84–93.

Kim, D. S., Hanifzadeh, M., & Kumar, A. (2018). Trend of biodiesel feedstock and its impact on biodiesel emission characteristics. *Environmental Progress & Sustainable Energy, 37*(1), 7–19.

Kirubakaran, M., & Selvan, V. A. M. (2018). A comprehensive review of low cost biodiesel production from waste chicken fat. *Renewable and Sustainable Energy Reviews, 82*, 390–401.

Knothe, G. (2010). Biodiesel and renewable diesel: A comparison. *Progress in Energy and Combustion Science, 36*(3), 364–373.

Kulkarni, M. G., & Dalai, A. K. (2006). Waste cooking oil – An economical source for biodiesel: A review. *Industrial & Engineering Chemistry Research, 45*(9), 2901–2913.

Leung, D., & Guo, Y. (2006). Transesterification of neat and used frying oil: Optimization for biodiesel production. *Fuel Processing Technology, 87*(10), 883–890.

Li, X., Xu, H., & Wu, Q. (2007). Large-scale biodiesel production from microalga *Chlorella protothecoides* through heterotrophic cultivation in bioreactors. *Biotechnology and Bioengineering, 98*(4), 764–771.

Lotero, E., Liu, Y., Lopez, D. E., Suwannakarn, K., Bruce, D. A., & Goodwin, J. G. (2005). Synthesis of biodiesel via acid catalysis. *Industrial & Engineering Chemistry Research, 44*(14), 5353–5363.

Lu, J., Nie, K., Xie, F., Wang, F., & Tan, T. (2007). Enzymatic synthesis of fatty acid methyl esters from lard with immobilized *Candida* sp., 99–125. *Process Biochemistry, 42*(9), 1367–1370.

Ma, F., & Hanna, M. A. (1999). Biodiesel production: A review. *Bioresource Technology, 70*(1), 1–15.

Maheswari, P., Haider, M. B., Yusuf, M., Klemeš, J. J., Bokhari, A., Beg, M., Al-Othman, A., Kumar, R., & Jaiswal, A. K. (2022). A review on latest trends in cleaner biodiesel production: Role of feedstock, production methods, and catalysts. *Journal of Cleaner Production, 131588.*

Mandari, V., & Devarai, S. K. (2021). Biodiesel production using homogeneous, heterogeneous, and enzyme catalysts via transesterification and esterification reactions: A critical review. *BioEnergy Research, 1–27.*

Molina Grima, E., Garcia Camacho, F., & Acien Fernandez, F. (1999). Production of EPA from *Phaeodactylum tricornutum*. In *Chemicals from Microalgae* (pp. 57–92).

Moser, B. R. (2009). Biodiesel production, properties, and feedstocks. *In Vitro Cellular & Developmental Biology – Plant, 45*(3), 229–266. https://doi.org/10.1007/s11627-009-9204-z.

Narwal, S. K., & Gupta, R. (2013). Biodiesel production by transesterification using immobilized lipase. *Biotechnology Letters, 35*, 479–490.

Neupane, D. (2022). Biofuels from renewable sources, a potential option for biodiesel production. *Bioengineering, 10*(1), 29.

Odetoye, T., Agu, J., & Ajala, E. (2021). Biodiesel production from poultry wastes: Waste chicken fat and eggshell. *Journal of Environmental Chemical Engineering, 9*(4), 105654.

Ogunkunle, O., & Ahmed, N. A. (2021). Overview of biodiesel combustion in mitigating the adverse impacts of engine emissions on the sustainable human–environment scenario. *Sustainability, 13*(10), 5465.

Öner, C., & Altun, Ş. (2009). Biodiesel production from inedible animal tallow and an experimental investigation of its use as alternative fuel in a direct injection diesel engine. *Applied Energy, 86*(10), 2114–2120.

Palmer, J. D., & Brigham, C. J. (2016). Feasibility of triacylglycerol production for biodiesel, utilizing *Rhodococcus opacus* as a biocatalyst and fishery waste as feedstock. *Renewable and Sustainable Energy Reviews, 56*, 922–928.

Palomo, J. M., Fernandez-Lorente, G., Mateo, C., Ortiz, C., Fernandez-Lafuente, R., & Guisan, J. M. (2002). Modulation of the enantioselectivity of lipases via controlled immobilization and medium engineering: Hydrolytic resolution of mandelic acid esters. *Enzyme and Microbial Technology, 31*(6), 775–783.

Parawira, W. (2009). Biotechnological production of biodiesel fuel using biocatalysed transesterification: A review. *Critical Reviews in Biotechnology, 29*(2), 82–93.

Pasha, M. K., Dai, L., Liu, D., Guo, M., & Du, W. (2021). An overview to process design, simulation and sustainability evaluation of biodiesel production. *Biotechnology for Biofuels, 14*(1), 1–23.

Phan, A. N., & Phan, T. M. (2008). Biodiesel production from waste cooking oils. *Fuel, 87*(17), 3490–3496. https://doi.org/10.1016/j.fuel.2008.07.008.

Poppe, J. K., Fernandez-Lafuente, R., Rodrigues, R. C., & Ayub, M. A. Z. (2015). Enzymatic reactors for biodiesel synthesis: Present status and future prospects. *Biotechnology Advances, 33*(5), 511–525.

Poppe, J. K., Matte, C. R., Peralba, M. D. C. R., Fernandez-Lafuente, R., Rodrigues, R. C., & Ayub, M. A. Z. (2015). Optimization of ethyl ester production from olive and palm oils using mixtures of immobilized lipases. *Applied Catalysis A: General, 490*, 50–56. https://doi.org/10.1016/j.apcata.2014.10.050.

Predojević, Z. J. (2008). The production of biodiesel from waste frying oils: A comparison of different purification steps. *Fuel, 87*(17–18), 3522–3528.

Ramadhas, A. S., Jayaraj, S., & Muraleedharan, C. (2005). Biodiesel production from high FFA rubber seed oil. *Fuel, 84*(4), 335–340. https://doi.org/10.1016/j.fuel.2004.09.016.

Raza, M., Ali, L., Inayat, A., Rocha-Meneses, L., Ahmed, S. F., Mofijur, M., Jamil, F., & Azimoh, C. L. (2022). Sustainability of biodiesel production using immobilized enzymes: A strategy to meet future bio-economy challenges. *International Journal of Energy Research, 46*(13), 19090–19108.

Robles-Medina, A., González-Moreno, P. A., Esteban-Cerdán, L., & Molina-Grima, E. (2009). Biocatalysis: Towards ever greener biodiesel production. *Biotechnology Advances, 27*(4), 398–408. https://doi.org/10.1016/j.biotechadv.2008.10.008.

Rocha, T. G., de L. Gomes, P. H., de Souza, M. C., Monteiro, R. R., & dos Santos, J. C. (2021). Lipase cocktail for optimized biodiesel production of free fatty acids from residual chicken oil. *Catalysis Letters, 151*, 1155–1166.

Rodrigues, J., Canet, A., Rivera, I., Osório, N., Sandoval, G., Valero, F., & Ferreira-Dias, S. (2016). Biodiesel production from crude jatropha oil catalyzed by non-commercial immobilized heterologous rhizopus oryzae and carica papaya lipases. Bioresource Technology, *213*, 88–95.

Rodrigues, R. C., Barbosa, O., Ortiz, C., Berenguer-Murcia, Á., Torres, R., & Fernandez-Lafuente, R. (2014). Amination of enzymes to improve biocatalyst performance: Coupling genetic modification and physicochemical tools. *RSC Advances, 4*(72), 38350–38374.

Saleh, J., Tremblay, A. Y., & Dubé, M. A. (2010). Glycerol removal from biodiesel using membrane separation technology. *Fuel, 89*(9), 2260–2266. https://doi.org/10.1016/j.fuel.2010.04.025.

Sangkharak, K., Mhaisawat, S., Rakkan, T., Paichid, N., & Yunu, T. (2020). Utilization of mixed chicken waste for biodiesel production using single and combination of immobilized lipase as a catalyst. *Biomass Conversion and Biorefinery*, 1–14.

Schmid, R. D., & Verger, R. (1998). Lipases: Interfacial enzymes with attractive applications. *Angewandte Chemie International Edition, 37*(12), 1608–1633.

Sharma, Y., Singh, B., & Upadhyay, S. (2008). Advancements in development and characterization of biodiesel: A review. *Fuel, 87*(12), 2355–2373.

Singh, A. R., Singh, S. K., & Jain, S. (2022). A review on bioenergy and biofuel production. *Materials Today: Proceedings, 49*, 510–516.

Singh, A., Smyth, B. M., & Murphy, J. D. (2010). A biofuel strategy for Ireland with an emphasis on production of biomethane and minimization of land-take. *Renewable and Sustainable Energy Reviews, 14*(1), 277–288.

Tacias Pascacio, V. G., Rosales Quintero, A., & Torrestiana Sánchez, B. (2016). Evaluación y caracterización de grasas y aceites residuales de cocina para la producción de biodiésel: Un caso de estudio. *Revista Internacional de Contaminación Ambiental, 32*(3), 303–313.

Tacias-Pascacio, V. G., Peirce, S., Torrestiana-Sanchez, B., Yates, M., Rosales-Quintero, A., Virgen-Ortíz, J. J., & Fernandez-Lafuente, R. (2016). Evaluation of different commercial hydrophobic supports for the immobilization of lipases: Tuning their stability, activity, and specificity. *RSC Advances, 6*(102), 100281–100294. https://doi.org/10.1039/C6RA21730C.

Tacias-Pascacio, V. G., Torrestiana-Sánchez, B., Dal Magro, L., Virgen-Ortíz, J. J., Suárez-Ruíz, F. J., Rodrigues, R. C., & Fernandez-Lafuente, R. (2019). Comparison of acid, basic and enzymatic catalysis on the production of biodiesel after RSM optimization. *Renewable Energy, 135*, 1–9. https://doi.org/10.1016/j.renene.2018.11.107.

Tacias-Pascacio, V. G., Virgen-Ortíz, J. J., Jiménez-Pérez, M., Yates, M., Torrestiana-Sanchez, B., Rosales-Quintero, A., & Fernandez-Lafuente, R. (2017). Evaluation of different lipase biocatalysts in the production of biodiesel from used cooking oil: Critical role of the immobilization support. *Fuel, 200*, 1–10. https://doi.org/10.1016/j.fuel.2017.03.054.

Talebian-Kiakalaieh, A., Amin, N. A. S., & Mazaheri, H. (2013). A review on novel processes of biodiesel production from waste cooking oil. *Applied Energy, 104*, 683–710.

Tayari, S., Abedi, R., & Rahi, A. (2020). Comparative assessment of engine performance and emissions fueled with three different biodiesel generations. *Renewable Energy, 147*, 1058–1069.

Toldrá-Reig, F., Mora, L., & Toldrá, F. (2020). Trends in biodiesel production from animal fat waste. *Applied Sciences, 10*(10), 3644.

Tomasevic, A. V., & Siler-Marinkovic, S. S. (2003). Methanolysis of used frying oil. *Fuel Processing Technology, 81*(1), 1–6. https://doi.org/10.1016/S0378-3820(02)00096-6.

Tongboriboon, K., Cheirsilp, B., & Aran, H. (2010). Mixed lipases for efficient enzymatic synthesis of biodiesel from used palm oil and ethanol in a solvent-free system. *Journal of Molecular Catalysis B: Enzymatic, 67*(1–2), 52–59.

Varanda, M. G., Pinto, G., & Martins, F. (2011). Life cycle analysis of biodiesel production. *Fuel Processing Technology, 92*(5), 1087–1094. https://doi.org/10.1016/j.fuproc.2011.01.003.

Vargas, M., Niehus, X., Casas-Godoy, L., & Sandoval, G. (2018). Lipases as biocatalyst for biodiesel production. In *Lipases and Phospholipases: Methods and Protocols*, 377–390.

Vasudevan, P. T., & Briggs, M. (2008). Biodiesel production—current state of the art and challenges. *Journal of Industrial Microbiology and Biotechnology, 35*(5), 421.

Yadav, A. K., Khan, M. E., Pal, A., & Dubey, A. M. (2016). Biodiesel production from Nerium oleander (Thevetia peruviana) oil through conventional and ultrasonic irradiation methods. *Energy Sources, Part A: Recovery, Utilization, and Environmental Effects, 38*(23), 3447–3452.

Yadav, A. K., Khan, M. E., Pal, A., & Dubey, A. M. (2017). Experimental investigations of performance and emissions characteristics of Kusum (Schleichera oleosa) biodiesel in a multi-cylinder transportation diesel engine. *Waste and Biomass Valorization, 8*, 1331–1341.

Yaqoob, H., Teoh, Y. H., Sher, F., Farooq, M. U., Jamil, M. A., Kausar, Z., Sabah, N. U., Shah, M. F., Rehman, H. Z. U., & Rehman, A. U. (2021). Potential of waste cooking oil biodiesel as renewable fuel in combustion engines: A review. *Energies, 14*(9), 2565.

Zhang, B., Weng, Y., Xu, H., & Mao, Z. (2012). Enzyme immobilization for biodiesel production. *Applied Microbiology and Biotechnology, 93*, 61–70.

Zhang, Y., Dubé, M. A., McLean, D. D., & Kates, M. (2003). Biodiesel production from waste cooking oil: 1. Process design and technological assessment. *Bioresource Technology, 89*(1), 1–16. https://doi.org/10.1016/S0960-8524(03)00040-3.

CHAPTER 11

Plant Extracts as Elicitors in Plants Against Diseases: Action Mode and Perspectives

MARCO ANTONIO TUCUCH-PÉREZ,[1] ROBERTO ARREDONDO-VALDÉS,[1] ELAN IÑAKY LAREDO ALCALÁ,[2] MÓNICA LIZETH CHÁVEZ GONZÁLEZ,[1] CRISTÓBAL NOÉ AGUILAR GONZÁLEZ,[1] FRANCISCO DANIEL HERNÁNDEZ CASTILLO,[3] AGUSTÍN HERNÁNDEZ JUAREZ,[3] EPIFANIO CASTRO DEL ÁNGEL,[3] and JULIA CECILIA ANGUIANO CABELLO[4]

[1]*School of Chemistry, Autonomous University of Coahuila, Saltillo, Coahuila, Mexico, ZC*

[2]*Research Center for the Conservation of Biodiversity and Ecology of Coahuila, Autonomous University of Coahuila, Cuatrociénegas, Coahuila, Mexico, ZC*

[3]*Department of Parasitology, Autonomous Agrarian University Antonio Narro, Saltillo, Coahuila, Mexico, ZC*

[4]*Health Sciences Area, La Salle Saltillo University, Saltillo, Coahuila, Mexico, ZC*

ABSTRACT

Food production is affected by phytopathogenic microorganisms that cause diseases in plants. Hence, it is necessary to control the causal agents of the diseases. The use of chemical products is the most common method. Nevertheless, the indiscriminate use of these products has environmental effects, causes problems in human health, and induces resistance in microorganisms

Sustainable Agriculture and Global Environmental Health: Toward the Efficient Management of Resources. Mónica L. Chávez González, Cristóbal Noé Aguilar, Juan Carlos Contreras-Esquivel, and A. K. Haghi (Eds.)

toward the active ingredients. Also, the necessity to change the food production paradigm to sustainable production has led to the development of natural alternatives to control plant diseases. Thus, plant extracts (PEs) emerge as a natural alternative. The activation of a plant's immune system by elicitors (Es) before an attack by phytopathogenic microorganisms is a mode of disease management; these Es activate local immune responses such as pattern-triggered immunity (PTI) or effector-triggered immunity (ETI); they can also activate systemic responses called systemic acquired resistance (SAR) or induced systemic resistance (ISR). These responses involve the production of substances capable of protecting the plants from infection. Plant extracts' phytochemical compounds (PCs) induce immune responses through several mechanisms. The increased use of PEs in sustainable agriculture makes it necessary to understand the action modes of the PEs in plant resistance induction, as well as the current perspectives in food production.

11.1 INTRODUCTION

Agriculture is the most critical activity for humanity. It is the main activity providing essential products such as food, fuel, energy, and wood. However, agricultural production is affected by phytosanitary problems such as insect plagues, weeds, and diseases caused by phytopathogens. In the case of diseases, there are records that these phytosanitary problems cause losses of 9.5 billion dollars worldwide, one of the causes that could affect the food safety of many countries (Conforti et al., 2018). The application of chemical products is the most used method to combat diseases. Nevertheless, the excessive and indiscriminate use of chemical products has led to the development of resistance by the causal agents toward the active ingredients (Rojas et al., 2017; Carmona & Sautua, 2017).

The demand for sustainable agriculture has increased in the last few years. Due to the paradigm change toward sustainable agriculture, farmers need to find natural alternatives with less environmental impact. Therefore, the development of alternatives is necessary to counter the impact of chemical products. Some legislations in many countries prohibit genetic crop improvement and assisted crossing. One alternative is the plant extracts (PEs) which contain phytochemical compounds (PCs) that confer antifungal, antibacterial, antiviral, insecticidal, antioxidant, and resistance inducer properties (Chandrashekhara et al., 2010; Jasso de Rodríguez et al., 2011; Burketova et al., 2015).

All properties previously mentioned are essential, but induced resistance (IR) has become the focus of study for researchers and companies due to the need for different approaches to controlling plant diseases. The plant's defense mechanism is an exciting option because it allows for reducing damage by phytopathogens. IR is also placed within the context of sustainable farming; the benefits outweigh the investment in this alternative (Walters et al., 2013; Cipollini & Heil, 2010).

Plants are naturally in contact with a vast spectrum of pathogenic microorganisms, but plants generally remain healthy due to defense mechanisms. These mechanisms include passive defenses such as structural barriers like waxes, cuticles, and cell walls, as well as active defense or induced resistance (IR). This type of resistance involves changes in metabolism triggered by the differential expression of genes (Laredo-Alcalá et al., 2017). IR stimulates the natural defenses of plants with elicitors that simulate an attack by phytopathogens, putting the plants in an alert state. When plants recognize a pathogen, the speed of their defenses often determines if they will resist an infection (Burketova et al., 2015; Llorens et al., 2017). Theoretically, plants possess genes to respond to pathogen aggression, showing that a complex network of hormonal signals controls the plant responses against the pathogen's attack (Laredo-Alcalá et al., 2017).

11.1.1 BASIS OF PLANT'S IMMUNITY

Plants have developed an efficient immunological system to respond to phytopathogen attacks. The plant defense activation begins when recognized molecules are called pathogen-associated molecular patterns (PAMPs) or damage-associated molecular patterns (DAMPs). These signals are recognized by receptors located in the cytoplasmic membrane called pattern recognition receptors (PRRs) which are formed of an extracellular Leucine-Rich Repeat (LRR) and an intracellular Kinase Domain (KD). PRRs interact with the protein Brassinosteroid insensitive 1-Associated kinase 1 (BAK1) to activate pattern-triggered immunity (PTI); PTI activates a massive amount of signals like ion fluxes leading to depolarization of the plasma membrane, production of reactive oxygen species (ROS), and activation of Mitogen-Activated Protein Kinases (MAPKs) (Jones & Dangl, 2006; Burketova et al., 2015). It has been revealed that these signals induce a significant transcriptional response in defense-related genes; the activation of these genes involves the production and accumulation of pathogenesis-related proteins

(PRPs), phytoalexins, lignin in the cellular wall, and ROS as executioners of pathogens (Tsuda & Katagiri, 2010; Burketova et al., 2015).

Like all organisms, pathogens evolve continually, adapting better mechanisms to infect their host; this can be corroborated through other plant defense strategies known as effector-triggered immunity (ETI). Pathogens develop proteins known as effectors, which are transferred into the plant cell, evading the PTI strategy. Similar to the plant's response, the production of receptors called Nucleotide-Binding Receptors (NBR) occurs to recognize the molecules called effectors. ETI generally occurs in cells near a pathogen in a process called the hypersensitive response (HR), providing evidence of dynamic co-evolution between microorganisms and plants (Dodds & Rathjen, 2010; Díaz, 2012; Tsuda & Katagiri, 2010).

In addition to the local defense mechanisms mentioned above, plants present a strategy that detects damage in a part of the plant and triggers a systemic response called induced resistance (IR), of which there are two types: systemic acquired resistance (SAR) and induced systemic resistance (ISR). Therefore, resistance activated in plants depends on biotic and abiotic factors known as Es. (Burketova et al., 2015; Mahesh et al., 2017) (Figure 11.1).

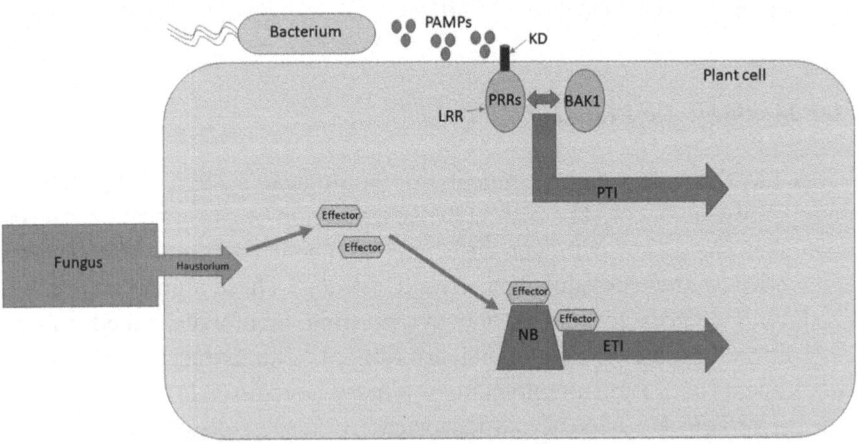

FIGURE 11.1 Schematic representation of defensive strategies PTI and ETI used by plants against phytopathogens.

11.1.2 SAR AND ISR

Phytohormones are chemical substances that regulate responses in plants, generally produced and transported in small amounts from one part to another of the plants. Some phytohormones are abscisic acid (ABA), brassinosteroids

(BR), ethylene (ET), auxins, gibberellins, salicylic acid (SA), and jasmonic acid (JA). With respect to IR, the most relevant phytohormones are SA and JA, which are involved in the immune response of plants against pathogens (Javid et al., 2011; Burketova et al., 2015).

SAR is an IR type of spread spectrum activated locally and systematically by necrotizing pathogens, conferring resistance to a massive variety of pathogens, not only the pathogen that activates it; once SAR is activated, it can last active for days or weeks, making it attractive for agriculture; SAR can be activated with synthetic or natural compounds (Laredo-Alcalá et al., 2017; Llorens et al., 2017). Studies have shown SAR depends on SA accumulation, SA pathway components like the methylated derivative of SA, diterpenoid dehydroabietic, free radicals nitric oxide (NO), ROS, as well as activation of Nonexpressor of Pathogenesis-related protein 1 (NPR1) (Gao et al., 2015). The SA pathway begins with SA synthesis, which occurs during PTI or ETI through acid chorismate. This is a metabolite obtained from the shikimic acid route; once present, SA interacts with NPR1, provoking structural change to the monomeric form and is translocated to the cell nucleus, activating the expression of Pathogenesis-Related genes (PR), thereby NPR1 and SA act as coactivators of PR (Llorens et al., 2017; Pieterse et al., 2012). Once PR is activated to initiate the PRPs production, which is accumulated in cytoplasm and vacuoles, and many of them have antimicrobial activity; there are 17 families of PRPs and each one presents different activities such as chitinases (PRPs3, 4, 8, and 11), glucanases (PRPs2), ribonucleases (PRPs10), peroxidases (PRPs9), defensins (PRPs12), and thionins (PRPs13) (Pieterse et al., 2012; Gao et al., 2015; Herrera et al., 2018).

On the other hand, the ISR is activated by soil pathogens able to colonize roots known as plant-growth-promoting rhizobacteria (PGPR), substances such as surfactants, chemical inducers, antibiotics, necrotizing pathogen wounds, or herbivores attack. ISR also is systemic protection of broad-spectrum that protects other parts of plants (Laredo-Alcalá et al., 2017; Burketova et al., 2015; Llorens et al., 2017). Unlike the SAR, ISR depends on JA and ET pathways; the JA biosynthesis pathway is known as the octadecanoic pathway, where many enzymes are implicated. The JA is formed from non-saturated linoleic acid released by phospholipids from the cell membrane. When a pathogen damages a plant, ET production is induced, and elicitors are generated conformed principally by elements from the destroyed cellular wall, ethylene response factors (ERF) are associated with enhancing defense response. Once JA and ET, the ethylene insensitive 3 (EIN3) integrates the JA and ET signaling pathway. EIN3 is linked to the ERF, ERF1, that acts as an activator of transcription of the same ERF1, the JA reinforces this reaction; as a result of ERF1 transcription, PRs such as PDF1.2, PR3, and PR4 are activated, the activation of these genes encodes the production of antimicrobial peptides known as defensins, protease inhibitors, enzymes involved in the biosynthesis of flavonoids, osmotines, and lipoxygenases, and different PRPs (Laredo-Alcalá et al., 2017; Shoresh et al., 2010; Herrera et al., 2018; Huang et al., 2016; Llorens et al., 2017) (Figure 11.2).

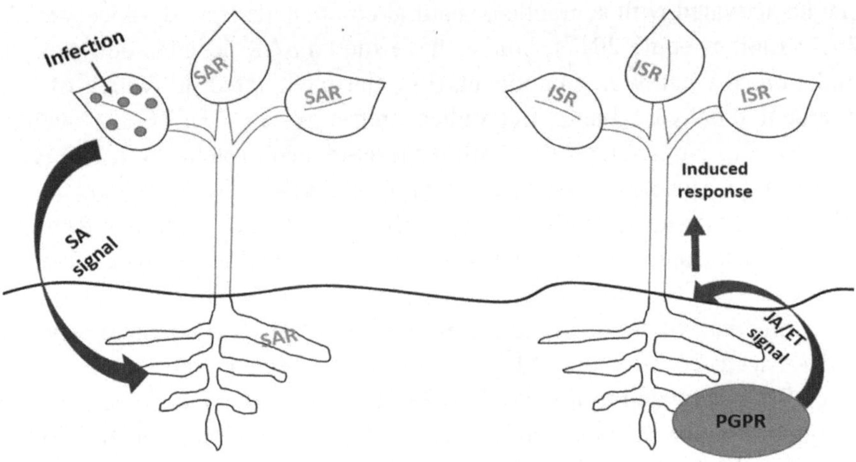

FIGURE 11.2 Schematic representation of SAR and ISR and their response mechanism.

11.2 NATURAL COMPOUNDS AS ELICITORS

Elicitors are compounds capable of stimulating plant defenses, acting on the plant, blocking or retarding the entry of pathogens, and limiting their activity in the plant. Depending on their origin, elicitors can be classified as abiotic and biotic (Ramirez-Estrada et al., 2016; Laredo-Alcalá et al., 2017). Abiotic elicitors generally have a non-biological origin such as physical treatments, which are mainly applied in postharvest treatment, such as heat treatment, low-temperature application, and UV irradiation (Terry & Joyce, 2004). Inorganic compounds also fall within this classification; Acibenzolar-S-methyl, known as benzothiadiazole (ASM/BTH), induces the SAR and encodes the production of lipoxygenases, sulfur-rich proteins, chitinases, and phytoalexins; or imidacloprid (IMID) which activates the SAR inducing the activation of gene PR2 (Thakur & Sohal, 2013; Navarro et al., 2017; Llorens et al., 2017).

On the other hand, biotic elicitors have biological origins and are compounds derived from pathogens or plants (Patel & Krishnamurthy, 2013).

Microorganisms interact with plants constantly, triggering many responses in their metabolism, such as the activation of immunity; a massive amount of microbial compounds can stimulate the plant defenses; one of these is the glucans produced by *Phytophthora* spp. capable of inducing phytoalexins production; likewise, syringolides from *Pseudomonas syringae* activate the hypersensitive response; not only bacterial elicitors induce the plant defenses but also fungal elicitors such as the sphinganine analog mycotoxins from *Fusarium moniliforme* that program the cell death (Thakur & Sohal, 2013). Finally, the compounds obtained from plants such as the phytohormones like SA, JA, and hexanoic acid (HA); JA regulates protein production through the octadecanoic pathway, whereas the SA and JA both can be applied exogenously to induce metabolic changes in plant defenses. HA is a monocarboxylic acid of short-chain. Among its effects is the expression of the PGIP gene which encodes polygalacturonase-inhibiting proteins involved in the deactivation of fungal polygalacturonases and further enhances the JA pathway (Llorens et al., 2017; Patel & Krishnamurthy, 2013).

The perspectives of the elicitors' use are promising, especially the biotic elicitors for which research for their application in agriculture has been increasing due to the current problem with chemical compounds. An optimum application of elicitors must consider factors like the elicitor's concentration, time of exposure, nutrients, and plant growth; a change in these features causes different effects on plant metabolism (Llorens et al., 2017; Patel & Krishnamurthy, 2013).

11.2.1 PES AS ELICITORS IN PLANTS

PEs are one of the most natural plant products for disease management, which contain several PCs. Generally, PEs are obtained from dried plants subjected to many processes, such as pulverization, filtration, and evaporation using various solvents (Wang et al., 2004). Recently, PEs have garnered interest worldwide because they have antimicrobial properties that affect phytopathogens' growth, acting as a curative alternative and as elicitors activating the plant defenses, providing a protective effect (Malo et al., 2017). These properties are conferred by PCs produced within the plant metabolism, such as phenolic compounds or their oxygen-substituted derivatives, terpenoids, carbohydrates, and vitamins (Altemimi et al., 2017; Calo et al., 2015; Gurjar et al., 2012; Hernández et al., 2015; Wink, 2011).

Regarding the curative effect, the principal action modes are related to the direct interaction between the microorganisms and the PCs. Some action modes include the inactivation of ROS present in fungal mitochondria that act in the fungal hyphae development by the flavonoids or the one presented by the phenolic compounds, which suggests that the number of hydroxyl groups on the phenolic ring is related to the degree of toxicity toward microorganisms, altering cellular membrane permeability (Da Cruz et al., 2013; Gurjar et al., 2012; Hintz et al., 2015; Warris & Ballou, 2019). On the other hand, the effect as elicitors involves turning on the plant defense mechanisms, increasing the activity of PRPs, Defense-Related Enzymes (DREs), secondary metabolites such as phytoalexins, antioxidants, and phytohormones, e.g., SA, JA, ET, and ABA (Khoa et al., 2017; Pretali et al., 2016).

Thus, several plant extracts are capable of acting as elicitors inducing plant defense responses. Malo et al. (2017) reported the application of extracts from *Capsicum annuum, Ruta graveolens*, and *Syzygium aromaticum* in tomato and tobacco plants in vitro inoculated with *Oidium* spp. They observed that SA accumulation triggered the activation of Pathogenesis-Related genes PR1, PR2, and PR3, inducing the production of glucanases, chitinases, and necrosis induction. Other species have been tested as elicitors and presented favorable results, such as the extract from *Kalanchoe pinnata* in rice against *Xanthomonas oryzae*, resulting in the production of DREs and antioxidant enzymes, namely peroxidase (POX), catalase (CAT), polyphenol oxidase (PPO), and phenylalanine ammonia-lyase (PAL) (Khoa et al., 2017). *Chrysanthemum indicum* has also been reported to elicit activity; aqueous extract enhances the activity of catalase (CAT) and peroxidase (POD), which are protective enzymes in *Atractylodis macrocephala* with *Sclerotium rolfsii* problems (Du et al., 2016). Phaseolus vulgaris is a common crop with problems of *P. syringae*; to control it, Duran-Flores & Heil (2014) applied leaf extract from the same crop *P. vulgaris* and induced the formation of ROS and extrafloral nectar as a response of the JA pathway. On the other hand, Draz et al. (2019) treated wheat plants with extracts from *Lawsonia inermis, Acalypha wilkesiana, Melia azedarach, Lantana camara,* and *Punica granatum* for *Puccinia triticina* control. These extracts decreased the disease severity and increased the content of chlorophyll a and b, phenolic compounds, and also increased the activity of POX and PPO.

Postharvest treatments to induce resistance in fruits are physical treatments; nevertheless, the resistance inducer capacity of PEs in postharvest crops has also been tested. The apple is an essential crop with many post-harvest diseases. Due to this problem, Gholamnezhad (2019) tested *Azadirachta indica, Lavandula spica,* and *Mentha pulegium* extracts in apple fruits with *Botrytis cinerea* fungus; the extracts increased the activity of POX, PAL, β-1,3-glucanase, and PPO. The effect was present until 12 days after the application of the treatments. The seeds can also be treated to induce resistance against diseases, benefiting the plant since seedling. Regarding this, Chandrashekhara et al. (2010) treated *Pennisetum glaucum* seeds inoculated with *Sclerospora graminicola* with *Viscum album* extract. The extract enhanced the seeds' parameters, having a high germination percentage, and the seedlings were more vigorous than the seeds without extract; furthermore, the extract activated the production of POX and PAL enzymes through the SAR pathway (Table 11.1).

11.2.2 ACTION MODE OF PES AS ELICITORS

The action mode to induce resistance in plants by the PEs has yet to be elucidated completely. It is probable that the mixture of diverse compounds acts synergistically as the trigger of plant responses (Burketova et al., 2015; Khoa et al., 2017). Nevertheless, researchers have tried to elucidate the possible action mode of the PEs to induce resistance in plants. Vitamins are metabolites produced in plants; hence, their presence in PEs is expected. Concerning induced resistance, Vitamins B such as thiamin and riboflavin have been proposed as compounds that induce resistance in plants. It suggests the role of thiamin in Ca^{2+} ions mobilization, which plays an essential role in HR due to the Ca^{2+} flow through the plasmatic membrane regulating the activity of protein kinase C, increasing the ROS (Ahn et al., 2005). Meanwhile, riboflavin was suggested as stimulating H_2O_2 production and lignin deposition formation, though it also promotes the expression of the 13-LOX gene encoding lipoxygenase production. Lipoxygenase plays an essential role in JA biosynthesis through the octadecanoid pathway (Taheri & Tarighi, 2010).

Within PCs produced in a plant's metabolism, the phenolic compounds are one of the more important due to their biological activity; a characteristic of these compounds is the presence of one aromatic ring with one hydroxyl substituent, one well known as the SA, which induces the activation of PR genes to encode PRPs, the HR, and triggers SAR (Liu et al., 2019). Flavonoids and other critical phenolic compounds are synthesized in all organisms of plants. Their ability to induce resistance is related to their capacity to stimulate the colonization of the roots by PGPR, triggering the mechanism of ISR in the plant's defense; researchers have also reported the accumulation of flavonoids to inhibit auxin transport, increasing resistance to phytopathogens (Mierziak et al., 2014; Dao et al., 2011).

TABLE 11.1 Different Types of Plant Extracts Used as Elicitors in Crops

Plant	Solvent	Crop	Pathogen	Response Induced	References
C. annum, R. graveolens, and *S. aromaticum*	Water	Tomato and tobacco	*Oidium* spp.	Activation of PR and production of PR1, PR2, and PR3.	Malo et al. (2017)
K. pinnata	Water	Rice	*X. oryzae*	Production of POX, CAT, PPO, and PAL.	Khoa et al. (2017)
C. indicum	Water	*A. maceo-cephalae*	*S. rolfsii*	Enhance CAT and POD.	Du et al. (2016)
P. vulgaris	Water + Tween 20	*P. vulgaris*	*P. syringae*	Formation of ROS and extrafloral nectar.	Duran-Flores & Heil (2014)
A. indica *L. spica* *M. pulegium*	Water and methanol	Postharvest apple fruit	*Botrytis cinerea*	Increase of the activity of POX, PAL, β-1,3-glucanase, and PPO.	Gholamnezhad (2019)
L. inermis, *A. wilkesiana*, *M. azedarach*, *L camara*, and *P. granatum*	Water	Wheat	*P. triticina*	Increase of chlorophyll a and b, and production of POX and PPO.	Draz et al. (2019)
V. album	Water	*Pennisetum glaucum*	*S. graminicola*	Production of POX and PAL, induction of SAR pathway.	Chandrashekhara et al. (2010)
Adathoda vasica	Methanol	Rice	*X. oryzae*	Increase in accumulation of POX, PAL, PPOe, and β-1,3-glucanase.	Govindappa et al. (2011)
Beta vulgaris	Ethanol	Potato plants	*Phytophthora infestans*	Activation of PR and production of PR1 and PR2.	Moushib et al. (2013)
Eucalyptus citriodora	Methanol	*Capsicum annuum*	*Fusarium oxysporum*	Production of POX, PPO, and PAL	Shafique et al. (2019)
Rosmarinus officinalis	Alcohol	Rice	*Magnaporthe oryzae*	Increase the activity of chitinase, β-1, 3-glucanase, lipoxygenase, PAL, and phenolic compounds, and increase in AS level.	Garcia et al. (2017)

TABLE 11.1 *(Continued)*

Plant	Solvent	Crop	Pathogen	Response Induced	References
Sophora flavescens (tuber), *Forsythia suspense* (tuber) and *Lonicera japonica.*	Water	Tobacco	*Phytophthora nicotianae and tobacco mosaic virus*	Activation of PR and production of peroxide dismutase, POX, and PAL.	Wang et al. (2018)
Agarista salicifolia	Methanol	*Ensete ventricosum*	*Xanthomonas campestris*	Increase of the activity of PAL, POX, PPO; and accumulation of total phenolics content and total soluble sugars.	Yemata & Fetene (2016).
Corymbia citriodora	Water	Cucumber	*Colletotrichum lagenarium*	Production of POX and β-1.3-glucanases.	Franzener et al. (2018)
Vitex negundo	Methanol	Rice	*X. oryzae*	Increase of the activity of PPO, POX, and b-1,3-glucanase.	Nisha et al. (2012)

Carbohydrate polymers such as polysaccharides or oligosaccharides have been reported as elicitors in plant defense; they are present in cell walls and thus are part of PAMPs; carbohydrates activate the PTI pathway by interacting with the PRRs. These receptors contain a ligand-binding ectodomain with lysine motifs, lectin motifs, epidermal growth factor, and leucine-rich repeats that allow them to recognize ligands, playing an essential role in interacting with carbohydrate ligands to activate defenses (Zheng et al., 2020). Some carbohydrates with eliciting activity include β-glucans, oligogalacturonides, and carrageenans; the responses induced include calcium influx, ROS burst, callose deposition, alkalinization of the medium, activation of MAPK, accumulation of SA, expression of defense-related genes, activation of defense enzymes, and production of phytoalexins (Zheng et al., 2020).

11.3 THE PERSPECTIVE OF PES AS ELICITORS IN SUSTAINABLE AGRICULTURE

Currently, sustainable farming is growing due to the environmental problems caused by chemical products; this has brought a paradigm change that involves a shift in production systems toward green technologies, which have less impact on the environment. As a consequence of this new paradigm, the market for natural and biological products is increasing. It is estimated that the demand for biopesticides has reached US $4 billion. The main reasons for choosing organic farming include better yields, resistance management, managing residue safety, biodegradability, and quick reentry of workers to crops. Most biopesticides can be used in organic production (Marrone, 2019).

The market for elicitors based on PEs is a relatively recent area that has begun to be exploited because the benefits and characteristics that PEs possess as elicitors have been demonstrated. An example is the capacity to play a dual role in plant defense, acting as elicitors and directly on the phytopathogens, representing an advantage in disease control; in addition to the above, biopesticides are better because they are designed with scientific studies allowing for a better manufacturing process and formulations (Burketova et al., 2015; Marrone, 2019). The PEs as elicitors can be applied with other biological control agents according to the EU Directive 2009/128/EC; this directive determines the rules for pesticide use. The elicitors based on PEs are mentioned as a component of integrated pest management (Burketova et al., 2015).

The elicitors are part of the natural and biological products that have also increased in agriculture due to the promising results presented by natural compounds, such as hexanoic acid, which decreased the incidence of *Alternaria alternata* in citrus trees. Chitosan is also a natural RI capable of activating the phenylpropanoid pathway or inducing the production of PRPs. Foliar application of chitosan in *Hordeum vulgare* reduced the infection by *Blumeria graminis. Reynoutria sacchalinensis* is a plant species that can cause resistance in cucumber and wheat against *Sphaerotheca fuliginea, B. graminis*, and *Sphaerotheca pannosa* when applied in extracts (Llorens et al., 2017; Burketova et al., 2015).

Within natural products, PEs have been demonstrated to be efficient in disease control through IR. In addition to the benefits mentioned above, this characteristic of natural products has turned PEs into an option for disease management. The increased need for disease management using PEs has spurred research and development by companies and centers. Currently, there are some commercial products with the ability to activate defenses in crops. *Reynoutria sacchalinensis* was one of the first plants to develop elicitors based on plant extracts because it induces the production of phenolic phytoalexins. This characteristic was utilized by the company Compo GmbH and later by KHH BioSci Inc. and Biofa AG with Milsana® to protect against powdery mildew in various crops (Siah et al., 2018). However, since 2009, this product has been reformulated by Marrone Bio Innovations with Regalia®, enhancing its ability to induce resistance in plants and broadening the spectrum of diseases it can combat (Su et al., 2012). *Macleaya cordata* is another plant exploited for its ability to induce resistance. Camas Technologies Inc. formulated a biofungicide based on an extract from this plant known as Qwel®, which increases polyphenolic phytoalexins and activates SAR. Stifénia® is a botanical fungicide based on *Trigonella foenum* extract formulated in France with eliciting capacity against grape oidium (Regnault-Roger, 2012). Another company focused on developing products based on plant extracts for disease management through IR is GreenCorp Biorganiks de México, with the products FullCoverO®, Sigatoxx®, and Virablock®. FullCoverO® is a resistance inducer with Camellia sinensis extract, which acts against many foliar fungi in several crops; Sigatoxx® contains diverse plant extracts to protect the banana crop against the fungus *Mycosphaerella fijiensis* and *Mycosphaerella musicola*. This product induces plant defense through the activation of mechanisms of systemic resistance and localized. Virablock® is a novel inducer of systemic and local resistance in plants that acts against Tomato Spotted Wilt Virus (TSWV), Tobacco Mosaic Virus (TMV), Cucumber Mosaic Virus (CMV), Pumpkin Mosaic Virus (SqMV), Papaya Ring Spot Virus (PRSV), and banana streak virus (BSV), in several vegetable crops (GreenCorp, 2020). Commercial products based on plant extracts induce resistance in plants against phytopathogens (Table 11.2).

TABLE 11.2 Commercial Products Based on Plant Extracts to Induce Resistance in Plants Against Phytopathogens

Product	Plant Extract	Crop	Pathogen	References
Milsana®	*R. sacchalinensis*	Ornamental plants, cucumber, tomato, and wheat.	*S. fuliginea*, *S. pannosa*, *Leveillula taurica*, and *Blumeria tritici*	Siah et al. (2018)
Regalia®	*R. sacchalinensis*	Eggplant, chili, tomato, potato, pepper, tobacco, cucumber, pumpkin, lemon, strawberry, and avocado.	*Alternaria solani*, *Botrytis cinerea*, *Colletotrichum gloesporioides*, *Colletotrichum fragariae*, *Erysiphe cichoracearum*, *Leveillula taurica*, *Podosphaera leucotricha*, *Sphaerotheca pannosa*, *Sphaerotheca fuliginea*, *Sphaerotheca macularis*, *Sphaerotheca humuli*, and *Uncinula necator*	Su et al. (2012)
Qwel®	*M. cordata*	Ornamental plants	*Sphaerotheca pannosa*, *Erysiphe* spp., *Oidium* spp., *Podosphaera* spp., *Microsphaera* spp., *Alternaria* spp., and *Septoria* spp.	Regnault-Roger (2012)
Stifénia®	*T. foenum*	Grape	*Oidium* spp.	Regnault-Roger (2012)
FullCoverO®	*Camellia sinensis*	Vegetables in general, fruit trees, and bananas.	Foliar fungus in general.	GreenCorp (2020)
Sigatoxx®	Diverse plant extracts	Banana	*M. fijiensis* and *M. musicola*	GreenCorp (2020)
Virablock®	Diverse plant extracts	Tomato, chili, potato, tobacco, cucumber, melon, watermelon, pumpkin, and banana.	*Banana streak virus* (*BSV*), *cucumber mosaic virus* (*VMV*), *papaya ring spot virus* (*PRSV*), *pumpkin mosaic virus* (*SqMV*), *tomato spotted wilt virus* (*TSWV*), and *tobacco mosaic virus* (*VMT*).	GreenCorp (2020)

11.4 CONCLUSION

The ISR is a mechanism of defense in plants that can be harnessed in agriculture to activate the immune system of plants, decreasing the possibility of infection from pathogens. In sustainable agricultural production, it is always desirable to prevent phytosanitary problems because it reduces costs and avoids unnecessary spending. Different pathways can be induced to activate plant defenses, with other products acting as elicitors. Within the context of sustainable agriculture, PEs are a natural option to decrease the impact of chemical products on the environment. Thus, PEs as elicitors bring benefits that can be used in sustainable agriculture. However, investigations about PEs as elicitors in disease management have considerably increased in research centers. In the market, there are a few compared with other biological products in agriculture.

Although diverse challenges exist to increase the production of elicitors based on PEs, the studies to improve the production process of elicitors based on PEs have increased. This can be corroborated by all the studies published in several scientific journals and the investments made by companies in this area. Coupled with this, new technologies such as genomics, proteomics, and metabolomics have allowed the development of new protocols and standards to enhance yield as well as the interest of industry in the PEs as elicitors (Abreu et al., 2012). So, the PEs are an excellent option to be used as elicitors in agriculture to change food production toward a sustainable product that will not affect the environment and will not compromise future generations due to all their properties and capacity to activate plant defenses. Although the available commercial products are currently limited, the new technologies are helpful in their production, so it is necessary to continue with related studies on the PEs.

KEYWORDS

- **elicitors**
- **food production**
- **genomics**
- **induced systemic resistance**
- **natural compounds**
- **plant defenses**
- **plant extracts**
- **plant immunity**
- **proteomics**
- **systemic acquired resistance**

REFERENCES

Abreu, A. C., McBain, A. J., & Simoes, M. (2012). Plants as sources of new antimicrobials and resistance-modifying agents. *Natural Product Reports, 29*(9), 1007–1021. doi: 10.1039/C2NP20035J.

Ahn, I. P., Kim, S., & Lee, Y. H. (2005). Vitamin B1 functions as an activator of plant disease resistance. *Plant Physiology, 138*(3), 1505–1515. doi: 10.1104/pp.104.058693.

Altemimi, A., Lakhssassi, N., Baharlouei, A., Watson, D., & Lightfoot, D. (2017). Phytochemicals: Extraction, isolation, and identification of bioactive compounds from plant extracts. *Plants, 6*(4), 42. doi: 10.3390/plants6040042.

Burketova, L., Trda, L., Ott, P. G., & Valentova, O. (2015). Bio-based resistance inducers for sustainable plant protection against pathogens. *Biotechnology Advances, 33*(6), 994–1004. doi: 10.1016/j.biotechadv.2015.01.004.

Calo, J. R., Crandall, P. G., O'Bryan, C. A., & Ricke, S. C. (2015). Essential oils as antimicrobials in food systems – A review. *Food Control, 54*, 111–119. doi: 10.1016/j.foodcont.2014.12.040.

Carmona, M., & Sautua, F. (2017). La problemática de la resistencia de hongos a fungicidas. Causas y efectos en cultivos extensivos. *Agronomía & Ambiente, 37*(1), 1–19. Retrieved from: http://agronomiayambiente.agro.uba.ar/index.php/AyA/article/view/60/59 (accessed on 28 May 2024).

Chandrashekhara, S., Niranjan-Raj, G., Manjunath, S., Deepak, H., & Shekar, S. (2010). Seed treatment with aqueous extract of Viscum album induces resistance to pearl millet downy mildew pathogen. *Journal of Plants Interactions, 5*(4), 283–291. doi: 10.1080/17429140903556539.

Cipollini, D., & Heil, M. (2010). Costs and benefits of induced resistance to herbivores and pathogens in plants. *Plant Science, 5*, 1–25. doi: 10.1079/PAVSNNR20105005.

Conforti, P., Ahmed, S., & Markova, G. (2018). Impact of disasters and crises on agriculture and food security, 2017. Available in: http://agris.fao.org/agris-search/search.do?recordID=XF2018001268 (accessed on 28 May 2024).

Da Cruz Cabral, L., Pinto, V. F., & Patriarca, A. (2013). Application of plant derived compounds to control fungal spoilage and mycotoxin production in foods. *International Journal of Food Microbiology, 166*(1), 1–14. doi: 10.1016/j.ijfoodmicro.2013.05.026.

Dao, T. T. H., Linthorst, H. J. M., & Verpoorte, R. (2011). Chalcone synthase and its functions in plant resistance. *Phytochemistry Reviews, 10*(3), 397. doi: 10.1007/s11101-011-9211-7.

Díaz, N. (2012). Resistencia sistémica adquirida mediada por el ácido salicílico. *Biotecnología en el Sector Agropecuario y Agroindustrial: BSAA, 10*(2), 257–267.

Dodds, P. N., & Rathjen, J. P. (2010). Plant immunity: towards an integrated view of plant–pathogen interactions. *Nature Reviews Genetics, 11*(8), 539–548. doi: 10.1038/nrg2812.

Draz, I. S., Elkhwaga, A. A., Elzaawely, A. A., El-Zahaby, H. M., & Ismail, A. W. A. (2019). Application of plant extracts as inducers to challenge leaf rust of wheat. *Egyptian Journal of Biological Pest Control, 29*(1), 1–8. doi: 10.1186/s41938-019-0109-9.

Du, N., Tian, W., Zheng, D., Zhang, X., & Qin, P. (2016). Extraction, purification and elicitor activities of polysaccharides from *Chrysanthemum indicum*. *International Journal of Biological Macromolecules, 82*, 347–354. doi: 10.1016/j.ijbiomac.2015.10.044.

Duran-Flores, D., & Heil, M. (2014). Damaged-self recognition in common bean (*Phaseolus vulgaris*) shows taxonomic specificity and triggers signaling via reactive oxygen species (ROS). *Frontiers in Plant Science, 5*, 585. doi: 10.3389/fpls.2014.00585.

Franzener, G., Schwan-Estrada, K. R. F., Moura, G. S., Kuhn, O. J., & Stangarlin, J. R. (2018). Induction of defense enzymes and control of anthracnose in cucumber by *Corymbia citriodora* aqueous extract. *Summa Phytopathologica, 44*(1), 10–16. doi: 10.1590/0100-5405/2218.

Gao, Q. M., Zhu, S., Kachroo, P., & Kachroo, A. (2015). Signal regulators of systemic acquired resistance. *Frontiers in Plant Science, 6*, 228. doi: 10.3389/fpls.2015.00228.

Garcia, M. L., Ferreira, N. L. O., Lima, R., Arruda, M. C. C. D. F., & Da, E. C. (2017). Standardized rosemary extract induces host plant defenses and suppresses rice leaf blast. *Journal of Medicinal Plants, 5*(5), 39–45. Retrieved from: http://www.plantsjournal.com/archives/?year=2017&vol=5&issue=5&part=A&ArticleId=684 (accessed on 28 May 2024).

Gholamnezhad, J. (2019). Effect of plant extracts on activity of some defense enzymes of apple fruit in interaction with *Botrytis cinerea*. *Journal of Integrative Agriculture, 18*(1), 115–123. doi: 10.1016/S2095-3119(18)62104-5.

Govindappa, M., Umesha, S., & Lokesh, S. (2011). *Adathoda vasica* leaf extract induces resistance in rice against bacterial leaf blight disease (*Xanthomonas oryzae* pv. *oryzae*). *International Journal of Plant Physiology and Biochemistry, 3*(1), 6–14. Retrieved from: https://academicjournals.org/journal/IJPPB/article-full-text-pdf/218B96B9946 (accessed on 28 May 2024).

GreenCorp. (2023). Biocontrol. Available from: https://greencorp.mx/producto/biocontrol/ (accessed on 28 May 2024).

Gurjar, M. S., Ali, S., Akhtar, M., & Singh, K. S. (2012). Efficacy of plant extracts in plant disease management. *Agricultural Science, 3*(3), 425. doi: 10.4236/as.2012.33050.

Hernández, T., García-Bores, A. M., Serrano, R., Ávila, G., Dávila, P., Cervantes, H., Peñalosa, I., Flores-Ortiz, C. M., & Lira, R. (2015). Fitoquímica y actividades biológicas de plantas de importancia en la medicina tradicional del Valle de Tehuacán-Cuicatlán. *TIP. Revista especializada En Ciencias Químico-Biológicas, 18*(2), 116–121. Retrieved from: http://www.scielo.org.mx/scielo.php?pid=S1405-888X2015000200116&script=sci_arttext&tlng=en (accessed on 28 May 2024).

Herrera, M., Portillo, D., Pulido, M. A., Diaz Tatis, P. A., & López Carrascal, C. E. (2018). Estudio de la expresión de genes que codifican para putativas proteínas PR en yuca (*Manihot esculenta* Crantz). *Acta Biológica Colombiana, 23*(3), 242–252. doi: 10.15446/abc.v23n3.70878.

Hintz, T., Matthews, K. K., & Di, R. (2015). The use of plant antimicrobial compounds for food preservation. *BioMed Research International*, 1–12. doi: 10.1155/2015/246264.

Huang, P. Y., Catinot, J., & Zimmerli, L. (2016). Ethylene response factors in Arabidopsis immunity. *Journal of Experimental Botany, 67*(5), 1231–1241. doi: 10.1093/jxb/erv518.

Jasso de Rodríguez, D. J., García, R. R., Castillo, F. H., González, C. A., Galindo, A. S., Quintanilla, J. V., & Zuccolotto, L. M. (2011). In vitro antifungal activity of extracts of Mexican Chihuahuan Desert plants against postharvest fruit fungi. *Industrial Crops and Products, 34*(1), 960–966. doi: 10.1016/j.indcrop.2011.03.001.

Javid, M. G., Sorooshzadeh, A., Moradi, F., Modarres Sanavy, S. A. M., & Allahdadi, I. (2011). The role of phytohormones in alleviating salt stress in crop plants. *Australian Journal of Crop Science, 5*(6), 726. Retrieved from: https://search.informit.com.au/documentSummary;dn=282135746215551;res=IELHSS (accessed on 28 May 2024).

Jones, J. D. G., & Dangl, J. L. (2006). The plant immune system. *Nature, 444*, 323–329. doi: 10.1038/nature05286.

Khoa, N. D., Xạ, T. V., & Hào, L. T. (2017). Disease-reducing effects of aqueous leaf extract of *Kalanchoe pinnata* on rice bacterial leaf blight caused by *Xanthomonas oryzae* pv. *oryzae* involve induced resistance. *Physiological and Molecular Plant Pathology, 100*, 57–66. doi: 10.1016/j.pmpp.2017.06.005.

Laredo-Alcalá, E. I., Martínez-Hernández, J. L., Iliná, A., Guillen-Cisneros, L., & Hernández-Castillo, F. D. (2017). Aplicación de ácido jasmónico como inductor de resistencia vegetal frente a patógenos. *Revista Mexicana de Ciencias Agrícolas, 8*(3), 673–683. doi: 10.29312/remexca.v8i3.40.

Liu, H., Kang, Y., Zhao, X., Liu, Y., Zhang, X., & Zhang, S. (2019). Effects of elicitation on bioactive compounds and biological activities of sprouts. *Journal of Functional Foods, 53*, 136–145. doi: 10.1016/j.jff.2018.12.019.

Llorens, E., García-Agustín, P., & Lapeña, L. (2017). Advances in induced resistance by natural compounds: towards new options for woody crop protection. *Scientia Agricola, 74*(1), 90–100. doi: 10.1590/1678-992x-2016-0012.

Mahesh, H. M., Murali, M., Pal, M. A. C., Melvin, P., & Sharada, M. S. (2017). Salicylic acid seed priming instigates defense mechanism by inducing PR-Proteins in *Solanum melongena* L. upon infection with *Verticillium dahliae* Kleb. *Plant Physiology and Biochemistry, 117*, 12–23. doi: 10.1016/j.plaphy.2017.05.012.

Malo, I., De Bastiani, M., Arevalo, P., & Bernacchia, G. (2017). Natural extracts from pepper, wild rue and clove can activate defenses against pathogens in tomato plants. *European Journal of Plant Pathology, 149*(1), 89–101. doi: 10.1007/s10658-017-1168-y.

Marrone, P. G. (2019). Pesticidal natural products–status and future potential. *Pest Management Science, 75*(9), 2325–2340. doi: 10.1002/ps.5433.

Mierziak, J., Kostyn, K., & Kulma, A. (2014). Flavonoids as important molecules of plant interactions with the environment. *Molecules, 19*(10), 16240–16265. doi: 10.3390/molecules191016240.

Moushib, L. I., Witzell, J., Lenman, M., Liljeroth, E., & Andreasson, E. (2013). Sugar beet extract induces defence against *Phytophthora infestans* in potato plants. *European Journal of Plant Pathology, 136*(2), 261–271. doi: 10.1007/s10658-012-0160-9.

Navarro, M. O., Simionato, A. S., Barazetti, A. R., dos Santos, I. M., Cely, M. V., Chryssafidis, A. L., & Andrade, G. (2017). Disease-induced resistance and plant immunization using microbes. In: Singh, D., Singh, H., Prabha, R. (Eds.), *Plant-Microbe Interactions in Agro-Ecological Perspectives*, pp. 447–465. Springer, Singapore. doi: 10.1007/978-981-10-5813-4_22.

Nisha, S., Revathi, K., Chandrasekaran, R., Kirubakaran, S. A., Sathish-Narayanan, S., Stout, M. J., & Senthil-Nathan, S. (2012). Effect of plant compounds on induced activities of defense-related enzymes and pathogenesis related protein in bacterial blight disease susceptible rice plant. *Physiological and Molecular Plant Pathology, 80*, 1–9. doi: 10.1016/j.pmpp.2012.07.001.

Patel, H., & Krishnamurthy, R. (2013). Elicitors in plant tissue culture. *Journal of Pharmacognosy and Phytochemistry, 2*(2), 60–65. Retrieved from: https://www.phytojournal.com/archives?year=2013&vol=2&issue=2&part=A&ArticleId=141 (accessed on 28 May 2024).

Pieterse, C. M., Van der Does, D., Zamioudis, C., Leon-Reyes, A., & Van Wees, S. C. (2012). Hormonal modulation of plant immunity. *Annual Review of Cell and Developmental Biology, 28*, 489–521. doi: 10.1146/annurev-cellbio-092910-154055.

Pretali, L., Bernardo, L., Butterfield, T. S., Trevisan, M., & Lucini, L. (2016). Botanical and biological pesticides elicit a similar induced systemic response in tomato (*Solanum lycopersicum*) secondary metabolism. *Phytochemistry, 130*, 56–63. doi: 10.1016/j.phytochem.2016.04.002.

Ramirez-Estrada, K., Vidal-Limon, H., Hidalgo, D., Moyano, E., Golenioswki, M., Cusidó, R. M., & Palazon, J. (2016). Elicitation, an effective strategy for the biotechnological production of bioactive high-added value compounds in plant cell factories. *Molecules, 21*(2), 182. doi: 10.3390/molecules21020182.

Regnault-Roger, C. (2012). Trends for commercialization of biocontrol agent (biopesticide) products. In: Mérillon, J., Ramawat, K. (Eds.), *Plant Defence: Biological Control.* Progress in Biological Control, Vol. 12, pp. 139–160. Springer, Dordrecht. doi: 10.1007/978-94-007-1933-0_6.

Rojas, J. G., Bermudez, A. R., Campos, O. C., Cazola, C. C., & Sandoval, G. P. (2017). Resistencia metabólica a insecticidas. *Revista Bio Ciencias, 4*(6), 16. doi: 10.15741/revbio.04.06.01.

Shafique, S., Shafique, S., Zameer, M., & Asif, M. (2019). Plant Defense System Activated in Chili Plants by Using Extracts from *Eucalyptus citriodora. Biocontrol Science, 24*(3), 137–144. doi: 10.4265/bio.24.137.

Shoresh, M., Harman, G. E., & Mastouri, F. (2010). Induced systemic resistance and plant responses to fungal biocontrol agents. *Annual Review of Phytopathology, 48*, 21–43. doi: 10.1146/annurev-phyto-073009-114450.

Siah, A., Magnin-Robert, M., Randoux, B., Choma, C., Rivière, C., Halama, P., & Reignault, P. (2018). Natural agents inducing plant resistance against pests and diseases. In: Mérillon, J. M., Riviere, C. (Eds.), *Natural Antimicrobial Agents*. Sustainable Development and Biodiversity, Vol. 19, pp. 121–159. Springer, Cham. doi: 10.1007/978-3-319-67045-4_6.

Su, H. (2012). Regalia® bioprotectant in plant disease management. *Outlooks on Pest Management, 23*(1), 30–34. doi: 10.1564/23feb09.

Taheri, P., & Tarighi, S. (2010). Riboflavin induces resistance in rice against *Rhizoctonia solani* via jasmonate-mediated priming of phenylpropanoid pathway. *Journal of Plant Physiology, 167*(3), 201–208. doi: 10.1016/j.jplph.2009.08.003.

Terry, L. A., & Joyce, D. C. (2004). Elicitors of induced disease resistance in postharvest horticultural crops: a brief review. *Postharvest Biology and Technology, 32*(1), 1–13. doi: 10.1016/j.postharvbio.2003.09.016.

Thakur, M., & Sohal, B. S. (2013). Role of elicitors in inducing resistance in plants against pathogen infection: a review. *ISRN Biochemistry, 2013*, 762412. doi: 10.1155/2013/762412.

Tsuda, K., & Katagiri, F. (2010). Comparing signaling mechanisms engaged in pattern-triggered and effector-triggered immunity. *Current Opinion in Plant Biology, 13*(4), 459–465. doi: 10.1016/j.pbi.2010.04.006.

Walters, D. R., Ratsep, J., & Havis, N. D. (2013). Controlling crop diseases using induced resistance: challenges for the future. *Journal of Experimental Botany, 64*(5), 1263–1280. doi: 10.1093/jxb/ert026.

Wang, R., Wang, S., Pan, W., Li, Q., Xia, Z., Guan, E., & Yi, Z. (2018). Strategy of tobacco plant against black shank and tobacco mosaic virus infection via induction of PR-1, PR-4 and PR-5 proteins assisted by medicinal plant extracts. *Physiological and Molecular Plant Pathology, 101*, 127–145. doi: 10.1016/j.pmpp.2017.07.002.

Wang, W., Ben-Daniel, B. H., & Cohen, Y. (2004). Control of plant diseases by extracts of *Inula viscosa. Phytopathology, 94*, 1042–1047. doi: 10.1094/PHYTO.2004.94.10.1042.

Warris, A., & Ballou, E. R. (2019). Oxidative responses and fungal infection biology. *Seminars in Cell & Developmental Biology, 89*, 34–46. doi: 10.1016/j.semcdb.2018.03.004.

Wink, M. (2011). *Annual Plant Reviews, Biochemistry of Plant Secondary Metabolism*. Wiley Blackwell. West Sussex, U.K.

Yemata, G., & Fetene, M. (2016). Induction of systemic resistance in *Ensete ventricosum* clones by the leaf extract of *Agarista salicifolia* against *Xanthomonas campestris* pv. *musacearum*. *Journal of Plant Pathology*, 429–440. Retrieved from: https://www.jstor.org/stable/44280485?seq=1#metadata_info_tab_contents (accessed on 28 May 2024).

Zheng, F., Chen, L., Zhang, P., Zhou, J., Lu, X., & Tian, W. (2020). Carbohydrate polymers exhibit great potential as effective elicitors in organic agriculture: A review. *Carbohydrate Polymers, 230*, 115637. doi: 10.1016/j.carbpol.2019.115637.

CHAPTER 12

Malanga and Its Sustainable Production

GILBER VELA-GUTIÉRREZ,[1] YANHSY HERNÁNDEZ PORTILLO,[2]
VEYMAR G. TACIAS-PASCACIO,[1] and ERIKA J. LÓPEZ-ZÚÑIGA[1]

[1]*Faculty of Nutrition Sciences and Food, University of Sciences and Arts of Chiapas, Mexico*

[2]*Direction of Research and Graduate Studies, University of Sciences and Arts of Chiapas, Mexico*

ABSTRACT

At present, despite the fact that first-world countries have a socioeconomically high lifestyle caused by scientific and technological advances that humanity has achieved, there is still a lag in access to a decent life in developing countries. Food and energy shortages are evident, causing health problems for the population due to nutritional deficiency and the use of conventional energy in inefficient and polluting production processes, limiting the achievement of sustainable development objectives, mainly "zero hunger," "health and welfare," and "affordable and non-polluting energy." A possible solution is functional foods processed by alternative methods implementing renewable and clean energies, which, in addition to their nutritional value, provide health benefits and reduce the risks of contracting certain diseases. In this sense, this work will address the production process of Malanga (Xanthosoma sagittifolium) and the use of solar thermal energy during the production chain, with emphasis on agricultural harvest, post-harvest treatment, solar thermal drying, and its derivatives.

12.1 INTRODUCTION

The edible aroids, predominantly *Xanthosoma sagittifolium* (popularly named taioba, tannia, or cocoyam) and *Colocasia esculenta* (taro), which are

Sustainable Agriculture and Global Environmental Health: Toward the Efficient Management of Resources.
Mónica L. Chávez González, Cristóbal Noé Aguilar, Juan Carlos Contreras-Esquivel, and A. K. Haghi (Eds.)

commonly called cocoyams, are consumed in many tropical and sub-tropical communities (Boakye et al., 2017). Xanthosoma sagittifolium is a dominant root and tuber crop globally (Boakye et al., 2018; De Almeida, Bernardes, Fonseca, Cristine, Amaya-Farfán, 2013). It is cultivated in many places in Oceania, Africa, and Asia, providing livelihoods for more than 400 million people. Africa is the leading producer, with West and Central Africa, especially Nigeria, Ghana, and Cameroon, contributing more than 60% of total African production, constituting an important source of energy for numerous populations in tropical and subtropical regions. Therefore, the importance of cocoyam for local food safety cannot be emphasized enough (Boakye et al., 2018; Markusse et al., 2018).

The crop is preferred over other root and tuber crops as it has highly digestible starch and is also rich in crude protein, calcium, phosphorus, pro-vitamin A, and B-complex vitamins (Sobowale et al., 2018). The tubers are commonly considered boiled vegetables resembling potatoes (*Solanum tuberosum* L., *Solanaceae*), but are also used for the production of soup thickener, porridge, flour, and beverages (Graf et al., 2018).

The nutritional quality of cocoyam compares favorably with cassava, potatoes, and yam (Abdulrashid & Agwunobi, 2012). The low cost of production, ease of cultivation, nutrient composition, and abundance of favorable climate support taioba cultivation in Latin America, and may make this plant a viable alternative for family farming (De Almeida, Bernardes, Fonseca, Cristine, Amaya-Farfán, 2013).

12.2 BACKGROUND

12.2.1 ORIGIN AND DISTRIBUTION

Cocoyams (*Colocasia esculenta* (L.) Schott and *Xanthosoma sagittifolium* (L.) Schott) are the dominant species of edible aroids, grown in tropical and subtropical countries.

Malanga corms of *Colocasia esculenta*, a starchy stem that is widely cultivated as a staple food in many parts of Africa, the Americas, the Pacific Islands, and Asia, rank fifth and sixth in world production and planting after cassava, potato, sweet potato, yam, and taro (Pérez et al., 2007; Morales & Santacruz, 2017). Corms of *X. sagittifolium*, known as taioba, tannia, or cocoyam, belong to the Araceae family and are native to tropical America (Vela-Gutiérrez, Velázquez, Tacias Vidal, León, & De La Cruz, 2021).

12.2.2 PLANT CHARACTERISTICS

The plant reaches a height of 1 to 3 m, without an aerial stem, with large leaves coming from the central stem (Figure 12.1(a)), has a primary starchy stem, which thickens as the plant grows, known as a corm (Figure 12.1(b)); the new leaves come out unrolling between the trunk of the leaves formed, the lateral leaves wither and dry up. The inflorescence sprouts between the leaves in a spadix provided with a spathe (small organ of the plant) white color of 12–15 cm, has female flowers in its lower portion, male in the upper, and sterile in the middle. The edible part is the corm, it can remain underground for up to 6 months (Onwueme, 1999).

FIGURE 12.1 Malanga (*Xanthosoma sagittifolium*) – (a) plant; and (b) plant with corm.

Source: Authors

The corms of *Colocasia esculenta* are ovoid-round in shape, the skin is dark brown, and the pulp is white and starchy (Figure 12.2(a)). This taro starch has multiple potential in the food industry. *Xanthosoma sagittifolium* produces corms similar to *C. esculenta*, with the difference that the pulp is white and has a purple pigmentation (Figure 12.2(b)).

The corms and leaves contain vitamin C, thiamine, riboflavin, minerals (calcium and phosphorus), starch, and several bioactive compounds. The phytochemicals and pharmacological potential of corms could improve health conditions by controlling some diseases. Table 12.1 shows the chemical composition of fresh malanga corms (Graf et al., 2018).

FIGURE 12.2 Malanga flesh – (a) *Colocasia esculenta*; and (b) *Xanthosoma sagittifolium*.

Source: Authors

TABLE 12.1 Proximate Chemical Composition of Fresh Corms of the Genera *Xanthosoma sagittifolium* (from Chiapas) and *Colocasia esculenta* (from Veracruz)

Parameter	Xanthosoma sagittifolium	Colocasia esculenta
Moisture	82.67 ± 0.28	64.90 ± 0.57
Crude protein (N × 6.25)	1.22 ± 0.07	2.89 ± 0.13
Fat	1.07 ± 0.10	0.82 ± 0.04
Crude fiber	0.88 ± 0.08	1.13 ± 0.04
Ash	0.33 ± 0.09	2.27 ± 0.11
Carbohydrates	13.83 ± 0.11	28.0 ± 0.55

Source: Vela-Gutiérrez et al. (2021).

12.2.3 PRODUCTION

The world production of both species (Colocasia and Xanthosoma) has increased in the last decades. Currently, the average world production is estimated to be ~11 million tons, of which 70% are produced in Africa alone, while ~21% and ~5% are produced in Asia and America, respectively (Calle, 2021; Nazario-Lezama et al., 2020).

In 2018 in Mexico, the area planted was 679 ha with a production of 35,190 t, with the main producing states being Veracruz, Tabasco, and Puebla. In Veracruz, about 620 ha are cultivated, with an average production of 77.43 t/ha (Servicio de Información Agropecuaria y Pesquera (SIAP), 2018), which has been the main producer. In the period from 2010 to 2016, there were 3,423 ha planted in Veracruz, with a production of 32,216 t, and an average yield of 56 t/ha. In Veracruz, the municipality of Actopan is the main producer, with 530 ha planted in 2017 (Servicio de Información Agropecuaria y Pesquera (SIAP), 2018). There are currently production reports in the states of Oaxaca and Chiapas.

Malanga produced in the state of Veracruz is generally exported to Canada and the United States of America (López, Arvizu, Asiain, Mayett, & Martínez, 2018). The economic and social importance of this product in Veracruz has encouraged production, causing an average annual growth rate of 28.16% during the period 2011–2017, as well as an increase in the area planted of 52.10% in the same period. (Salgado-Molina et al., 2023).

12.3 PRODUCTION SYSTEMS

12.3.1 PRODUCTION SYSTEMS IN MEXICO

According to a study developed by Nazario-Lezama et al. (2020), the state of Veracruz, Mexico, has the largest production of malanga, and specifically the municipality of Actopan, Veracruz has a planted agricultural area of 17,651 ha, of which 402 ha is malanga. Likewise, SIAP (Servicio de Información Agropecuaria y Pesquera (SIAP), 2018) indicated that this crop in this state represents 68.8% of the nationwide area planted with the crop. About 47,000 tons are produced in Mexico; Veracruz produces 43,500 tons, followed by Nayarit with 4,300 tons (Secretaría de Agricultura y Desarrollo Rural (SADR), 2023). Around 57% of producers are considered small-scale, ranging from 0.5–3.0 ha, 4.0–6.0 ha (14.3%), 7.0–9.0 ha (21.4%), and more than 10 ha (7.1%), with an average annual yield of 50 t ha^{-1}. The producers' experience in malanga production is 5–9 years (57.14%), 1–4 years (32.14%), and 10–15 years (10.71%).

12.3.2 PRODUCTION SYSTEMS IN CHIAPAS

In Chiapas, malanga production systems are generally small-scale (0,5 to 2,0 ha) or backyard gardens (Figure 12.3). The production obtained covers the local or regional market.

FIGURE 12.3 Backyard Gardens of Malanga.

Source: Authors

In an interview with Francisco Gutiérrez (2022), a small malanga producer from the community of Cuahutemoc in the municipality of San Fernando in Chiapas, Mexico, he said that several factors are involved in the cultivation and production of malanga; however, it is necessary to approach it from an integral and sustainable perspective.

Before describing malanga cultivation, we will contextualize sustainability and agroecology. Beckford, Campbell, & Barker (2011) define sustainable agriculture as a farming operation that limits the degradation and consumption of resources, expands local food production, boosts food safety, and secures rural income and quality of life along with the generation of income.

According to Azadi et al. (2011), Sustainable Food Production Systems (SFPS) provide healthy food to meet today's food needs while carrying on healthy ecosystems for generations to come with the least negative effects on the environment. A sustainable food system also encourages local production and distribution and makes nutritious food more accessible and cost-effective to all.

Several authors define agroecosystems as production units where a diversity of events and interactions of organisms (biotic and abiotic) occur. Likewise, an agroecosystem can be characterized by its adaptation and adjustment to the possibilities offered by nature, the purpose being food supply (Gómez, 2012). In this sense and according to Nazario-Lezama (2018), a malanga agroecosystem is a production system where there are different actions of transformation and value addition in the different phases that integrate it, from inputs to obtaining malanga corms, which can be used for self-consumption, exchange, and others for commercialization. Considering the above, the following stages in the cultivation of malanga are proposed.

12.3.3 MALANGA CULTIVATION

12.3.3.1 SOIL CHARACTERISTICS

Malanga plants grow well in fertile and moist soils, although they can adapt to different types of soils (Langeland et al., 2008). Under suitable conditions, the length of the growth cycle is 270 to 330 days. Plants produce a large amount of foliage between six and nine months; harvesting takes place at nine months for white malanga and 12 months for purple malanga (Vela Gutiérrez, n.d.).

12.3.3.2 SEEDING OR SEED

Two parts of the plant can be used, corms or propagules as vegetative material, depending on the availability of plant material (Figure 12.4). If corms are used, they should be small, without tears, and without the presence of fungi. If propagules are used, they should be conditioned by pruning the leaves and cutting part of the corm. The planting distance is 50 cm between plants and 80 cm between rows, with an average density of 25,000 plants per hectare.

12.3.3.3 CULTIVATION

Once planting is done, it is necessary to keep the soil of the crop moist; in the dry season, it is recommended to irrigate daily or at least every two days,

FIGURE 12.4 Dry sowing of malanga, using propagules as vegetative material.

Image credit: Martínez, Muñozcano, & Santoyo, 2011

and in the rainy season, it is irrigated unless it is not constant. It is feasible to establish a malanga crop with an irrigation system (Figure 12.5).

For example, the malanga (*Xanthosoma sagittifolium*) crop grown by Francisco Gutiérrez (2022) on his ranch in the municipality of San Fernando, Chiapas, Mexico, is moistened by wastewater from a system of springs and natural ponds where tilapia and banana are grown. In the interview, he also mentions that strict crop cleanliness and fertility are also required; a condition that is generated through the association of crops, including sheep breeding, which allows him to eliminate weeds among the crops, in addition to using animal feces (manure) as natural fertilizer.

Harvesting is done between 10 and 12 months of age. For this activity, the foliage of the plant must be removed 24–48 hours previously, the soil must be abundantly moistened, and the corms must be extracted. The corms are cleaned, and the roots are removed and washed thoroughly (Figure 12.6). These are distributed to the local market in nearby cities (Gutierrez, 2022).

FIGURE 12.5 Malanga cultivation with drip irrigation system in southern Sinaloa Mexico (Martínez, Muñozcano, & Santoyo, 2011).

Image credit: Martínez, Muñozcano, & Santoyo, 2011

FIGURE 12.6 Corms of malanga (*Xanthosoma sagittifolium*).

Source: Authors

12.4 PROCESSING AND PRESERVATION OF THE MALANGA

Due to its relatively high moisture content, cocoyam tubers are susceptible to deterioration after harvest. As such, cocoyam is further processed using different techniques to make it available in various forms including boiled, roasted, or fried forms and starch (Sobowale et al., 2018).

The malanga (*Xanthosoma sagittifolium*) is a functional food that is presented as an alternative to help meet the sustainable development goals (SDGs), specifically "zero hunger" and "good health and well-being." Its nutritional value lies in its high content of starch (30–85% dry base), protein (1.4–7%), as well as being a good source of fiber (0.6–0.8%), vitamin A, C, calcium, and phosphorus (Arici, Kahraman, Özülkü, Metin, & Toker, 2016). According to Cruz-Ordóñez (2017), this food can be processed by various methods that prevent the growth of unwanted microorganisms (pathogens). For this, it is necessary to follow the recommendations established in NOM 251-SSA1-2009 (Secretaría de Salud, 2009) to avoid Foodborne Diseases (ETAS) or food poisoning (Vela Gutiérrez, n.d.).

According to Kapcum et al. (2022), cooked malanga presents significant differences from the raw product, mainly when boiling, steaming, baking, microwaving, and frying. All these methods decrease the oxalate content, the total phenolic content, and the antioxidant activity of the malanga. The frying process is the one that presents the greatest advantage in the retention of compounds that promote health and the reduction of oxalate; however, it can also present contraindications to human health due to the high-fat content.

In this process, it is possible to implement renewable energies for heat generation. A clear example is biofuels such as biogas, which could be generated from residual biomass through anaerobic digestion. According to Laibuta (2014) and Mandala (1999), this biofuel is a gas mixture, predominantly methane (CH_4) and carbon dioxide (CO_2), approximately between 60% and 40%, respectively, in addition to traces such as hydrogen sulfide (H_2S). This fuel could be used for cooking malanga in rural areas where this food is produced. For this, it is necessary to design or adapt the cooking systems. Mainly, the characteristics of the burners used by the stoves and grills must be considered. Determining the space between the flame and the container (pot) is important, as reported by Orhorhoro (2018) and Jiménez (2016). The biogas stove can reach an efficiency between 60% and 69%.

Returning to the processing methods of malanga, it is important to remember that the starch content of this plant is between 30% and 85% (dry basis). Therefore, it can replace other foods that are used for the production of flour, such as cereals, legumes, tubers, etc. For the preparation of this fine powder, Vela-Gutiérrez (n.d.) establishes the following process for its preparation:

1. Reception and inspection of malanga corms, considering weight, size, and state of maturity.
2. Washed and disinfected with a commercial antiseptic solution (manufacturer's specifications).
3. Peeled and portioned corns.
4. Cut into thin slices with a peeler or mandolin (thickness approx. 3 mm).
5. Rest in 10% citric acid solution for a period of 15 min.
6. Distribute in trays with aluminum foil at a temperature of 62±1°C for a period of 24 hours, using an air velocity of 1 m/s.
7. It is ground in a motorized mill for dry grains, it can also be done in an industrial blender.
8. Sift with number 120 (it can be done using a strainer or mesh), and vacuum pack to improve flour conservation.

In the process established by Vela-Gutiérrez (n.d.), point six refers to the dehydration of the product. This process can be carried out using technologies that use traditional fuels for heat generation, such as LP gas or electric ovens. Likewise, it is possible to reduce the humidity in the raw material (malanga) through eco-friendly technologies, such as solar dryers.

Solar dryers are technologies that utilize solar radiation to reduce the humidity of a solid (e.g., agricultural product). Through thermal processes, they increase the heat of a drying agent (air-fluid), which is then transferred to the product being dried (food) (Conde, 2020). As mentioned by Hernandez (2015), this device consists of a collector where solar radiation is concentrated and the fluid is dispersed (drying agent), as well as a drying chamber where the drying agent interacts with the product being dried to dehydrate it. The classification of this technology is very broad and depends on many variables. Considering the energy capture, solar dryers can be divided into three types: direct, indirect, and mixed. According to Leon (2002), air circulation can also be used to distinguish between devices: forced circulation or natural convection (Figure 12.7).

The use of this technology for drying malanga has been studied by various researchers (Martínez-Pereyra et al., 2018). They studied the effect of airspeed on the kinetics of drying in a thin layer of malanga. For this,

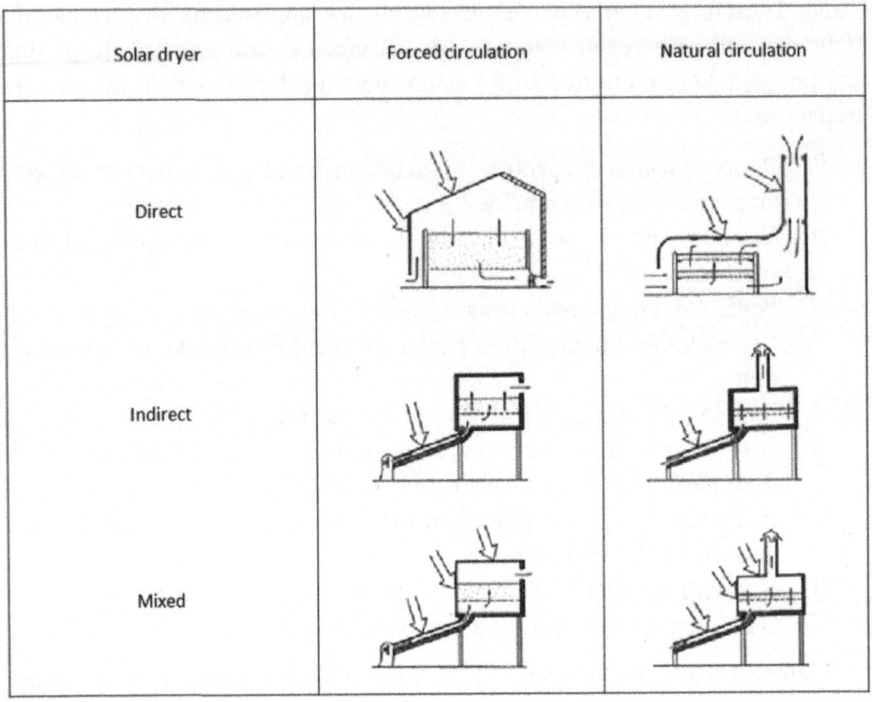

FIGURE 12.7 Solar dryer classification.

Source: Reprinted/Adapted from Valdés et al., 2017

they achieved an indirect solar dryer at 0.16, 0.24, and 0.32 m/s, where they appreciated that the air speed of drying has a significant impact on drying time and speed. In addition, according to the author, these parameters are affected by the variation in solar radiation, decreasing as it decreases. Moheno-Barrueta (2021) presented a similar problem in his study carried out on an indirect solar dryer where he experimented with the dehydration of bananas and malanga. To reduce the problem of intermittent solar radiation, it is possible to use energy vectors to store the necessary heat in the process. In addition, the development of hybrid devices is proposed, either with traditional fuels or the use of another source of renewable energy (Conde, 2020). The inventor used solar thermal energy and the combustion of biomass to heat the working fluid.

After the product is dehydrated, ground, and sieved, it can be used in different processes for food preparation, such as boxed and integral bread added with malanga flour (Vela Gutiérrez, n.d.).

12.5 OTHERS USES

The malanga plant consists of leaves, a stem, and a tuber, but the use is generally limited to the leaves, while the other parts are discarded. This generates environmental and inefficiency problems. These discarded parts have great potential. The stem is mainly composed of lignocellulose, and the tuber has a considerable percentage of starch. The average starch content in the cocoyam is 25% w/w, which can be considered as a high value compared with other starchy crops such as potatoes (15%) and cassava (18%) (Serna-Loaiza, Martínez, Pisarenko, & Cardona-Alzate, 2018).

12.6 CONCLUSION

In the future, food should be based strictly on the nutritional requirements of each person or individual, or generically at the population level, or perhaps even further towards the treatment or prevention of certain diseases; in this sense, functional foods will have a high impact, those foods that, in addition to providing nutrients to the organism, have bioactive ingredients or principles that have been scientifically proven to have an effect on health.

Malanga corms have exceptional nutrients and bioactive ingredients that make it easy to design and formulate functional foods. In this sense, processing and/or preservation methods can have an infinite number of applications for the cultivation of malanga and its ingredients in particular, but it is necessary to propose strategies for sustainable production. Regarding the integral use of the malanga plant, there is still much to be studied, including the use of each of its components; for example, for the formulation of food for human consumption, as well as for animals, or the development of biodegradable and/or edible packaging or containers.

The use of renewable energies for the processing of Malanga is feasible mainly in rural areas, where various processes such as cooking or dehydrating it can be carried out. In order for the process to be more efficient, the use of energy vectors to store heat is recommended, as well as the implementation of hybrid devices where more than one energy source can be used, such as solar thermal and biomass.

KEYWORDS

- **bioactive ingredients**
- **biomass**
- **cultivation**
- **functional foods**
- **hybrid devices**
- **Malanga**
- **soil characteristics**
- **Xanthosoma sagittifolium**

REFERENCES

Abdulrashid, M., & Agwunobi, L. N. (2012). Tannia (Xanthosoma sagittifolium) cocoyam as dietary substitute for maize in broiler chicken. *Greener Journal of Agricultural Sciences*, *2*, 167–171.

Arici, M., Kahraman, B., Özülkü, G., Metin, R., & Toker, O. (2016). Physicochemical and nutritional properties of taro (Colocasia esculenta L. Schott) flour as affected by drying temperature and air velocity. *LWT – Food Science and Technology, 74*, 434–440. https://doi.org/10.1016/j.lwt.2016.08.006.

Azadi, H., Schoonbeek, S., Mahmoudi, H., Derudder, B., De Maeyer, P., & Witlox, F. (2011). Organic agriculture and sustainable food production system: Main potentials. *Agriculture, Ecosystems & Environment*, 144(1), 92–94.

Beckford, C., Campbell, D., & Barker, D. (2011). Sustainable food production systems and food security: Economic and environmental imperatives in yam cultivation in Trelawny, Jamaica. *Sustainability*, 3(3), 541–561. https://doi.org/10.3390/su3030541.

Boakye, A. A., Gudjónsdóttir, M., Skytte, J. L. S., Chronakis, I., Wireko-Manu, F. D., & Oduro, I. (2017). Characteristics of Xanthosoma sagittifolium roots during cooking, using physicochemical analysis, uniaxial compression, multispectral imaging and low field NMR spectroscopy. *Journal of Food Science and Technology*, *54*, 2670–2683. Advance online publication. https://doi.org/10.1007/s13197-017-2704-7.

Boakye, A. A., Wireko-Manu, F. D., Oduro, I., Ellis, W. O., Gudjónsdóttir, M., & Chronakis, I. S. (2018). Utilizing cocoyam (Xanthosoma sagittifolium) for food and nutrition security: A review. *Food Science & Nutrition*. https://doi.org/10.1002/fsn3.602.

Calle, D. J. (2021). Properties of malanga flours and their use in pastes and gluten-free breads [Doctoral dissertation, Universidad Politecnica de Valencia].

Conde, O. (2020). Development of an automated biomass rotary solar dryer for the dehydration of coffee beans [Master's thesis, Universidad de Ciencias y Artes de Chiapas].

Cruz Ordóñez, M. A., Palacio-Pola, G., Márquez-Montes, R., & Vela Gutiérrez, G. (2017). Technologies for the conservation of malanga corms (Xanthosoma sagittifolium). In J. R.

Velázquez Martínez & A. Corzo Sosa (Eds.), *Contributions to Food Sciences*. Universidad Juárez Autónoma de Tabasco.

De Almeida, J. E., Bernardes, M. E., Fonseca, R. H., Cristine, V. E., Amaya-Farfán, J. (2013). Taioba (Xanthosoma sagittifolium) leaves: Nutrient composition and physiological effects on healthy rats. *Journal of Food Science*. https://doi.org/10.1111/1750-3841.12301.

Falade, K. O., & Okafor, C. A. (2015). Physical, functional, and pasting properties of flours from corms of two Cocoyam (Colocasia esculenta and Xanthosoma sagittifolium) cultivars. *Journal of Food Science and Technology*, *52*, 3440–3448. https://doi.org/10.1007/s13197-014-1368-9.

Gómez, A. S. (2012). Agroecosystems. In *Spanish Millennium Ecosystem Assessment*. Universidad de Alcalá, Spain. Retrieved from: http://www.ecomilenio.es/wp-content/uploads/2012/03/17-Agroecosistemas-web.pdf (accessed on 28 May 2024).

Graf, B. L., Zhang, L., Corradini, M. G., Kuhna, P., Newman, S. S., Salbaumc, J. M., & Raskin, I. (2018). Physicochemical differences between malanga (Xanthosoma sagittifolium) and potato (Solanum tuberosum) tubers are associated with differential effects on the gut microbiome. *Journal of Functional Foods*, *45*, 268–276. https://doi.org/10.1016/j.jff.2018.04.032.

Gutierrez, F. (2022). Interview with a small malanga producer in the community of Cuahutémoc, municipality of San Fernando, Chiapas, Mexico.

Hernández, J. M. (2015). Implementación y evaluación de un secador solar para queso en la comunidad UNESCO, Cintalapa, Chiapas y pruebas de secado de queso fresco en chiapaneco [Master's thesis, Universidad de Ciencias y Artes de Chiapas].

Jimenez, G. (2016). Biogas combustion efficiency analysis of a Bolivian burner. *Revista de Investigación e Innovación Agropecuaria y Recursos Naturales, 3*, 1016.

Kapcum, C., Pasada, K., Kantiwong, P., Sroysang, B., Phiwtawee, J., Suphantharika, P., Belur, P. D., Agoo, E. M. G., Janairo, J. I. B., & Wongsagonsup, R. (2022). Effects of different cooking methods on chemical compositions, in vitro starch digestibility and antioxidant activity of taro (Colocasia esculenta) corms. *International Journal of Food Science & Technology, 57*, 5144–5154. https://doi.org/10.1111/ijfs.15823.

Laibuta, R. (2014). Maximizing the adiabatic flame temperature of biogas using hydrogen [Master's thesis, The University of Sheffield].

Langeland, et al. (2008). *Identification and Biology of Nonnative Plants in Florida's Natural Areas* (2nd edn.) [University of Florida IFAS Publication SP 257].

Leon, M. A., Kumar, S., & Bhattacharya, S. C. (2002). A comprehensive procedure for performance evaluation of solar dryers. *Renewable & Sustainable Energy Reviews, 6*, 367–393.

López, S. Y., Arvizu, B. E., Asiain, H. A., Mayett, M. Y., & Martínez, F. J. L. (2018). Competitive analysis of malanga production activity: An approach based on Michael Porter's theory. *Revista Internacional Administracion & Finanzas*, *8*(16). https://doi.org/10.23913/ride.v8i16.366.

Mandala, T., & Kiran, B. (1999). Determination of the quality of biogas by flame temperature measurement. *Energy Conversion and Management, 40*, 1225–1228. https://doi.org/10.1016/S0196-8904(99)00009-6.

Markusse, D., Nguimbou, R. M., Aboubakar, X., Njintang, Y. N., Scher, J., & Mbofung, F. C. M. (2018). Production, physicochemical and sensory characterization of cocoyam mixed flours and pastes (achu). *Journal of Food Measurement and Characterization*. https://doi.org/10.1007/s11694-018-9738-z.

Martínez, A. C. O., Muñozcano, R. M., & Santoyo, J. J. A. (2011). Technological package for the establishment of malanga. *Fundación Produce Sinaloa*, 1–22.

Martínez-Pereyra, G., Ramírez-Morales, E., Rojas-Blanco, L., Córdova Bautista, Y., Gonzales-Solano, M., & Pérez-Hernández, G. (2018). Thin-layer drying of taro (Colocasia esculenta) in a solar dryer at different velocities. *Journal of Energy, Engineering Optimization and Sustainability, 2*(1), 77–92. https://doi.org/10.19136/jeeos.a2n1.2696.

Moheno-Barrueta, M., Tzuc, O. M., Martínez-Pereyra, G., Cardoso-Fernández, V., Rojas-Blanco, L., Ramírez-Morales, E., Pérez-Hernández, G., & Bassam, A. (2021). Experimental evaluation and theoretical optimization of an indirect solar dryer with forced ventilation under tropical climate by an inverse artificial neural network. *Applied Sciences, 11*, 7616. https://doi.org/10.3390/app111676.

Morales, V., & Santacruz, S. (2017). Uso de películas comestibles a base de carboximetilcelulosa y goma xantana para la disminución de absorción de grasa de malanga frita (Xanthosoma sagittifolium). *Revista Politécnica*, *40*(1), 1–6.

Nazario-Lezama, N. (2018). Análisis de la cadena de valor de malanga (Colocasia esculenta L.) en Veracruz, México [Master's thesis, Colegio de Posgraduados]. Retrieved from: http://colposdigital.colpos.mx:8080/jspui/bitstream/10521/3525/1/Nazario_Lezama_Nancy%20_MC%20_Agroecosistemas_2018.pdf (accessed on 28 May 2024).

Nazario-Lezama, N., Arvizu-Barrón, E., Mayett-Moreno, Y., Álvarez-Ávila, M., & García-Pérez, E. (2020). Production and marketing of taro (Colocasia esculenta (L.) Shott) in Actopan, Veracruz, Mexico: Value chain perspective. *Agroproductividad*, *13*(5), 59–64. https://doi.org/10.32854/agrop.vi.1660.

Onwueme, I. (1999). Taro cultivation in Asia and the Pacific. Bangkok: Food and Agriculture Organization (FAO) of the United Nations Regional office for Asia and the Pacific.

Orhorhoro, E. K. (2018). Design and construction of an improved biogas stove. *Arid Zone Journal of Engineering, Technology and Environment, 14*, 325–335.

Pérez, E. E., Gutiérrez, M. E., Pacheco de Delahaye, E., Tovar, J., & Lares, M. (2007). Production and characterization of Xanthosoma sagittifolium and Colocasia esculenta flours. *Journal of Sensory Studies*. https://doi.org/10.1111/j.1750-3841.2007.00420.x.

Salgado-Molina, A., Alvarez-Avila, M. C., Asiain, Hoyos, A., Platas-Rosado, D. E., Figuera-Rodríguez, K. A., & Ducrocq, C. (2023). Taro (Colocasia esculenta (L.) SCHOTT) value chain in Veracruz, Mexico. *Agro Productividad*. https://doi.org/10.32854/agrop.v16i2.2290.

Secretaría de Agricultura y Desarrollo Rural (SADR). (2023). Agriculture promotes malanga cultivation for the benefit of producers in the southeast of the country. Retrieved from: https://www.gob.mx/agricultura/prensa/promueve-agricultura-cultivo-de-malanga-en-beneficio-de-productores-del-sur-sureste-del-pais?idiom=es (accessed on 28 May 2024).

Secretaría de Salud. (2009). NOM 251-SSA1-2009. Norma Oficial Mexicana: Prácticas de higiene para el proceso de alimentos, bebidas o suplementos alimenticios. Retrieved from: https://www.dof.gob.mx/normasOficiales/3980/salud/salud.htm (accessed on 28 May 2024).

Serna-Loaiza, S., Martínez, A., Pisarenko, Y., & Cardona-Alzate, C. A. (2018). Integral use of plants and their residues: The case of cocoyam (Xanthosoma sagittifolium) conversion through biorefineries at small scale. *Environmental Science and Pollution Research, 25, 35949*–35959. https://doi.org/10.1007/s11356-018-2313-7.

Servicio de Información Agropecuaria y Pesquera (SIAP). (2018). Statistical Yearbook of Agricultural Production. Retrieved from: https://nube.siap.gob.mx/cierreagricola/ (accessed on 28 May 2024).

Sobowale, S. S., Animashaun, O. H., Mulaba-Bafubiandi, A. F., Abidoye, T. S., Kewuyemi, Y. O., & Adebo, O. A. (2018). Process optimization of extrusion variables and its effect on properties of extruded cocoyam (Xanthosoma sagittifolium) noodles. *Food Science & Nutrition*. https://doi.org/10.1002/fsn3.786.

Valdés, C. F., Chejne Janna, F., Marrugo, G., Gomez, C., Montoya, J., Macias, R., Londoño, C., De La Cruz, J., & Ocampo, A. (2017). La gasificación, alternativa de generación de energía y productos con alto valor agregado para la industria. Arenas (1st edn.). Universidad Nacional de Colombia.

Vela Gutiérrez, G. (n.d.). *Processing and Preservation of Malanga Corms* (1st ed.). Montebello UNICACH.

Vela-Gutiérrez, G., Velázquez, L. A. A., Tacias P. V. G., Vidal, L. D. G., León, G. E., & De La Cruz, M. J. (2021). Effect of heat treatment on oxalate and hydrocyanic acid levels of malanga corms of two cultivars (Xanthosoma sagittifolium and Colocasia esculenta) in a murine model. *Journal of Food Science and Technology, 59,* 220–227. https://doi.org/10.1007/s13197-021-05004-9.

CHAPTER 13

Valorization of Agro-Industrial Wastes for Ultrasound-Assisted Extraction of Pectin: A Review

CARLOS N. CANO-GONZALEZ,[1] MARÍA DE LA LUZ HERRERA-ESTRADA,[2] JUAN CARLOS CONTRERAS-ESQUIVEL,[2] LLUVIA DE ABRIL ALEXANDRA SORIANO-MELGAR,[2] and ROCIO YANELI AGUIRRE-LOREDO[1]

[1]*Centro de Investigación en Química Aplicada, Blvd. Enrique Reyna Hermosillo 140, San José de los Cerritos, Saltillo, Coahuila, Mexico*

[2]*Facultad de Ciencias Químicas, Universidad Autónoma de Coahuila, Blvd. Venustiano Carranza y José Cárdenas s/n. Republica Oriente, Saltillo, Coahuila, Mexico*

ABSTRACT

Agro-industrial wastes and by-products are excellent candidates for extracting valuable components such as pectin, providing sustainable processes and environmental protection. Pectin is a polysaccharide found in the plant wall and has applications in the health industry. This polysaccharide is essential due to its rheological properties as a stabilizer and biological properties such as anti-inflammatory, antioxidant, and prebiotic. The function depends directly on the chemical structure of the polysaccharide, which includes the content of uronic acids, neutral sugars, degree of esterification, and molecular weight. The extraction method is related to the efficiency and quality of the product, which depends on the parameters of the process and the properties of the plant matrix. This review presents information on using different alternative plant sources for pectin extraction. The ultrasound-assisted extraction

Sustainable Agriculture and Global Environmental Health: Toward the Efficient Management of Resources. Mónica L. Chávez González, Cristóbal Noé Aguilar, Juan Carlos Contreras-Esquivel, and A. K. Haghi (Eds.)

factors, as well as the characterization of the polysaccharide by different methods.

13.1 INTRODUCTION

Agro-industrial wastes have been presented as an attractive source of by-products. Due to the great demand for food production, waste has increased, generating environmental pollution. These wastes contain compounds with biological activities with applications in different fields such as food, pharmaceuticals, cosmetics, etc. (Jayanthi et al., 2023). The polysaccharides in the plant wall, such as pectin, have focused on their extraction, characterization, and application. Pectin presents three regions that depend on the conformation of the monosaccharides: Homogalacturonan, Rhamnogalacturonan I, and Rhamnogalacturonan II. Pectin gives physicochemical properties related to the polysaccharide's structural composition. The quality and application of the pectin depend on the galacturonic acid (GalA) content, degree of esterification (DE), and extraction yield (Singh & Tingirikari, 2021; Shivamathi et al., 2022).

The extraction of the pectin has been carried out by different methodologies, including conventional, microwave, ultrasound, and enzymatic. The extraction yield and the physicochemical properties of pectin depend on factors such as extraction time, power, pH, extracting agent, temperature, frequency, and solid-liquid ratio (Filho, Coelho, Muniz, & Barbosa, 2022). Depending on the extraction method, the factors that can be applied vary. Ultrasound-assisted extraction generates a degradation of the plant wall to release pectin by acoustic cavitation. This method is environmentally friendly and presents extraction yields and quality pectin composition. Common sources of pectin are citrus peel and apple, but the update focuses on alternate plant sources such as grape peel and pomace, passion fruit, dragon fruit, date, fig, witch, pineapple, etc. (Santos et al., 2020; Petkowicz & Williams, 2020).

The characterization of the pectin extracted from different plant sources by ultrasound implies the determination of the monosaccharides that make up the chemical structure of the polysaccharide. The galacturonic acid content must be $\geq$ 65% of the chemical structure of the polysaccharide, and by the degree of esterification, it is presented as a low degree (<50%) and a high degree (>50%). The characterization can be carried out by colorimetric and instrumental methods that can be general or specific (Petkowicz & Williams,

2020; Marić et al., 2018). This review presents information on the most critical ultrasonic extraction factors for yield and characteristics of quality pectin from different plant sources. Comparison of yield, galacturonic acid content, and degree of pectin esterification from common and alternative plant sources. In addition, the characterization methods involved in pectin.

13.2 AGRO-INDUSTRIAL WASTES APPROACH

The food industry has presented a large amount of waste generated by vegetables that can cause environmental problems. For this reason, it has focused on using agro-industrial waste as a sustainable method to mitigate the environmental impact. Some works present information on using various plant sources to obtain pectin (Petkowicz & Williams, 2020). The agri-food sector is focused on solving crucial challenges. The fulfillment of the demand for food consumption is not simply based on the production chain but also faces the problem of food loss. The increase in food production is related to the rise in the world population; in less than three decades, a population of 9,000 million is expected. Approximately 50% of fresh products worldwide are lost in postharvest due to physical, chemical, and biological factors during transport and storage (Muñoz-Almagro et al., 2021).

Vegetable processing industries generate many by-products, mainly peels, seeds, and stems. The disposal of these wastes presents an environmental problem. One of the alternatives offered in the last decade is obtaining bio compounds from agro-industrial waste. To date, works on extracting important compounds in different areas of health have been presented. FAO statisticians guarantee that vegetable production will increase; therefore, the incidence of residues will be gradual (Jayanthi et al., 2023). Currently, the food industry presents an approach regarding food safety and environmental contamination due to the large amount of waste generated by food that causes environmental problems. It is estimated that 33% of the food produced is wasted worldwide, with a tendency to increase in the coming years, according to the UN and the FAO. The use of agro-industrial waste in a sustainable way with the minimum adverse effect on the environment has generated areas of knowledge for the recovery of by-products. Agro-industrial residues present compounds with bioactivities that can be used in products in food, cosmetics, medicine, and bioremediation with benefits for living beings (Jayanthi et al., 2023; Petkowicz & Williams, 2020).

13.3 BRANCHED HETEROPOLYSACCHARIDE: PECTIN

The plant cell wall comprises different polysaccharides, including hemicellulose, cellulose, and pectin, with a structural function and other macromolecules, such as proteins, enzymes, and phenolic compounds. Pectin is found in dicotyledonous plants and a few monocotyledons in more significant presence in the middle lamina; the highest concentration gradient goes from the primary wall to the plasma membrane with structural function (Zioga, Tsouko, Maina, Koutinas, Mandala, & Evageliou, 2022).

Pectin is a heteropolysaccharide made up mainly of galacturonic acid as a skeleton for the molecular structure. This polysaccharide is approved as a food additive by the FAO. It is considered dietary fiber and presents benefits in improving and controlling cholesterol levels and caloric deficit. In addition, partially hydrolyzed pectin can modulate the immune system and biostimulant activity in plants. Pectin extraction presents two main steps: hydrolysis of protopectin to pectin using acids and ethanol precipitation (Santos et al., 2020; Petkowicz & Williams, 2020). Pectins can be classified by their degree of esterification into two types: high methoxyl (> 50%) and low methoxyl (< 50%). It is a complex branched polysaccharide composed of three regions: Homogalacturonan (HG) comprises a linear chain of D-glucuronic acids. Rhamnogalacturonan I (RG I) includes alternating arabinose and D-galacturonic acid units with arabinan, arabinogalactan, and galactan. Rhamnogalacturonan II (RG II) contains D-galacturonic acids and complex branches (Singh & Tingirikari, 2021). The most recent data show that pectins are composed of at least 17 types of monosaccharides in different proportions. D-GalA is the main one in the chain, followed by D-galactose or L-arabinose. It comprises a linear chain of α-D-GalA units linked by (1,4) bonds, in which some of the C-6 carboxyl groups are methyl/acetyl esterified. In this context, the degree of esterification is the percentage of esterified galacturonic acid units out of the total number of galacturonic acid units in the polysaccharide. The degree of pectin esterification is essential for the functional properties in the plant cell wall and affects its application as a texture-modifying agent (Marić et al., 2018).

The primary sources of pectin extraction are citrus and apple. However, the focus on only a few vegetables is a problem due to the shortage of these. Therefore, alternative plant sources that meet the characteristics of quality pectin for use as an additive in the pharmaceutical, cosmetic, and food industries, among others, are sought (Patience, Schieppati, & Boffito, 2021).

For applying pectin, the galacturonic acid content, degree of esterification, molecular weight, and extraction yield, among others, are considered. The physicochemical properties of pectins indicate the most appropriate alternate sources for their production. Recently, research has been directed at residues such as shells, pomace, stems, leaves, roots, etc., from alternate plant sources such as grapes (Spinei & Oroian, 2023), passion fruit (Santos et al., 2020; Sucheta, Misra, & Yadav, 2020), carrots (Sucheta, Misra, & Yadav, 2020), sunflower (Ezzati, Ayaseh, Ghanbarzadeh, & Heshmati, 2020), dragon fruit (Nguyen & Pirak, 2019), Gac (Tran, Saifullah, Nguyen, Nguyen, & Vuong, 2021), date (Djaoud et al., 2022), fig (Gharibzahedi, Smith, & Guo, 2019), custard apple (Shivamathi et al., 2019), chayote (Ke, Jiang, Shen, Wu, Liu, & Zhang, 2020), pineapple (Shivamathi et al., 2022), sisal (Yang, Wang, Hu, Xiao, & Wu, 2018).

Pectin has presented a considerable increase in the market in recent years, taking into account the challenge of plant sources and the physicochemical properties of pectin. In 2050, the market will reach approximately 2 billion dollars. Due to the interest in the areas of health, it is essential to cover each of the stages of pectin production. The demand for pectin production is related to the generation of gelled products that cause an annual increase in pectin production of 4–5% (Shivamathi et al., 2022' Filho, Coelho, Muniz, & Barbosa, 2022).

13.3.1 ULTRASONIC-ASSISTED EXTRACTION OF PECTIN

Extraction by the conventional method involves using some acids and solvents, which presents a concern for the environment, in addition to pectin depolymerization, equipment damage, and costs. Therefore, alternative extraction technologies that offer efficiency and a positive relationship with the environment have been implemented. The extraction technologies applied are microwaves, ultrasound, supercritical fluids, and enzymes which involve advantages in the extraction process and quality characteristics for the pectin (Zioga, Tsouko, Maina, Koutinas, Mandala, & Evageliou, 2022).

Ultrasound-assisted extraction is an ecological and efficient method for extracting plant metabolites, presenting more excellent selectivity. The high-frequency waves generated by ultrasound cause acoustic cavitation in the extracting agent. This phenomenon induces fractures in the plant wall, facilitating the interaction of the metabolite with the extracting agent and increasing the mass transfer rate without compromising the stability of the product. In addition, among its advantages is the low consumption of energy and solvent, which indicates a higher performance and efficiency in the extraction process that leads to acceptable product quality (Shivamathi et al., 2022; Patience, Schieppati, & Boffito, 2021).

The UAE is mainly characterized by determining the controllable extraction conditions, frequency, and wavelength amplitude, to increase the solid-liquid extraction performance. The cavitation process exerted by ultrasound occurs in two steps: the first step is expansion, where the bubbles are formed and expand, and the second step is compression, where the bubbles collapse due to the negative pressure on the resistance liquid to traction (Fernández-Delgado et al., 2023). Some factors can cause pectin degradation, such as the internal ones, taking power and frequency involved, and the external ones, which are the temperature, application time, and type of solvent. Power density is an important factor for the yield and quality of extracted pectin. Therefore, applying high power and prolonged degrees can generate depolymerization and low esterification in addition to the different content of monosaccharides of the pectin (Ling, Ramaswamy, Lyng, Gao, & Wang, 2023).

To contribute to the conservation of resources and use of plant residues, economically more viable extraction methods have sought to achieve better quality pectin with physicochemical and functional properties consistent with its application. Therefore, several alternative sustainable approaches have emerged, which are faster to extract pectin and other natural compounds from biological materials. Pectin extraction consists of multi-level physicochemical processes influenced by various factors, such as temperature, extraction time, type of solvent, solid-liquid ratio, acid concentration, pH, etc. (Filho, Coelho, Muniz, & Barbosa, 2022). The acids implemented in the extraction of pectin from different plant matrices are citric acid (Spinei & Oroian, 2023), hydrochloric acid (Shivamathi et al., 2019), acetic acid (Sucheta, Misra, & Yadav, 2020), and nitric acid (Patience, Schieppati, & Boffito, 2021). The temperature applied in the extraction oscillates between 70 and 100°C, and an acid pH of a range of 1.0 to 4.0, presenting better extraction results at a pH less than 2.0. The power that is applied ranges from 70 to 800 W for a time of 10 to 90 min. The solid-liquid ratio ranges from 1:10 to 1:100 depending on the characteristics of the plant matrix (Santos et al., 2020; Petkowicz & Williams, 2020). Table 13.1 shows the extraction conditions applying ultrasound in different plant sources.

TABLE 13.1 Parameters for the Ultrasound-Assisted Extraction of Pectin

Time (min)	Power (W)	Temperature (°C)	Frequency (kHz)	LRS (g/mL)	Solvent	pH	References
60	200	NR	25	1:10	Citric acid	1.8	Spinei & Oroian (2023)
60	140	75	37	1:10	Citric acid	2.0	Minjares-Fuentes et al. (2014)
28.3	130	61.9	20	1:17	NR	1.2	Moorthy et al. (2015)
27.95	800	66.7	20	1:50	Citric acid	1.5	Wang, Ma, Xu, Cao, Jiang, Ding, Ye, & Liu (2015)
51.79	800	60	20	1:50	Hydrochloric acid	1.5	Xu, Zhang, Bailina, Ge, Ding, Ye, & Liu (2014)
25	200	70	24	1:50	Hydrochloric acid	1.5	Bagherian, Zokaee Ashtiani, Fouladitajar, & Mohtashamy (2011)
30	250	85	20	1:30	Nitric acid	2.0	Freitas de Oliveira et al. (2016)
24	500	40	20	1:30	Citric acid	2.0	Panwar, Panesar, & Chopra (2023)
30	70	NR	20	1:30	Citric acid	1.8	Dranca, Vargas, & Oroian (2020)
20	70	NR	20	1:10	Citric acid	1.8	Dranca, Vargas, & Oroian (2020)
20	86	60	20	1:100	Citrate buffer	5.2	Encalada et al. (2019)
30	400	33	25	1:20	NR	NR	Ezzati, Ayaseh, Ghanbarzadeh, & Heshmati (2020)
10	150	30	20	1:20	Citric acid	1.5	Hundie (2020)
22	155	NR	NR	1:22.5	Citric acid	1.5	Hundie (2020)
40	NR	70	NR	1:50	NR	NR	Ke, Jiang, Shen, Wu, Liu, & Zhang (2020)
30	550	70	37	1:100	Acetic acid	2.5	Sucheta, Misra, & Yadav (2020)
27	323	NR	20	1:15	Citric acid	3.2	Maran et al. (2017)
60	NR	70	NR	1:30	Citric acid	2.0	Nguyen & Pirak (2019)
18	NR	63	20	1:23	Hydrochloric acid	2.3	Shivamathi et al. (2019)
35	200	60	50	1:50	Citric acid	1.5	Tran, Saifullah, Nguyen, Nguyen, & Vuong (2021)
15	500	80	20	1:40	Citric acid	2.5	Wang, Huang, Fan, Zhao, Hu, Xu, Pan, & Liu (2016)

TABLE 13.1 *(Continued)*

Time (min)	Power (W)	Temperature (°C)	Frequency (kHz)	LRS (g/mL)	Solvent	pH	References
70	80	65	132	1:12	Citric acid	2.0	Zaid et al. (2020)
60	160	50	40	1:18	Citric acid	2.0	Villamil-Galindo & Piagentini (2022)
90	NR	NR	20	1:20	NR	1.0	Fernández-Delgado et al. (2023)
90	350	60	45	1:10	Nitric acid	1.5	Djaoud et al. (2022)
25	70	70	24	1:15	Citric acid	1.4	Gharibzahedi, Smith, & Guo (2019)
30	NR	67	NR	1:50	Citric acid	NR	Santos et al. (2020)
10	200	NR	20	1:15	Citric acid	1.5	Asgari, Labbafi, Khodaiyan, Kazemi, & Hosseini (2020)
30	400	60	20	1:20	Citric acid	2.5	Calvete-Torre et al. (2021)
24	NR	60	20	1:15	NR	1.6	Moorthy, Maran, Ilakya, Anitha, Sabarima, & Priya (2017)
30	497	85	20	1:30	Nitric acid	2.0	Guandalini, Rodrigues, & Marczak (2019)
60	500	40	NR	1:30	Nitric acid	2.0	Patience, Schieppati, & Boffito (2021)
22	NR	71	20	1:15	Hydrochloric acid	1.0	Shivamathi et al. (2022)
60	450	70	20	1:15	Ammonium oxalate	4.0	Yang, Wang, Hu, Xiao, & Wu (2018)

NR: Not reported.

The structural characterization of pectin is an essential factor in the quality and application of the product. The yield, galacturonic acid content, and degree of esterification present the physicochemical properties of pectin. The application of this polysaccharide depends on its rheological and biological properties. According to the Joint FAO/WHO Expert Committee on Food Additives and the European Commission, the pectin sample must not contain less than 65% GalA content to be considered food-grade pectin (Panwar, Panesar, & Chopra, 2023). Some pectin extraction works using response or factorial surfaces do not present the galacturonic acid content or the degree of esterification. The plant sources where the design has been applied are *Musa balbisiana* (Maran et al., 2017), pomegranate peel (Moorthy et al.,

2015), passion fruit peel (Santos et al., 2020), jackfruit peel (Moorthy, Maran, Ilakya, Anitha, Sabarima, & Priya, 2017), orange peel (Patience, Schieppati, & Boffito, 2021). The use of surfactants has been presented in works to support extracting pectins, such as Tween80, PEG4000, and Saponin. Compared with water, the extraction yields with the surfactants give a more significant response in the release of pectin (Fernández-Delgado et al., 2023). The structural characteristics that determine the application of pectin are the galacturonic acid content and the degree of esterification, indicating biological and rheological properties for implementation in different areas of health. Ultrasound-assisted extraction in different plant sources is attractive due to the characteristics and yield of pectin (Table 13.2).

TABLE 13.2 Agro-Industrial Waste for Pectin Extraction and its Characterization

Type of Waste	Extraction Yield (%)	Galacturonic Acid (%)	Degree of Esterification (%)	References
Sour orange peel	28.07	65.3	Low	Hosseini, Khodaiyan, Kazemi, & Najari (2019)
Orange peel	26.87	68.25	High	Hundie (2020)
Orange peel	11.0	NR	NR	Patience, Schieppati, & Boffito (2021)
Grapefruit peel	27.34	50.03	High	Wang, Ma, Xu, Cao, Jiang, Ding, Ye, & Liu (2015)
Grapefruit peel	26.74	NR	NR	Xu, Zhang, Bailina, Ge, Ding, Ye, & Liu (2014)
Grapefruit peel	17.92	68.21	High	Bagherian, Zokaee Ashtiani, Fouladitajar, & Mohtashamy (2011)
Citrus limetta peel	28.73	79.60	High	Panwar, Panesar, & Chopra (2023)
Apple pomace	9.18	98.12	High	Dranca & Oroian (2019)
Apple pomace	9.18	92.83	High	Dranca, Vargas, & Oroian (2020)
Apple pomace	7.9	66.7	High	Calvete-Torre et al. (2021)
Granny Smith apple peel	6.38	NR	High	Villamil-Galindo & Piagentini (2022)
Carrot roots	20.6	39.0	Low	Encalada et al. (2019)
Sunflower heads and stems	11.15	72.94	Low	Ezzati, Ayaseh, Ghanbarzadeh, & Heshmati (2020)
Grape pomace	8.94	78.64	High	Spinei & Oroian (2023)
Grape pomace	32.35	97.00	High	Minjares-Fuentes et al. (2014)
Pomegranate peel	24.18	NR	NR	Moorthy et al. (2015)

TABLE 13.2 *(Continued)*

Type of Waste	Extraction Yield (%)	Galacturonic Acid (%)	Degree of Esterification (%)	References
Chayote	6.19	57.25	Low	Ke, Jiang, Shen, Wu, Liu, & Zhang (2020)
Black carrot pomace	8	17.03	Low	Sucheta, Misra, & Yadav (2020)
Musa Balbisiana	8.99	NR	NR	Maran et al. (2017)
Dragon fruit peel	19.48	66.55	Low	Nguyen & Pirak (2019)
Custard apple peel	8.93	70.24	High	Shivamathi et al. (2019)
Gac pulp	53.8	NR	Low	Tran, Saifullah, Nguyen, Nguyen, & Vuong (2021)
Mango peel	17.15	53.35	High	Wang, Huang, Fan, Zhao, Hu, Xu, Pan, & Liu (2016)
Dragon fruit peel	30.11	44.34	High	Zaid et al. (2020)
Sugar beet pulp	24.6	39.1	High	Fernández-Delgado et al. (2023)
Algerian dates	7.6	54.6	Low	Djaoud et al. (2022)
Fig skin	8.79	66.12	Low	Gharibzahedi, Smith, & Guo (2019)
Passion fruit peels	51.0	NR	High	Freitas de Oliveira et al. (2016)
Passion fruit peels	12.67	76.29	High	Freitas de Oliveira et al. (2016)
The green husk of walnuts	12.78	69.44	High	Asgari, Labbafi, Khodaiyan, Kazemi, & Hosseini (2020)
Artocarpus heterophyllus	14.5	NR	NR	Moorthy, Maran, Ilakya, Anitha, Sabarima, & Priya (2017)
Fruit peel				
Mango peel	8.6	70.0	High	Guandalini, Rodrigues, & Marczak (2019)
Pineapple peel	16.24	75.6	High	Shivamathi et al. (2022)
Sisal	11.9	62.28	Low	Yang, Wang, Hu, Xiao, & Wu (2018)

NR: Not reported.

13.3.2 COMMON PLANT SOURCES IN PECTIN EXTRACTION

Among the citrus fruits are orange, lemon, and grapefruit peel. One of the factors with the most significant impact on ultrasound-assisted extraction is acidic pH. The highest extraction yield of pectin from sour orange peel was at pH 1.5. But this factor intervenes negatively with the DE, which was 6.77%. GalA content was 65.3%, proving that the backbone of pectin is homogalacturonan. The following monosaccharides present a lower rate of galactose, rhamnose, arabinose, and xylose, which were 25.1%, 2.8%,

2.5%, and 2.0%, respectively (Hosseini, Khodaiyan, Kazemi, & Najari, 2019). In the work of Hundie (2020), orange peel established 63.13% DE, and GalA content was 68.25%. Grapefruit peel (*Citrus paradisi* Macf.) GalA content was 50.03%. GalA appears as the main monomer after glucose, galactose, rhamnose, and arabinose, with 13.99%, 13.84%, 7.32%, and 4.43%, respectively. It also presents traces of monosaccharides that are not present in the chemical structure of pectin and DE 65.52% (Wang, Ma, Xu, Cao, Jiang, Ding, Ye, & Liu, 2015). Turkish grapefruit pectin, by applying sonication, shows GalA content was 68.21% and DE 76.12%. Treatment in combination with ultrasound and microwave presented an increase in the rate of GalA from 74.25%. Also, the percentage increase in performance to 33.88%. The variety of both technologies gives better advantages in the extraction and composition characteristics of the pectin (Bagherian, Zokaee Ashtiani, Fouladitajar, & Mohtashamy, 2011). The pectin of the sweet lemon peel (*Citrus limetta*) presents a composition of monosaccharides where it predominates with a GalA content of 79.60% and DE 55.29% (Panwar, Panesar, & Chopra, 2023).

One of the critical factors in the content of GalA in the pectin of apple pomace (*Malus domestica* 'Falticeni') is the pH in the extraction process. The GalA content decreases when the pH increases, as does the pectin yield—one of the also essential aspects in the structural composition of the polysaccharide. The critical factors for the GalA content were the low pH, increased amplitude, and extraction time, obtaining pectin with 98.12% GalA and a DE of 76.71% (Dranca & Oroian, 2019). In another work by Dranca et al. (2020), the GalA content of apple pectin for microwave and ultrasound extraction was 90.60% and 92.83%, respectively. Extraction by alternative technologies presents a higher GalA content in the chemical structure of pectin and DE of 77.0%. The GalA content was 66.7% in the pectin of the apple pomace. For neutral sugars glucose, arabinose, rhamnose, galactose, and fucose, it was 10.3%, 8.5%, 5.0%, 3.6%, and 2.8%, respectively. The application of ultrasound presents a higher GalA content, giving a higher region of HG. The DE ranges from 50–88%, depending on the treatment and the apple variety (Calvete-Torre et al., 2021).

13.3.3 ALTERNATIVE PLANT SOURCES FOR PECTIN EXTRACTION

In the grape pomace (*Vitis vinifera* var. Feteasca Neagra), the GalA content and the degree of pectin esterification were 80.24% and 81.07%, respectively (Spinei & Oroian, 2023). In the Cabernet Sauvignon variety, GalA occurs as the main monosaccharide of pectin, with a percentage of 95–97%. Monosaccharides such as rhamnose, arabinose, and galactose are found in less than 2%. This may be due to the removal of the side chains by the power of the ultrasound and the extracting agent (citric acid)—the DE of 64.1% (Minjares-Fuentes et al., 2014). Carrot root pectin has low GalA content (39%) and DE (37.2%). One of the critical factors for the extraction and the high content of GalA is the acidic pH (Encalada et al., 2019). In another study, black carrot pomace pectin has low GalA and DE content of 17.03% and 34.3%, respectively. The extraction technologies can affect the chemical structure of the pectin by decreasing the percentage of GalA and DE (Sucheta, Misra, & Yadav, 2020). The GalA content of the passion fruit peel was high. One of the critical factors for increasing this monosaccharide in the chemical structure of pectin was the temperature used in the extraction. There is a more significant presence of GalA content at lower temperatures presenting 76.29% (Freitas de Oliveira et al., 2016). The GalA content in sunflower pectin was 72.94%. Neutral sugars may have a more significant presence of galactose, rhamnose, and arabinose with a rate of 9.85%, 8.01%, and 2.86%, respectively. The pectin regions comprise 70.09% homogalacturonan and 23.58% rhamnogalacturonan I, and DE of 34.06% (Ezzati, Ayaseh, Ghanbarzadeh, & Heshmati, 2020). The GalA content of coyote pectin was 57.25%, with a low DE (17.6%) (Ke, Jiang, Shen, Wu, Liu, & Zhang, 2020). The amount of GalA in the pectin from the dragon fruit peel was 66.55% and low DE 36.41% under the best extraction conditions. The best extraction conditions in yield present modifications in the chemical structure of the pectin, both in the percentage of uronics and in the DE (Shivamathi et al., 2019). The optimization in extracting pectin from the cherimoya peel presents a chemical structure with a GalA content higher than 70.24% and a DE higher than 70% (Shivamathi et al., 2019). The pectin extraction yield of Guc pulp was high, but the DE was around 5%. Pectin has a low degree of methylation (Tran, Saifullah, Nguyen, Nguyen, & Vuong, 2021). The GalA content of the pectin from the mango peel was low to present quality pectin. GalA content was 53.35%, and the DE is higher than 86% (Wang, Huang, Fan, Zhao, Hu, Xu, Pan, & Liu, 2016). The percentage of GalA and DE depends on the extraction factors, mainly pH, time, temperature, and solid-liquid ratio. Dragon fruit pectin has a GalA content of 44.34%. The neutral sugars were 27.32% galactose, 17.93% mannose, 7.65% rhamnose, and 2.8% glucose. The DE was 56.10% classified as a high DE (Zaid et al., 2020). The acid extraction solution can hydrolyze the insoluble pectin and make it soluble, causing depolymerization and modifying its chemical structure and molecular weight. The pectin of the date peel presents a low GalA content of 54.6%, as well as the DE, which was 25%. The neutral sugars with the highest presence were glucose, arabinose, galactose, mannose, and rhamnose, with concentrations of 13.3, 10.3, 9.9, 7.0, and 4.9%, respectively (Djaoud et al., 2022). The fig peel pectin has an acceptable GalA content of 66.12% with a low DE of 39.42%. The combination of ultrasound-microwaves reduces an increase in extraction yield of 11.71%. The GalA content was 76.85%, higher than ultrasound alone, while the DE dropped to 32.41%. The proportion of the monosaccharide's glucuronic acid, GalA, glucose, fructose, arabinose, galactose, rhamnose, and mannose for the extraction by ultrasound and ultrasound-microwave was 0.2:3.2:0.6:0.7:0.3:1.1:0.2:0.2 and 0.2:3.4:0.7 :0.4:0.5:1.0:0.2:0.1, respectively (Asgari, Labbafi, Khodaiyan, Kazemi, & Hosseini, 2020). The GalA content of the pectin of the green walnut shell was 69.44%. The characteristic of the pectin was established as a high DE with 56.21%. The GalA content in pectin is essential in different properties as a functional emulsifier and antioxidant activity. The DE corresponds to the ratio of esterified GalA carboxyl groups to methyl and acetyl groups (Asgari, Labbafi, Khodaiyan, Kazemi, & Hosseini, 2020). The GalA content was 70%, and a DE of 58% of the pectin from the mango peel. In previous studies, some parameters were found to increase the content of GalA, highlighting the size of the shell particles, blanching, and the state of maturity. In addition, to the conditions applied in the extraction method, the pectin will present different characteristics (Guandalini, Rodrigues, & Marczak, 2019). The sisal waste presents pectin with a GalA content of 62.28% and a DE of 44.35%. The neutral sugars with the most significant presence were glucose, rhamnose, mannose, galactose, and arabinose, with a concentration of 23.45, 2.90, 0.48, 0.22, and 0.09%, respectively. The high glucose content may be due to the partial hydrolysis of cellulose and xyloglucan in the plant wall (Yang, Wang, Hu, Xiao, & Wu, 2018). The GalA content of the pectin from the pineapple peel was 75.6% and DE 58% (Shivamathi et al., 2022).

13.3.4 PHYSICOCHEMICAL CHARACTERIZATION METHODS OF PECTIN

The extraction methods and the vegetable source determine the physicochemical factors that the pectin presents. Therefore, its structural and physicochemical characterization is essential after the polysaccharide extraction. The methods involved include colorimeters or instruments that show information such as monosaccharide composition, degree of esterification (acetyl, methyl), molecular weight, particle size, thermal stability, crystallinity, etc. (Ling, Ramaswamy, Lyng, Gao, & Wang, 2023). Table 13.3 shows the physicochemical characterization methods of pectin.

TABLE 13.3 Physicochemical Characterization Methods of Pectin

Methods	Determination	References
Colorimetric Methods		
m-hydroxydiphenyl	Uronic acids content	Blumenkrantz & Asboe-Hansen (1973)
Hydroxylamine	Acetyl content	Hestrin (1949)
Titration	Degree of methyl-esterification	Santos et al. (2020)
Instrumental Methods		
GC-MS	Monosaccharides	Petkowicz & Williams (2020)
	Degree of esterification	Petkowicz & Williams (2020)
HPAEC-PAD	Uronic acids content	Nagel et al. (2014)
	Neutral sugar content	Nagel et al. (2014)
FT-IR	Degree of methyl-esterification	Santos et al. (2020)
NMR	Chemical structure	Santos et al. (2020)
TGA	Thermal stability	Panwar, Panesar, & Chopra (2023)
XRD	Crystallinity or amorphous nature	Tran, Saifullah, Nguyen, Nguyen, & Vuong (2021)
HPSEC	Molecular weight	Djaoud et al. (2022)
SEM	Morphological analysis	Tran, Saifullah, Nguyen, Nguyen, & Vuong (2021)
DLS	Particle size	Sucheta, Misra, & Yadav (2020)
IEC	Purification	Shivamathi et al. (2022)

The determination of uronic acids uses the m-hydroxydiphenyl technique. About 1.2 mL of $H_2SO_4/Na_2B_4O_7$ (0.075 M) are added to the samples. This stage is carried out in an ice bath, left for 10 min. Samples are shaken and heated at 100°C for 5 min. They are then cooled with cold water and added 20 μL of MHDF (0.15%) in NaOH solution (0.5%). GalA is used as a standard. Samples were analyzed in a spectrophotometer at 520 nm (Blumenkrantz & Asboe-Hansen, 1973).

The alkaline hydroxylamine reagent (2 M) determines the acetyl content. Hydrochloric acid is added to adjust the pH to 1.2, then 1 mL of ferric chloride solution (0.37 M) is added. Samples are analyzed at 540 nm. Penta-O-acetyl-β-D-galactopyranose is used as a standard (Hestrin, 1949).

The titration method supports the determination of methyl esterification. The pectin is added to water and left under depression at 40°C, 150 rpm for 2 h. The solution is titrated with 0.1 N NaOH using the phenolphthalein indicator. Then add 10 mL of 0.1 N NaOH to the solution and stir for two h for saponification of pectin-esterified carboxylic acids. Then 0.1 N HCl is added to the solution and titrated with 0.1 N NaOH using the phenolphthalein indicator (Santos et al., 2020).

The degree of methyl esterification can be determined by the instrumental method using FT-IR of the signal areas at 1,750 cm^{-1} for the methyl esterified carboxylic group (COO-R) and 1,630 cm^{-1} for the carboxylic ion (COO^-). The average of the areas of the 1,750 cm^{-1} peaks over the sum of the area of the 1,750 and 1,630 cm^{-1} peaks, the data obtained are plotted concerning their degree of esterification. The samples are interpolated on the calibration curve. The FT-IR method helps to establish the structural behavior of the pectin by the degree of esterification. FT-IR analysis is fast, does not destroy the sample, and can be applied to determine pectin quality (Petkowicz & Williams, 2020).

Gas chromatography with mass spectrometry (GC-MS) for the determination of monosaccharides. The pectin is hydrolyzed with TFA (2 M) at 120°C for two hours, $NaBH_4$ is used for reduction and then acetylated with pyridine acetic anhydride. The derivatization generates alditol acetates which are determined using a carrier gas (helium) for dragging the sample and a capillary column (DB-225) for separation and identification (Petkowicz & Williams, 2020).

High-resolution anion exchange chromatography with pulsed amperometric detection (HPAEC-PAD) for the determination of uronic acids. The pectin is hydrolyzed with TFA (2 M). Samples were analyzed using the AminoTrap column (3 × 30 mm), keeping the CarboPac PA20 column (3 × 30 mm) and CarboPac PA20 column (3 × 150 mm) connected in series. The eluents used were ultrapure water for A, sodium acetate (170 mM) in NaOH (100 mM) for B, and NaOH (500 mM) for C. It uses a gradient profile of eluents with a flow rate of 0.2 mL/min (Santos et al., 2020).

Nuclear magnetic resonance (NMR) analysis determines the chemical structure of pectin. The sample is prepared at 1.0% in distilled water and capped in a standard NMR tube. The spectra determine H^1 and C^{13} with a total of 64 transient sweeps with a data acquisition time of 4 s (Santos et al., 2020).

The thermal stability of the pectin is performed by thermogravimetric analysis (TGA). The sample is analyzed in an inert atmosphere with nitrogen in a temperature range of 50 to 600°C at a heating rate of 10°C/min. Weight loss regions are presented in 3. The first region offers a loss of approximately 10% due to the sample's water loss. The second region offers a loss greater than 50%, corresponding to the appearance of the pyrrole polysaccharide. The third region is slower and corresponds to the deformation of the main chain of the polysaccharide (Panwar, Panesar, & Chopra, 2023).

X-ray diffraction (XRD) allows for determining the crystallinity or amorphous nature of the pectin. This property of polysaccharides depends on the type and bonds of the monomer, and the structure is linear or branched. Samples are scanned at 25°C in a 2θ diffraction angle range of 5° to 50° with a given scan step (0.01°) and count time (2 s per step) (Tran, Saifullah, Nguyen, Nguyen, & Vuong, 2021).

The molecular weight of pectin is estimated by High-Pressure Size Exclusion Chromatography (HPSEC). For the determination of the relative molecular weight, a TSK-Gel wash column (6.0 mm × 400 mm) was obtained connected to two G5000 PWXL columns in series (7.8 mm × 300 mm, 10 μm) to G2500 PWXL (7.8 mm × 300 mm, 6 μm). The eluent as mobile phase is 0.01 M ammonium acetate with a flow of 0.5 mL/min for 50 min (Djaoud et al., 2022).

The pectin particle size is a physical factor for the different types of analysis dependent on surface contact. The study is carried out using Dynamic Light Scattering (DLS) equipment. The sample was solubilized in deionized water, applying suppression at 80°C, 500 rpm for 30 min. The sample is subjected to a laser emitting a wavelength of 633 nm at 25°C (Sucheta, Misra, & Yadav, 2020).

Ion exchange chromatography (IEC) makes generating a total or partial purification of pectin possible. The pectin is dissolved in distilled water with a 15 mg/mL concentration. The DEAE-cellulose column (25 x 2.4 cm) is used for the physicochemical properties of pectin. Eluent with NaCl with a rising gradient of 0.1–0.5 mol/L and a 1 mL/min flow rate. The fractions obtained are analyzed by different methods to determine the purity of the pectin (Shivamathi et al., 2022).

13.4 CONCLUSION

Pectin is a polysaccharide in the cell wall that can be obtained from agro-industrial residues. It is made up mainly of galacturonic acid and a smaller percentage of neutral sugars. The chemical structure of the polysaccharide determines the physicochemical and biological properties. The extraction methods applied depend on the efficiency and quality of the polysaccharide, considering the extraction yield, the uronic content, and the degree of esterification. The ultrasound-assisted extraction method presents advantages for generating value-added by-products in alternative plant sources. Also, identify controllable factors to indicate efficient extraction and pectin quality.

KEYWORDS

- **agro-industrial residues**
- **characterization-methods**
- **degree-esterification**
- **homogalacturonan**
- **pectin**
- **polysaccharide**
- **uronic-acid**
- **value-added by-products**

REFERENCES

Asgari, K., Labbafi, M., Khodaiyan, F., Kazemi, M., & Hosseini, S. S. (2020). High-methylated pectin from walnut processing wastes as a potential resource: Ultrasound assisted extraction and physicochemical, structural and functional analysis. *International Journal of Biological Macromolecules, 152*, 1274–1282. https://doi.org/10.1016/j.ijbiomac.2019.10.224.

Bagherian, H., Zokaee Ashtiani, F., Fouladitajar, A., & Mohtashamy, M. (2011). Comparisons between conventional, microwave- and ultrasound-assisted methods for extraction of pectin from grapefruit. *Chemical Engineering and Processing: Process Intensification, 50*, 1237–1243. https://doi.org/10.1016/J.CEP.2011.08.002.

Blumenkrantz, N., & Asboe-Hansen, G. (1973). New method for quantitative determination of uronic acids. *Analytical Biochemistry, 54*, 484–489. https://doi.org/10.1016/0003-2697(73)90377-1.

Calvete-Torre, I., Muñoz-Almagro, N., Pacheco, M. T., Antón, M. J., Dapena, E., Ruiz, L., Margolles, A., Villamiel, M., & Moreno, F. J. (2021). Apple pomaces derived from mono-varietal Asturian ciders production are potential source of pectins with appealing functional properties. *Carbohydrate Polymers, 264*. https://doi.org/10.1016/j.carbpol.2021.117980.

Djaoud, K., Muñoz-Almagro, N., Benítez, V., Martín-Cabrejas, M. Á., Madani, K., Boulekbache-Makhlouf, L., & Villamiel, M. (2022). New valorization approach of Algerian dates (*Phoenix dactylifera* L.) by ultrasound pectin extraction: Physicochemical, techno-functional, antioxidant and antidiabetic properties. *International Journal of Biological Macromolecules, 212*, 337–347. https://doi.org/10.1016/j.ijbiomac.2022.05.115.

Dranca, F., & Oroian, M. (2019). Ultrasound-Assisted Extraction of Pectin from *Malus domestica* 'Fălticeni' Apple Pomace. *Processes, 7(8), 488*. https://doi.org/10.3390/PR7080488.

Dranca, F., Vargas, M., & Oroian, M. (2020). Physicochemical properties of pectin from *Malus domestica* 'Fălticeni' apple pomace as affected by non-conventional extraction techniques. *Food Hydrocolloids, 100, 105383*. https://doi.org/10.1016/j.foodhyd.2019.105383.

Encalada, A. M. I., Pérez, C. D., Flores, S. K., Rossetti, L., Fissore, E. N., & Rojas, A. M. (2019). Antioxidant pectin enriched fractions obtained from discarded carrots (*Daucus carota* L.) by ultrasound-enzyme assisted extraction. *Food Chemistry, 289*, 453–460. https://doi.org/10.1016/j.foodchem.2019.03.078.

Ezzati, S., Ayaseh, A., Ghanbarzadeh, B., & Heshmati, M. K. (2020). Pectin from sunflower by-product: Optimization of ultrasound-assisted extraction, characterization, and functional analysis. *International Journal of Biological Macromolecules, 165*, 776–786. https://doi.org/10.1016/j.ijbiomac.2020.09.205.

Fernández-Delgado, M., del Amo-Mateos, E., Coca, M., López-Linares, J. C., García-Cubero, M. T., & Lucas, S. (2023). Enhancement of industrial pectin production from sugar beet pulp by the integration of surfactants in ultrasound-assisted extraction followed by diafiltration/ultrafiltration. *Industrial Crops and Products, 194,* 116304. https://doi.org/10.1016/j.indcrop.2023.116304.

Filho, L. B. S., Coelho, R. C., Muniz, E. C., & Barbosa, H. de S. (2022). Optimization of pectin extraction using response surface methodology: A bibliometric analysis. *Carbohydrate Polymer Technologies and Applications, 4,* 100229. https://doi.org/10.1016/j.carpta.2022.100229.

Freitas de Oliveira, C., Giordani, D., Lutckemier, R., Gurak, P. D., Cladera-Olivera, F., & Ferreira Marczak, L. D. (2016). Extraction of pectin from passion fruit peel assisted by ultrasound. *LWT – Food Science and Technology, 71*, 110–115. https://doi.org/10.1016/J.LWT.2016.03.027.

Gharibzahedi, S. M. T., Smith, B., & Guo, Y. (2019). Pectin extraction from common fig skin by different methods: The physicochemical, rheological, functional, and structural evaluations. *International Journal of Biological Macromolecules, 136*, 275–283. https://doi.org/10.1016/j.ijbiomac.2019.06.040.

Guandalini, B. B. V., Rodrigues, N. P., & Marczak, L. D. F. (2019). Sequential extraction of phenolics and pectin from mango peel assisted by ultrasound. *Food Research International, 119*, 455–461. https://doi.org/10.1016/j.foodres.2018.12.011.

Hestrin, S. (1949). The reaction of acetylcholine and other carboxylic acid derivatives with hydroxylamine, and its analytical application. *Journal of Biological Chemistry, 180*, 249–261. https://doi.org/10.1016/S0021-9258(18)56740-5.

Hosseini, S. S., Khodaiyan, F., Kazemi, M., & Najari, Z. (2019). Optimization and characterization of pectin extracted from sour orange peel by ultrasound assisted method. *International Journal of Biological Macromolecules, 125*, 621–629. https://doi.org/10.1016/j.ijbiomac.2018.12.096.

Hundie, K. B. (2020). Ultrasound-Assisted Optimization of Pectin Extraction from Orange Peel Using Response Surface Methodology (RSM) and Artificial Neural Network (ANN). *International Journal of Applied Science and Engineering, 8,* 102231. https://doi.org/10.30954/2322-0465.2.2020.1.

Jayanthi Antonisamy, A., Marimuthu, S., Malayandi, S., Rajendran, K., Lin, Y. C., Andaluri, G., Lee, S. L., & Ponnusamy, V. K. (2023). Sustainable approaches on industrial food wastes to value-added products – A review on extraction methods, characterizations, and its biomedical applications. *Environmental Research, 217,* 114758. https://doi.org/10.1016/j.envres.2022.114758.

Ke, J., Jiang, G., Shen, G., Wu, H., Liu, Y., & Zhang, Z. (2020). Optimization, characterization, and rheological behavior study of pectin extracted from chayote (*Sechium edule*) using ultrasound assisted method. *International Journal of Biological Macromolecules, 147,* 688–698. https://doi.org/10.1016/j.ijbiomac.2020.01.055.

Ling, B., Ramaswamy, H. S., Lyng, J. G., Gao, J., & Wang, S. (2023). Roles of physical fields in the extraction of pectin from plant food wastes and byproducts: A systematic review. *Food Research International, 164,* 112343. https://doi.org/10.1016/j.foodres.2022.112343.

Maran, J. P., Priya, B., Al-Dhabi, N. A., Ponmurugan, K., Moorthy, I. G., & Sivarajasekar, N. (2017). Ultrasound assisted citric acid mediated pectin extraction from industrial waste of *Musa balbisiana*. *Ultrasonics Sonochemistry, 35*, 204–209. https://doi.org/10.1016/j.ultsonch.2016.09.019.

Marić, M., Grassino, A. N., Zhu, Z., Barba, F. J., Brnčić, M., & Rimac Brnčić, S. (2018). An overview of the traditional and innovative approaches for pectin extraction from plant food wastes and by-products: Ultrasound-, microwaves-, and enzyme-assisted extraction. *Trends in Food Science & Technology, 76*, 28–37. https://doi.org/10.1016/j.tifs.2018.03.022.

Minjares-Fuentes, R., Femenia, A., Garau, M. C., Meza-Velázquez, J. A., Simal, S., & Rosselló, C. (2014). Ultrasound-assisted extraction of pectins from grape pomace using citric acid: A response surface methodology approach. *Carbohydrate Polymers, 106,* 179–189. https://doi.org/10.1016/j.carbpol.2014.02.013.

Moorthy, I. G., Maran, J. P., Ilakya, S., Anitha, S. L., Sabarima, S. P., & Priya, B. (2017). Ultrasound assisted extraction of pectin from waste *Artocarpus heterophyllus* fruit peel. *Ultrasonics Sonochemistry, 34*, 525–530. https://doi.org/10.1016/j.ultsonch.2016.06.015.

Moorthy, I. G., Maran, J. P., Surya, S. M., Naganyashree, S., & Shivamathi, C. S. (2015). Response surface optimization of ultrasound assisted extraction of pectin from pomegranate peel. *International Journal of Biological Macromolecules, 72*, 1323–1328. https://doi.org/10.1016/j.ijbiomac.2014.10.037.

Muñoz-Almagro, N., Ruiz-Torralba, A., Méndez-Albiñana, P., Guerra-Hernández, E., García-Villanova, B., Moreno, R., Villamiel, M., & Montilla, A. (2021). Berry fruits as source of pectin: Conventional and non-conventional extraction techniques. *International Journal of Biological Macromolecules, 186*, 962–974. https://doi.org/10.1016/j.ijbiomac.2021.07.016.

Nagel, A., Neidhart, S., Anders, T., Elstner, P., Korhummel, S., Sulzer, T., Wulfkuhler, S., Winkler, C., Qadri, S., Rentschler, C., Pholpipattanapong, N., Wuthisomboon, J., Endress, H. U., Sruamsiri, P., & Carle, R. (2014). Improved processes for the conversion of mango peel into storable starting material for the recovery of functional co-products. *Industrial Crops and Products, 61*, 92–105. https://doi.org/10.1016/J.INDCROP.2014.06.034.

Nguyen, B. M. N., & Pirak, T. (2019). Physicochemical properties and antioxidant activities of white dragon fruit peel pectin extracted with conventional and ultrasound-assisted extraction. *Cogent Food & Agriculture, 5*. https://doi.org/10.1080/23311932.2019.1633076.

Panwar, D., Panesar, P. S., & Chopra, H. K. (2023). Ultrasound-assisted extraction of pectin from *Citrus limetta* peels: Optimization, characterization, and its comparison with commercial pectin. *Food Bioscience, 51,* 102231. https://doi.org/10.1016/j.fbio.2022.102231.

Patience, N. A., Schieppati, D., & Boffito, D. C. (2021). Continuous and pulsed ultrasound pectin extraction from navel orange peels. *Ultrasonics Sonochemistry, 73*, 105480. https://doi.org/10.1016/j.ultsonch.2021.105480.

Petkowicz, C. L. O., & Williams, P. A. (2020). Pectins from food waste: Characterization and functional properties of a pectin extracted from broccoli stalk. *Food Hydrocolloids, 107,* 105930. https://doi.org/10.1016/j.foodhyd.2020.105930.

Santos, E. E., Amaro, R. C., Bustamante, C. C. C., Guerra, M. H. A., Soares, L. C., & Froes, R. E. S. (2020). Extraction of pectin from agroindustrial residue with an ecofriendly solvent: Use of FTIR and chemometrics to differentiate pectins according to degree of methyl esterification. *Food Hydrocolloids, 107,* 105921. https://doi.org/10.1016/j.foodhyd.2020.105921.

Shivamathi, C. S., Gunaseelan, S., Soosai, M. R., Vignesh, N. S., Varalakshmi, P., Kumar, R. S., Karthikumar, S., Kumar, R. V., Baskar, R., Rigby, S. P., Syed, A., Elgorban, A. M., & Moorthy, I. M. G. (2022). Process optimization and characterization of pectin derived from underexploited pineapple peel biowaste as a value-added product. *Food Hydrocolloids, 123,* 107141. https://doi.org/10.1016/j.foodhyd.2021.107141.

Shivamathi, C. S., Moorthy, I. G., Kumar, R. V., Soosai, M. R., Maran, J. P., Kumar, R. S., & Varalakshmi, P. (2019). Optimization of ultrasound assisted extraction of pectin from custard apple peel: Potential and new source. *Carbohydrate Polymers, 225,* 115240. https://doi.org/10.1016/j.carbpol.2019.115240.

Singh, R. P., & Tingirikari, J. M. R. (2021). Agro waste derived pectin poly and oligosaccharides: Synthesis and functional characterization. *Biocatalysis and Agricultural Biotechnology, 31,* 101910. https://doi.org/10.1016/j.bcab.2021.101910.

Spinei, M., & Oroian, M. (2023). Structural, functional and physicochemical properties of pectin from grape pomace as affected by different extraction techniques. *International Journal of Biological Macromolecules, 224*, 739–753. https://doi.org/10.1016/j.ijbiomac.2022.10.162.

Sucheta, Misra, N. N., & Yadav, S. K. (2020). Extraction of pectin from black carrot pomace using intermittent microwave, ultrasound and conventional heating: Kinetics, characterization and process economics. *Food Hydrocolloids, 102*, 105592. https://doi.org/10.1016/j.foodhyd.2019.105592.

Tran, T. T. B., Saifullah, M., Nguyen, N. H., Nguyen, M. H., & Vuong, Q. V. (2021). Comparison of ultrasound-assisted and conventional extraction for recovery of pectin from Gac (*Momordica cochinchinensis*) pulp. *Future Foods, 4,* 100074. https://doi.org/10.1016/j.fufo.2021.100074.

Villamil-Galindo, E., & Piagentini, A. M. (2022). Sequential ultrasound-assisted extraction of pectin and phenolic compounds for the valorisation of 'Granny Smith' apple peel. *Food Bioscience, 49*, 101958. https://doi.org/10.1016/j.fbio.2022.101958.

Wang, M., Huang, B., Fan, C., Zhao, K., Hu, H., Xu, X., Pan, S., & Liu, F. (2016). Characterization and functional properties of mango peel pectin extracted by ultrasound assisted citric acid. *International Journal of Biological Macromolecules, 91*, 794–803. https://doi.org/10.1016/J.IJBIOMAC.2016.06.011.

Wang, W., Ma, X., Xu, Y., Cao, Y., Jiang, Z., Ding, T., Ye, X., & Liu, D. (2015). Ultrasound-assisted heating extraction of pectin from grapefruit peel: Optimization and comparison with the conventional method. *Food Chemistry, 178*, 106–114. https://doi.org/10.1016/j.foodchem.2015.01.080.

Xu, Y., Zhang, L., Bailina, Y., Ge, Z., Ding, T., Ye, X., & Liu, D. (2014). Effects of ultrasound and/or heating on the extraction of pectin from grapefruit peel. *Journal of Food Engineering, 126*, 72–81. https://doi.org/10.1016/j.jfoodeng.2013.11.004.

Yang, Y., Wang, Z., Hu, D., Xiao, K., & Wu, J. Y. (2018). Efficient extraction of pectin from sisal waste by combined enzymatic and ultrasonic process. *Food Hydrocolloids, 79*, 189–196. https://doi.org/10.1016/j.foodhyd.2017.11.051.

Zaid, R. M., Mishra, P., Siti Noredyani, A. R., Tabassum, S., Ab Wahid, Z., & Mimi Sakinah, A. M. (2020). Proximate characteristics and statistical optimization of ultrasound-assisted extraction of high-methoxyl-pectin from *Hylocereus polyrhizus* peels. *Food and Bioproducts Processing, 123*, 134–149. https://doi.org/10.1016/j.fbp.2020.06.011.

Zioga, M., Tsouko, E., Maina, S., Koutinas, A., Mandala, I., & Evageliou, V. (2022). Physicochemical and rheological characteristics of pectin extracted from renewable orange peel employing conventional and green technologies. *Food Hydrocolloids, 132*, 107887. https://doi.org/10.1016/j.foodhyd.2022.107887.

Index

B

C

I

M

N

S